THE MEDICAL-
PHARMACEUTICAL
KILLING
MACHINE

FACING FACTS COULD SAVE YOUR LIFE

CHILDREN'S HEALTH DEFENSE

Skyhorse Publishing

All Rights Reserved. No part of this book may be reproduced in any manner without the express written consent of the publisher, except in the case of brief excerpts in critical reviews or articles. All inquiries should be addressed to Skyhorse Publishing, 307 West 36th Street, 11th Floor, New York, NY 10018.

Skyhorse Publishing books may be purchased in bulk at special discounts for sales promotion, corporate gifts, fund-raising, or educational purposes. Special editions can also be created to specifications. For details, contact the Special Sales Department, Skyhorse Publishing, 307 West 36th Street, 11th Floor, New York, NY 10018 or info@skyhorsepublishing.com.

Skyhorse® and Skyhorse Publishing® are registered trademarks of Skyhorse Publishing, Inc.®, a Delaware corporation.

Visit our website at www.skyhorsepublishing.com.

10 9 8 7 6 5 4 3 2 1

Library of Congress Cataloging-in-Publication Data is available on file.

Print ISBN: 978-1-64821-129-4
eBook ISBN: 978-1-64821-130-0

Cover design by Children's Health Defense
Cover photo by Lisa F. Young

Printed in the United States of America

CONTENTS

List of Acronyms   vii

Introduction   xi

CHAPTER ONE: FROM QUACKERY TO CRIMINALITY   1

The "Messianic" Benjamin Rush   1

A "Patently Criminal" Model   4

Iatrogenocide Takes Center Stage   7

CHAPTER TWO: THE MEDICAL CASSANDRAS   9

Ivan Illich: Iatrogenesis   10

Dr. Robert Mendelsohn: Hazardous Medicine   17

Dr. Barbara Starfield: Implicating the Medical System   26

Gary Null: Gruesome Statistics   29

Dr. Peter Breggin: The Psychopharmaceutical Complex   38

Jon Rappoport and Celia Farber: The Spin Machine   51

Sasha Latypova and Katherine Watt: Legalizing Democide   65

**CHAPTER THREE: TWENTY-FIRST CENTURY VACCINE
TECHNOLOGIES**   77

Modified RNA   78

Lipid Nanoparticles (LNPs)   85

Russian Roulette Batches   90

"Adulteration" and "Contamination"   94

CHAPTER FOUR: ASSISTED SUICIDE AND EUTHANASIA   99

The Trendsetters   100

"Not So Simple"   105

Killing Kids and Young Adults, Too   108
Freedom to Die—But Not to Live?   110
Slippery Slope   111

CHAPTER FIVE: DEADLY MEDICINE: HOW DO THEY GET AWAY WITH IT?   **117**
Consolidation of Medical Authority   117
The Rise of the Modern Pharmaceutical Industry   131
The Sidelining and Weaponization of Nutrition   137
The Evidence-Based Medicine Juggernaut   141
Medical Gaslighting   147
The Creation of a Global Enforcement Infrastructure   155

CHAPTER SIX: WHY DO THEY DO IT? MONEY, PRESTIGE, AND CONTROL   **163**
Customers for Life   164
Power, Prestige, and Perks   178
Depopulation and Central Control?   184

CHAPTER SEVEN: LIFESAVING FACTS FOR YOU AND THOSE YOU LOVE   **193**
Plummeting Trust   194
New Threats on the Horizon   196
Refocusing on Health, Not Medicine   198

Endnotes   203

LIST OF ACRONYMS

AAP	American Academy of Pediatrics
ACAMAID	American Clinicians Academy on Medical Aid in Dying
ACIP	Advisory Committee on Immunization Practices (CDC)
ADHD	Attention-deficit/hyperactivity disorder
AMA	American Medical Association
AONL	American Organization for Nursing Leadership
API	Active pharmaceutical ingredient
ARPA-H	Advanced Research Projects Agency for Health
ASD	Autism spectrum disorder
BARDA	Biomedical Advanced Research and Development Authority
BIS	Bank for International Settlements
CARPA	Complement activation-related pseudoallergy
CDC	Centers for Disease Control and Prevention
cGMP	Current good manufacturing practices
CHD	Children's Health Defense
CICP	Countermeasures Injury Compensation Program
D4CE	Doctors for COVID Ethics
DARPA	Defense Advanced Research Projects Agency
DEA	Drug Enforcement Administration
DHS	Department of Homeland Security
DNR	Do not resuscitate orders
DOD	Department of Defense
DOJ	Department of Justice
DSM	Diagnostic and Statistical Manual of Mental Disorders
EAS	Euthanasia and physician-assisted suicide
EBM	Evidence-based medicine

EMR	Electronic medical record
EPA	Environmental Protection Agency
EUA	Emergency use authorization
FDA	Food and Drug Administration
FND	Functional neurological disorder
FNIH	Foundation for the National Institutes of Health
FOIA	Freedom of Information Act
GAO	Government Accountability Office
HHS	Department of Health and Human Services
HPV	Human papillomavirus vaccine
HUD	Department of Housing and Urban Development
ICU	Intensive care unit
IHR	International Health Regulations
LNP	Lipid nanoparticle
MAID	Medical Assistance in Dying
MMR	Measles, mumps, rubella vaccine
NAS	Neonatal abstinence syndrome
NIAID	National Institute of Allergy and Infectious Diseases
NIH	National Institutes of Health
NVIC	National Vaccine Information Center
OPPR	Office of Pandemic Preparedness and Response
OTA	Office of Technology Assessment (government office from 1974–1995)
OTA	Other Transaction Agreement (DOD funding mechanism)
OTC	Over-the-counter
OWS	Operation Warp Speed
PCR	Polymerase chain reaction
PDR	Physicians' Desk Reference
PEG	Polyethylene glycol
PHE	Public health emergency
PHEIC	Public health emergency of international concern
PhRMA	Pharmaceutical Research and Manufacturers of America
POCUS	Point-of-care ultrasound
PPPR	Pandemic prevention, preparedness, and response
PREP Act	Public Readiness and Emergency Preparedness Act

QALY	Quality-adjusted life year
RCT	Randomized controlled trial
RSV	Respiratory syncytial virus
SADS	Sudden adult death syndrome
SIDS	Sudden infant death syndrome
SSNI	Serotonin and norepinephrine reuptake inhibitor
SSRI	Selective serotonin reuptake inhibitor
UN	United Nations
VAERS	Vaccine Adverse Event Reporting System
VICP	National Vaccine Injury Compensation Program
VITT	Vaccine-induced immune thrombocytopenia and thrombosis
VRBPAC	Vaccines and Related Biological Products Advisory Committee (FDA)
WHO	World Health Organization

INTRODUCTION

The turbulent events surrounding the 2020 declaration of the COVID-19 "pandemic" have attracted attention as never before to the potential for medicine to be both lethal and malevolent.[1] This is a lesson that Americans could have learned decades ago. Notably, when 40 million Americans were maneuvered into accepting a neurologically risky vaccine during and subsequent to the 1976 swine flu fiasco,[2] it prefigured the recurrent fact that epidemics and pandemics (real or engineered)[3] and the amped-up fear that accompanies them offer a convenient pretext for staging unfriendly medical assaults on credulous populations.

The spooky, "supercharged" biopharmaceutical technologies[4] ushered onto the world stage under cover of COVID—and the scale of their implementation—suggest that the latest crop of medical-pharmaceutical interventions may represent a new level of medical skulduggery. At the same time, it cannot be denied that even before 2020, medical history was replete with examples of dangerous interventions that poisoned, injured, or killed. In the view of writer Jon Rappoport—who has outlined many examples of medicine's "grotesque" track record—the "medical cartel destroying millions of lives is nothing new."[5]

At our current juncture, there is much we can learn from medical history—and from 20th- and 21st-century doctors and writers who have sounded the alarm about medicine's treacherous waters. Ivan Illich, a theologian, philosopher, sociologist, and historian who was one of the most well-known critics of medicine in the last century, popularized the term iatrogenesis (meaning "doctor-caused") in the 1970s in seminal writings that called attention to "injury done to patients by ineffective, toxic, and unsafe treatments."[6] Illich famously proclaimed that the medical establishment had become "a major threat to health," adding, "The disabling impact of professional control over medicine has reached the proportions of an epidemic."

In more recent times, political economist Dr. Toby Rogers has put forth the term iatrogenocide "the mass killing of a population by scientific and medical professionals"—as an even more apt descriptor of what medicine seems to be up to.[7] Distressingly, Rogers notes,[8] the pharmaceutical poisoning of the population—which often begins in utero[9]—has become a major engine of the U.S. economy, creating perverse and entrenched incentives to propagate medical harm. He somberly observes, "We appear rich (in terms of dollars and cents) but, because we are pursuing such a catastrophic economic model, we are actually desperately poor (in terms of health and happiness) as a nation."[10]

In April 2023, father and patients' rights advocate Scott Schara filed a landmark lawsuit against a Wisconsin hospital, alleging purposeful medical battery by doctors and nurses against his 19-year-old daughter Grace, resulting in her wrongful death.[11] Commenting on the Schara case and many others like it, Andrew Lohse of the communications firm Overton & Associates suggested in a September 2023 press release that "medical murder" is eclipsing heart disease and cancer and "becoming America's #1 cause of death"; differentiating "medical murder" from "medical malpractice," and with the Schara case in mind, Lohse characterizes it as "a degree of negligence and recklessness," which he says "can only be identified as intentional."[12]

Even so, four years into the world events launched by COVID, many people still have a hard time believing that governments and corporations "would really do something this diabolical"; Rogers' answer is, "*of course they would*. Genociders gonna genocide" [italics in original].[13] Rappoport shares a similar perspective, asserting that "doctors, public health agencies, other government leaders, and mainstream journalists are fully aware" that death is often caused by "medicine."[14] Sadly, citizens who do not understand or accept the reality of medicine's perils will find it impossible to duck the threats.

The goal of this book is to help readers contextualize current events and more clearly understand the lengthy history of "medical weapons of mass destruction,"[15] so they can protect themselves and their family and friends.

Chapter One, "From Quackery to Criminality," tells the story of Founding Father Dr. Benjamin Rush and posits a historical continuity between some of Rush's practices and beliefs and those of his modern-day medical brethren.

Chapter Two, "The Medical Cassandras," reviews the cautionary reportage of nine articulate medical skeptics—Ivan Illich, Dr. Robert Mendelsohn, Dr. Barbara Starfield, Gary Null, Dr. Peter Breggin, Jon Rappoport, Celia Farber, Sasha Latypova, and Katherine Watt. Writing in the 20th and 21st centuries, each in their own way has tried to warn the public about medicine's life-threatening underbelly. Their warnings, taken together, create a powerful portrait of widespread harm.

Chapter Three, "Twenty-First Century Vaccine Technologies," highlights the mRNA vaccine technology inaugurated with COVID. Among other troubling features, it has become apparent that this new mechanism for iatrogenesis inflicts novel forms of toxicity, not all of which are understood or readily identifiable.

Chapter Four, "Assisted Suicide and Euthanasia," discusses the global proliferation of policies and propaganda promoting assisted suicide and euthanasia, including for babies, children, individuals with autism, and those with mental illness. According to the critiques outlined in the chapter, some of the policymakers espousing a rhetoric of compassion actually may have a less benevolent agenda.

Chapter Five, "Deadly Medicine: How Do They Get Away With It?" dives deeper into some of the mechanisms that medicine and pharma use to facilitate and perpetuate medical harm.

Those who entertain the possibility that some of the harm caused by medicine may be intentional understandably wonder *why*. Chapter Six, "Why Do They Do It? Money, Prestige, and Control," considers potential answers to that question.

Finally, Chapter Seven, "Lifesaving Facts for You and Those You Love," explains why it is so important to acknowledge the reality of the medical-pharmaceutical killing machine, whether the damage results from carelessness or a plan. Only then will it be possible to avoid the threats posed by the current medical model and take steps toward building a different model that prioritizes life and genuine health.

CHAPTER ONE

FROM QUACKERY TO CRIMINALITY

The medicinal use of mercury offers a long-running example of medically induced harm. Although centuries of whistleblowers have warned that dosing patients with it constitutes reckless quackery—and the U.S. government presently places mercury at number three on its "Substance Priority List," right under arsenic and lead[1]—the heavy metal has figured prominently in the "medical armamentarium" from as far back as the sixth century BC through the present day.[2,3]

In his important book *Evidence of Harm*, author David Kirby exposed the pharmaceutical industry's controversial practice of including mercury preservatives in vaccines.[4] Pointedly using the word "criminal," Kirby wrote in the foreword to another book about mercury (*The Age of Autism* by Dan Olmsted and Mark Blaxill) that the "blind belief in a known poison" has been "misguided, immoral, and in some cases, patently criminal."[5]

The "Messianic" Benjamin Rush

In many ways, the medical practices and beliefs of U.S. Founding Father, physician, and University of Pennsylvania medical school professor Benjamin Rush may have set the stage for modern medicine's stubborn adherence to dangerous protocols—despite clinical evidence of harm— and its silver-bullet fascination with vaccines "as substitutes for right

1

living," as Eleanor McBean put it in her 1957 book *The Poisoned Needle: Suppressed Facts About Vaccination*.[6]

The reportedly "messianic" and "uncompromising" Rush's late-1700s stock-in-trade was a radical protocol involving bloodletting and purging with—what else?—mercury,[7] a practice that medical historians later dubbed "heroic medicine."[8] Rush had his own proprietary brand of laxative called "Thunderclappers," consisting of approximately 60% mercury chloride (also called calomel), which he promoted as "a purgative of explosive power."[9,10] As Rush honed his clinical methods, he passed them on to a phalanx of enthusiastic students and disciples during yellow fever epidemics in Philadelphia, where he would bleed and purge up to 100 patients a day.[11] Although use of calomel was not uncommon among doctors of that era,[12] Rush prescribed up to 10 times more than his medical peers and also recommended the removal of huge amounts of patients' blood, erroneously believing that the blood would replenish itself in a matter of a day or two. "A patient's failure to respond to this disastrous therapy," one historian wrote in 2004, "won [the patient] only another round of bleeding and purging."[13] In another modern writer's colorful description, "So much blood was spilled in the front yard that the site became malodorous and buzzed with flies."[14]

No less a figure than George Washington underwent a rapid and gruesome death after Rush protégé Dr. Elisha Dick (and two other Johnny-on-the-spot physicians) poisoned Washington with mercury and removed 40% of the beleaguered general's total blood volume—a quantity that, to this day, "continues to amaze and appall laymen and physicians alike."[15] From many historians' point of view, Washington's doctors caused his death, a death that may well have changed the course of history.[16]

Rush was enthusiastic about promoting his "heroic medicine" protocol, "proclaim[ing] the success of his cure to the public and his medical colleagues" in newspapers, advertisements, and brochures, and even "harangu[ing] people in the streets."[17] In addition, he was an early and explicit proponent of smallpox vaccination. In 1803, he joined with 30 other Philadelphia doctors in signing a public notice "expressing their confidence in vaccination and recommending it for general use."[18] Significantly, smallpox vaccination represented a turning point in the "medicalization of the general public" in both early nineteenth-century America and Europe, and a boon for the burgeoning medical profession:

> Since the late eighteenth century, doctors had intensified their efforts to win government support for their plans to bring the whole population under medical control. . . . Thus Jenner's method of cowpox vaccination presented medical practitioners with **a new chance to increase their prestige and influence** on public health affairs [bold added]. Doctors also foresaw an increase in their income through vaccination fees and hoped to establish themselves, with the help of the vaccine, among those classes of the population who had not consulted doctors before.[19]

From 1813 to 1822, the young U.S. government appointed James Smith as the nation's "federal vaccine agent," charging him with "maintaining a supply of the smallpox vaccine and distributing it nationwide"; Smith had been a student of Rush's at the University of Pennsylvania and was a fellow member of the "well-educated medical elite."[20] Although other physicians of the day argued that smallpox vaccination was both dangerous and ineffective, then—as now—defenders of the practice prevailed by using "more or less perverted statistics," with one doctor urging his "professional brethren to be slow to publish fatal cases of small-pox after vaccination" and others passing off vaccine-induced fatalities as some other disease.[21]

Reflecting on Rush's medical legacy, U.S. Army medical officer P.M. Ashburn made remarks in 1929 that highlight one of the many reasons why Rush's cautionary tale is still pertinent today. Ashburn wrote that by virtue of Rush's "social and professional prominence, his position as teacher and his facile pen," the Philadelphia physician "was more potent in propagation and long perpetuation of medical errors than any man of his day," thereby "blacken[ing] the record of medicine."[22] This observation illustrates how social prestige—coupled with "unyielding devotion to dogma"—often helps practitioners of dangerous medicine beat back their critics.[23]

In Rush's time, those critics included fellow physician Elisha Barlett, who opined about Rush's medical theories, "In the whole vast compass of medical literature, there cannot be found an equal number of pages containing a greater amount of utter nonsense and unqualified absurdities,"[24] as well as feisty British journalist and pamphleteer William Cobbett, who dared to publish tracts asserting that Rush's yellow fever treatments were both ineffective and dangerous—and "a perversion of nature's healing powers."[25] In response, Rush sued Cobbett for libel and won, in "one of the largest libel awards in American history at the time."[26]

One of Cobbett's fascinating observations—which reverberates uncannily in the COVID era—was that extreme fear (in this instance, of yellow fever) made members of the public far more willing to subject themselves to Rush's "experiments" than they otherwise might have been. Cobbett wrote:

> [Rush] seized, with uncommon alacrity and address, the occasion presented by the Yellow Fever, the fearful ravages of which were peculiarly calculated to dispose the minds of the panick-struck people to the tolerance, and even to the admiration, of experiments, which, at any other time, they would have rejected with disdain.[27]

Interestingly, after Rush's libel victory, Cobbett exacted a modicum of revenge by assembling data from municipal records (acknowledged today as "an epidemiological tour de force"), which pointed to a 56% mortality rate among Rush's yellow fever patients that contrasted starkly with the physician's own claim of a greater than 90% survival rate.[28] When word of those dismal statistics got out to the public, Rush's medical practice suffered. Undaunted, Rush went on to become Treasurer of the U.S. Mint under President John Adams. As the author of America's first psychiatric textbook, he is also revered today as "the father of American psychiatry." Rush proposed the same general treatments for madness that he favored for physical ailments, supplemented by straitjackets and other "modes of punishments" for tough cases.[29]

For his part, in 1800, a disgusted Cobbett returned to London, where he continued to hold medicine's feet to the fire, including condemning smallpox vaccination as "quackery."[30]

A "Patently Criminal" Model

Some modern medical historians are willing to go so far as to characterize medicine, in periods and places like 18th-century America, as "deplorable,"[31] and to suggest that back then, "a doctor was just as likely to kill you as save you."[32] Most, however, frame medical barbarity as a thing of the past. Shielded by high-end machines, complex drug technologies, glossy scientific publications, and lingo like "rigorous" and "evidence-based," the current medical-pharmaceutical-regulatory establishment and its hagiographers would have the public believe that "safe and effective" now rules the day.[33]

There is ample evidence to show that pledges of safety often are either disingenuous or false, and there are indications that Kirby's description of the medical model as sometimes "patently criminal" was squarely on the mark. At the level of individual medical practitioners, law firms specialized in malpractice note that if a doctor "appears to be indifferent to patients' well-being or safety," that indifference can be grounds for criminal liability.[34] A search of the word "criminal" on the website of *Medpage Today* (a conventional news service that is generally protective of medicine's reputation) brings up countless articles about doctors and other health care providers running "pill mill" operations, carrying out fraud, taking kickbacks, tampering with drugs, faking data, sexually assaulting or abusing patients, and engaging in other types of "unprofessional" and unethical conduct. The site's "Investigative Roundups" feature stories (often formulated as questions to soften the impact) with titles like "Columbia protected predator doc?", "Psychiatrist held patients against their will?", "$15K surgery shakedown?" or "Doc pushed unneeded surgery?"[35] Other *Medpage Today* headlines flamboyantly bandy about words like "deadly," "loophole," "games," "tactics," "unethical," and "secretive."

Sometimes, individuals who defend the medical status quo blame whichever reports of misbehavior manage to surface (many do not) on "a few bad apples."[36] Others, such as Harvard scientist and patient safety advocate Lucian Leape, do the reverse, shifting the blame from "bad people" to nebulous "bad systems;"[37] Leape suggests that a cycle of disrespect is "learned, tolerated, and reinforced in the hierarchical hospital culture."[38] The fact is, however, that medical harms flow from both individuals and institutions. Most health care providers operate in broader organizational and corporate contexts—and it is policymakers and decision-makers at those levels who often give medical-pharmaceutical corruption and criminality a green light. This is illustrated by the phenomenon (for which there is even an academic field of study) called "clinicide," defined as serial medical killers responsible for "the unnatural death of multiple patients in the course of treatment;"[39] not infrequently, the killers' host institutions countenance or "enable" this clinicide by choosing to ignore red flags.[40]

As an extension of the "bad apples" argument, some upholders of the status quo point to the fines that the U.S. Department of Justice (DOJ) routinely levies on hospitals and pharmaceutical corporations, suggesting that these are an adequate mechanism to catch and punish players engaged in malfeasance.[41,42] However, given that medical-pharmaceutical culprits not infrequently are criminal recidivists and that the fines generally amount to "little more than a slap on the wrist,"[43,44] it is fair to ask "whether such a monetary punitive system really does much to prevent bad behavior."[45]

Moreover, DOJ rarely prosecutes or holds corporate leaders accountable, despite having a "powerful legal tool" at its disposal to go after the executives at the helm of medical misconduct; it has done so only 13 times since the year 2000.[46] Instead, many signs point to a wink-and-a-nod sub rosa understanding between the various parties, with the penalties doing nothing to prevent future harms but instead furnishing a generous flow of kickbacks that prosecutors and regulators can funnel into various sectors of the federal budget (see **Illegal But Profitable**).[47] In fact, under the False Claims Act, the U.S. Department of Health and Human Services (HHS) gets a 20 to 1 return on every dollar it "invest[s] in prosecutions and investigations."[48]

Illegal but Profitable

In 2018, the nonprofit consumer advocacy organization Public Citizen published a report summarizing 27 years of pharmaceutical industry criminal and civil penalties. The report concluded:

> To our knowledge, a parent company has never been excluded
> from participation in Medicare and Medicaid for illegal

activities, which endanger the public health and deplete tax-payer-funded programs. Criminal prosecutions of executives leading companies engaged in these illegal activities have been extremely rare. Much larger penalties and successful prosecutions of company executives that oversee systemic fraud, including jail sentences if appropriate, are necessary to deter future unlawful behavior. Otherwise, these illegal but profitable activities will continue to be part of companies' business model.[49]

Iatrogenocide Takes Center Stage

Even before COVID, available data indicated that 20th- and 21st-century Western medicine had failed to improve health in any meaningful way, instead trading off the industrial-age diseases of yore for modern chronic disease epidemics, many or most with iatrogenic causes or contributors. Unfortunately, recent events suggest that medicine—forging an unhealthy partnership with government—may now be more dangerous than it has ever been.

Until 2020, the Americans who were most concerned about medical risks and medical criminality belonged to groups already adversely affected, such as those injured by vaccines or opioids. However, with the advent of life-threatening COVID "countermeasures" and lethal protocols in U.S. hospitals and in other countries such as the UK,[50,51,52] medical-pharmaceutical gangsterism—seemingly occurring with government cognizance—has begun attracting more widespread notice. When governments began parlaying the dubious health "emergency" into an excuse to authorize and mandate the COVID vaccines and boosters[53]—and proceeded full tilt even when unprecedented injuries and deaths immediately began piling up[54]—some segments of the public saw the contours of an officially sanctioned medical crime.

As Holocaust survivor and human rights activist Vera Sharav communicated in her docuseries *Never Again Is Now Global*,[55] medical coercion and the suspension of constitutional freedoms have never led anywhere good. Unfortunately, history shows that governments intent on "state repression, brutality and genocide" can usually count on the readiness of some doctors to serve as accomplices, even if their complicity has the potential to turn them into "mass murderers on an exponential scale."[56]

CHAPTER TWO

THE MEDICAL
CASSANDRAS

In Greek mythology's Trojan War saga, Cassandra is the daughter of Priam, king of Troy. As recounted by Greek tragedian Aeschylus, Apollo gifts Cassandra with the power of prophecy but, after she rejects the god's romantic overtures, Apollo ordains that "her prophecies should never be believed."[1] As a result, no one pays attention to her warnings about the fall of Troy, with disastrous consequences.

Chapter One's discussion of Dr. Benjamin Rush and his journalistic nemesis William Cobbett, who unhesitatingly denounced the trendy practices of bloodletting and smallpox vaccination, shows that medicine has always had its share of "Cassandras." In fact, though a "heroic account" of medicine still predominates—propelled by "mesmerizing ideals such as evolution and progress"—literature and fiction often have shone a light on a parallel history of "uncertainty, incredulity, and contempt . . . toward medicine in Western culture."[2]

Medical historian Andrea Carlino traces this "antimedical literary tradition" back to at least Greco-Roman times. He observes that it has persisted "almost without interruption across the centuries until today," with medieval and early modern writers like Italy's Petrarch (14th century) and France's Montaigne (16th century) and Molière (17th century) giving colorful "literary respectability to some of the most popular beliefs about medical deceits, defects and physicians' deplorable habits."[3] In Petrarch's cynical view (summarized by Carlino), doctors:

- Deceive patients "with false promises and illusory expectations"
- Take risks "with their patients' suffering bodies"
- Commit "errors and abuses [that] are not prosecuted"
- Are "always granted immunity"

In this chapter, we will see what some of the most eloquent medical skeptics of modern times have had to say—in somber nonfiction writings—about the medical cartel and its kissing cousin, the pharmaceutical industry, now metastasized into the *bio*pharmaceutical industry. (This may be a suitable place to remind readers that pharmaceutical companies provide "outsized funding of medical schools, medical textbooks, and medical associations," and spend inordinate sums on "the legacy and online digital media, as well as U.S. lawmakers at the state and federal level.")[4] The data and observations assembled by these decades of indignant and high-integrity faultfinders show that not only do "deceits," "defects," "deplorable habits," and "false promises" still abound, they seem to be core traits for much of Western medicine. Although these modern writers often have achieved a significant readership, the general public's reluctance to hear their message has often had tragic results not so very different from the Trojans'.

Ivan Illich: Iatrogenesis
"The medical establishment has become a major threat to health."
—Ivan Illich

Since the 1970s when Illich (1926–2002) popularized the term "iatrogenesis," overwhelming evidence of harm has forced modern medicine to grudgingly acknowledge iatrogenesis as a major problem and even publish books and studies about it. In fact, iatrogenicity has become its own clinical discipline, with one 2018 book devoting 29 chapters to iatrogenesis exclusively in cardiovascular medicine![5]

Accounts in the popular press and the scientific literature define iatrogenesis as encompassing an alarmingly wide range of bad outcomes. As one example, a study published in *JAMA Internal Medicine* in January 2024 reported that diagnostic errors were both "common" and "harmful," affecting nearly one in four hospitalized adults who ended up in intensive care or died.[6] The researchers acknowledged that the incidence of errors and harms was "higher than expected."[7] Table 1 summarizes some of the iatrogenic problems reported in the medical literature.

Table 1. What Does Iatrogenesis Look Like?	
Iatrogenic Problem	**Examples/Additional Comments**
Diagnostic errors	Misdiagnosis estimated to cause death or permanent disability in 795,000 Americans annually (*BMJ Quality & Safety*, July 2023)[8]
Medication "side effects"	Both short-term and longer-term effects
Medication "errors"	For example, wrong dose or wrong medication given
Anesthesia complications	For example, nerve damage, paralysis, blood clots, death
Other surgery complications	For example, hemorrhage, infection, thrombosis, pulmonary
"Traumatic stress"[9]	Notably in pediatrics
"E-iatrogenesis"[10]	Harms related to health information technology
"Organizational iatrogenesis"[11]	Problems flowing from decision errors
"Cascade iatrogenesis"[12]	The "serial development of multiple medical complications . . . set in motion by a seemingly innocuous first event"
Various and sundry	"[S]lips of the scalpel, lapses like mixing up lab results, faulty decision-making, inadequate training, evasion of known safety practices, miscommunication, equipment failures, and many more"[13]

When iatrogenesis makes it into the news—which it does infrequently but dramatically—reporters try to soften the blow by promoting the notion that iatrogenic outcomes are generally unintentional,[14] that is, the result of "medical oversights or mistakes."[15] Even with this relatively benign framing of the problem, however, reporters admit that the "ease with which medical errors can occur is striking"[16] (see **Iatrogenesis: "Widespread, Frequent, Massive, and Continuous"**).

Iatrogenesis: "Widespread, Frequent, Massive, and Continuous"

While characterizing iatrogenic harms as the consequence of "medical error or accident," a 2021 article in *STAT* was surprisingly frank about both the magnitude of the problem and the efforts taken to hide it:

> "This summer, surgeons . . . transplanted a donor kidney into the wrong patient. . . . The most surprising thing about the story is not that a serious medical error occurred, but that it found its way into the news. . . . [Iatrogenic harm is] so widespread, so frequent, so massive, and so continuous that it rarely makes headlines. And unlike a plane crash or a building collapse, **the vast majority of iatrogenic deaths can be kept under wraps**—and they are" [bold added].[17]

STAT also noted that "existing incentives push the wrong way," stating, "Because iatrogenic harm requires additional medical care, errors bring more revenue into the organization." Conversely, any organization that decides to invest in "system redesigns" to *prevent* harms is likely to "be rewarded by seeing its income fall"!

Over the course of his colorful life, Illich studied natural science, history, and art history; had a lengthy stint as a renegade Catholic priest; founded a think tank; and was an early advocate of homeschooling. However, his role as medical gadfly may be his most enduring legacy—in part because he was willing to cut straight to the chase and was not inclined to accept medicine's feeble apologies for iatrogenic outcomes.

Illich's critique of medicalized healthcare extended well beyond the clinical sphere to encompass social and cultural dimensions.[18] Even so, his scathing remarks about "clinical iatrogenesis"—defined as "all clinical conditions for which remedies, physicians, or hospitals are the . . . 'sickening' agents"—were bad enough (see **Running the Hospital Gauntlet: Then and Now**).

Running the Hospital Gauntlet: Then and Now

In the 1970s, according to Illich, mining and high-rise construction were the only two industries to surpass hospitals in terms of the frequency of reported "accidents." Citing government data, Illich summarized the dangers of hospitalization at the time as follows:

- Seven percent of **all patients** (about one in 14) suffered "compensable injuries" while hospitalized.
- One in five patients admitted to **university research hospitals** ended up with an "iatrogenic disease," which, moreover, proved fatal for one in 30; Illich noted the particular hazards of university hospitals.
- One in 50 hospitalized **children** experienced iatrogenic outcomes that required further treatment.
- About half of iatrogenic injuries could be accounted for by **drug complications**; other frequent sources of harm included unnecessary diagnostic procedures and actions taken by doctors to avoid accusations of malpractice.

Fast-forwarding to 2012, a nursing document reported that the incidence of hospital admissions due to adverse drug events had not, as of that time, "decreased in the past 20 years and the absolute numbers may have increased."[19,20] The report cited ongoing and concerning rates of iatrogenic harm among the hospitalized, including:

- Iatrogenic complications in anywhere from 2% to 36% of **hospital patients**

- Especially high rates of iatrogenic complications in **intensive care unit (ICU) patients**, with 6.5% ending up permanently disabled and from 4% to 14% ending up dead
- Disproportionate risks in **seniors** (age 65 and up), with "twice as many diagnostic complications, two and one half times as many medication reactions, [and] four times as many therapeutic mishaps"

Still more recently, researchers analyzed medical records data for patients hospitalized in 2018 in a dozen Massachusetts hospitals.[21] The study defined iatrogenic events as adverse drug events, surgical or other procedural events, patient-care events, health care-associated infections, and, in seven cases, death. The researchers found that nearly one in four patients (24%) had experienced "at least one adverse event," and a third of those "had a severity level of serious . . . or higher."

In his book *Limits to Medicine* (originally titled *Medical Nemesis*), Illich boldly asserted that doctors' effectiveness was "an illusion."[22] Although some might find his claims overstated, he believed doctors' contributions during epidemics to be especially fabled. He wrote:

> The study of the evolution of disease patterns provides evidence that during the last century doctors have affected epidemics no more profoundly than did priests during earlier times. Epidemics came and went, imprecated by both but touched by neither. They are not modified any more decisively by the rituals performed in medical clinics than by those customary at religious shrines.

Warning the public not to be "passive," Illich made a number of points that have not lost their relevance over the ensuing decades:

- Considering broader influences on health, Illich observed that "the environment is the **primary determinant** of the state of general health of any population."

- According to Illich, most of the decline in mortality before 1965 was due to **improved nutrition** and had nothing to do with "the professional practice of physicians"; he also highlighted undernutrition and poisons and mutagens in food as "a new kind of malnutrition" to which most doctors remain oblivious. (With the wave of 21st-century synthetic foods that is now upon us, the latter remarks seem particularly forward-looking.)[23]

- If medical care were merely "futile" but "otherwise harmless," that would be bad enough, said Illich, but he deemed the situation to be far worse, with serious "pain, dysfunction, disability, and anguish resulting from technical medical intervention" constituting "one of the most **rapidly spreading epidemics** of our time."

- Illich was skeptical that "new devices, approaches, and organizational arrangements" could correct the hazardous health care system, suggesting that such innovations tend to become "pathogens" in their own right, creating "a self-reinforcing **iatrogenic loop**."

- Illich also chillingly prefigured the rise of technocratic medicine,[24] calling attention to the shift from the doctor as caring artisan ("exercising a skill on personally known individuals") to the doctor as impersonal **technician** who applies "scientific rules to classes of patients." The "inevitable" result of an "engineering model" of health, Illich suggested, would be "managed maintenance of life on high levels of sublethal illness"—a status quo that could quite easily describe the large proportion of poisoned and medication-reliant Americans who limp along with debilitating chronic illness.[25]

- Perhaps even more frighteningly, Illich insisted that incompetence, greed, and laziness were responsible for relatively few of the harms caused by modern doctors. Instead, he alleged, "most of the damage . . . occurs in the **ordinary practice** of well-trained men and women who have learned to bow to prevailing professional judgment and procedure, even though they know (or could and should know) what damage they do."

During Illich's lifetime, many sought to dismiss his warnings about modern medicine as "extreme," but insiders such as Richard Smith—long-time editor-in-chief of the influential journal *The BMJ* (formerly the *British Medical Journal*)—later deemed Illich's writings on medicine and health to have had "something of a prophetic quality."[26] (Smith himself has been a persistent thorn in the side of the health care industry, suggesting in a 2021 *BMJ* blog that we should assume all health research "to be untrustworthy until there is some evidence to the contrary.")[27] Illich presciently condemned the ever more widespread use of powerful drugs and the corresponding explosion in side effects—including both obvious reactions and more "subtle kinds of poisoning." In this regard, he underscored the potential for addiction and pointed not only to risks such as taking wrong, old, contaminated, or counterfeit drugs, but also to the problem of taking multiple drugs "in dangerous combinations," a phenomenon now referred to as "polypharmacy."

Recent studies show that polypharmacy increases the risks of hospitalization and all-cause death.[28] In the UK, life expectancy began stalling noticeably around 2010, coinciding with a period of skyrocketing prescription drug use—with the life expectancy slowdown "being one of the most significant in the world's leading economies."[29] U.S. studies confirm out-of-control polypharmacy as a "21st century iatrogenic epidemic."[30] Between 1988 and 2010, for example—driven by increased use of heart drugs and antidepressants—the proportion of American seniors (> 65 years) taking five or more medications tripled (going from 12.8% to 39%).[31] In 2019, KFF (formerly known as the Kaiser Family Foundation) reported that over half (54%) of seniors and one-third (32%) of middle-aged adults (50–64 years old) regularly take at least four prescription drugs, as do 7% to 13% of adults under age 50[32] (see **Chapter Six** for more on this topic). The 2018 National Ambulatory Medical Care Survey found that two in five patients leave an office visit with three or more medications or prescriptions in hand; 3% of respondents even answered "yes" to the top response category (15 or more medications)![33]

Drawing attention to the personal dimensions of the polypharmacy problem, a grieving mother whose 21-year-old son died of drug interactions wrote in 2023 that "the layering of multiple medications on top of one another" often occurs "without regard to what other doctors have already prescribed or the potential interactions between the drugs."[34] The mother's clear conclusion: Overprescribing is a "scourge."

One of Illich's most visionary statements should give pause in light of the war that government officials have waged on health—and health freedom—in the COVID era:

> [T]he medical monopoly over health care has **expanded without checks** and has encroached on our liberty with regard to our own bodies [bold added]. Society has transferred to physicians the exclusive right to determine what constitutes sickness, who is or might become sick, and what shall be done to such people.

Illich also foresaw the massive gaslighting that now seems to be par for the course when medicine—including vaccination—injures or kills someone.[35,36] He warned that we should not be surprised when the medical-pharmaceutical complex blames the victim for the damage it causes.

Dr. Robert Mendelsohn: Hazardous Medicine

"Murder is a clear and present danger."
—Dr. Robert Mendelsohn (with reference to the dangers
of hospitals)

Ivan Illich formulated his critiques of modern medicine from an outsider's vantage point, but pediatrician and medical school faculty member Dr. Robert Mendelsohn (1926–1988) unabashedly practiced "medical heresy" from the inside. At a time when suppression of dissenting voices had not yet escalated to the level of the rampant censorship that operates today, Mendelsohn even published a syndicated newspaper column for 12 years running (titled "The People's Doctor"), as well as appearing on television shows with celebrity hosts like Phil Donahue and Joan Rivers.[37] Mendelsohn's half dozen books—with titles like *Confessions of a Medical Heretic* (1979)[38] and *How to Raise a Healthy Child . . . in Spite of Your Doctor* (1984)[39]—sold hundreds of thousands of copies, offering sometimes humorous and often shocking descriptions of "bad medicine" as well as practical advice on how to steer clear of it.

Mendelsohn's insights were the result of observations painfully accrued from his own practice. In the early part of his career when he practiced medicine conventionally, his patients would later return with iatrogenic diseases; ultimately, his conclusion was that his interventions were doing more harm than good.[40] Like many of modern medicine's detractors,[41] Mendelsohn was not particularly sanguine about the prospects for better medicine. In the introduction to *How to Raise a Healthy Child*, for example, he wrote that doctors emerge from medical school "with their heads so stuffed with institutionalized foolishness that there is no room left for common sense." (Thirty years later, medical interns could still be seen posting comments on student forums along the lines of, "Don't go into medicine if critical thinking is important to you.")[42]

Spelling out some of the scarier history and implications in *Medical Heretic*, Mendelsohn asserted that "doctors have throughout the ages embraced the wrong ideas." Citing a variety of dangerous medical practices, past and present, his conclusion was that one "could make a case that medicine has always been hazardous to the majority of patients." He also observed, "Doctors almost always get more reward and recognition for *intervening* than for non intervening"; he wryly noted that "first, do no harm" had morphed into "First Do *Something*" [italics in original].

Like Illich, Mendelsohn was leery about doctors' power to "define or manipulate the limits of health and disease any way [they] choos[e]." In 2002, *The BMJ's* Richard Smith and Australian journalist Ray Moynihan agreed, writing in an article titled "Too much medicine?" that "the concept of what is and what is not a disease is extremely slippery."[43] Commenting on the profession's growing propensity to medicalize normal life processes such as "birth, ageing, sexuality, unhappiness, and death," they alerted readers that a "boost to status, influence, and income . . . comes when new territory is defined as medical," also pointing out that "[a]dvances in genetics open up the possibility of defining almost all of us as sick."

Mendelsohn cited the inconstancy of hypertension definitions as one blatant example of doctors' control over definitions of health and illness.

In 2017, as if illustrating this very point, medical trade groups revised their definition of high blood pressure, arbitrarily lowering the threshold from 140/90 to 130/80.[44] In one fell swoop, they moved 14% more Americans from the "healthy" column to the "hypertensive" column. Coincidentally or not, a study for the 2017–2022 period later showed a 25% increase in prescriptions for statin drugs,[45] and especially the riskier "high-intensity" statins[46] promoted for both cholesterol-lowering and blood-pressure-lowering effects.[47] As one of the most prescribed medicines in history, powerful statins happen to be the very definition of an iatrogenic intervention; since their inception, not only have they not delivered on their promise of improving heart health, but they have caused a wide range of horrific side effects that "mimic the effects of aging"[48]—including "brain fog, joint pain, kidney injury, impaired liver function, heart failure, and even dementia."[49]

A central—and bold—premise of the *Medical Heretic* chapter titled "Miraculous Mayhem" is that there is no such thing as a completely safe drug (see **Dr. Mendelsohn's List of "Pharmaceutical Backfires"**). In fact, citing official statistics, Mendelsohn observed that doctor-prescribed drugs were bigger killers than illegal street drugs. A 2016 story in *U.S. News & World Report* confirmed that almost five times as many people "die each year as a result of taking medications as prescribed" as perish from illegal drugs or overdoses, making prescription drugs America's fourth leading cause of death, according to official rankings.[50] Another news story has suggested that "the traditional distinction between illegal 'street drugs' and legal 'therapeutic prescription drugs' [has] become so blurred as to be almost nonexistent."[51]

In a statement likely to speak to any person who has ever been injured by a drug or vaccine, Mendelsohn cautioned:

> Like a game of Russian Roulette, for the person who gets the loaded chamber, the risk is 100 percent. But *unlike* that game, for the person taking a drug *no chamber is entirely empty* [italics in original]. Every drug stresses and hurts [the] body in some way.

Dr. Mendelsohn's List of "Pharmaceutical Backfires"

Briefly outlining "pharmaceutical backfires" from the late 1890s through the early 1960s, Dr. Mendelsohn underscored that such incidents were not exceptions: "drug disasters like these are going on every day."

1890: Tuberculin—Dr. Robert Koch injected tuberculosis (TB) patients with a TB "remedy" (tuberculin) of "greatly exaggerated" healing power.[52] In an 1891 trial, the recovery rate was no better in treated than untreated patients, and 1% to 4% of treated subjects died.[53] Adding interesting historical context to Koch's mixed motivations, renowned Max Planck Institute immunologist Stefan H.E. Kaufmann wrote in a 2001 paper that Koch was under professional pressure to present "spectacular" results at an International Congress of Medicine but also had personal reasons to spin tuberculin as a breakthrough: "He had fallen in love with the 17-y-old beauty Hedwig Freiberg and wanted to divorce his wife Emma, who had demanded a hefty financial settlement in return."[54] According to Kaufmann, when Koch was 50 and Hedwig was 21, the "substantial payments" that Koch received for tuberculin finally allowed him to divorce his first wife and remarry.

1928: Thorotrast—Doctors began using "an apparently innocent contrast medium" to take x-rays of organs such as the liver and spleen; within a couple of decades, it became evident that radioactive Thorotrast, even in small doses, "irradiated such organs for a lifelong period" and caused cancer.[55]

1937: Sulfanilamide—Harold Watkins, chief chemist at the S.E. Massengill Co., developed a liquid version of the popular antibiotic sulfanilamide, discovering that while it would not dissolve in water, it readily dissolved in diethylene glycol (a deadly poison better known as antifreeze).[56] The chemist designed the raspberry-flavored "Elixir Sulfanilamide" to appeal to pediatric patients. Roughly six gallons of the "elixir" killed 107 people, mostly children,[57] who suffered symptoms characteristic of kidney failure for as long as three weeks, including "intense and unrelenting pain."[58] University of Chicago pharmacology graduate student Frances Kelsey was part of the team that identified diethylene glycol as the culprit (see "1959: Thalidomide" below).[59] The next year, the 1938 Food, Drug, and Cosmetic Act increased the authority of the Food and Drug Administration (FDA) to regulate drugs. Watkins ended up committing suicide, but the Massengill company denied any wrongdoing.

1955: Salk Polio Vaccine—When the U.S. rolled out the fast-tracked "inactivated" polio vaccine developed by Jonas Salk, at least 220,000 children received a famously defective batch that killed 10 and left tens of thousands with permanent muscle weakness or paralysis.[60] The number of "polio" cases also rose immediately and dramatically, with some health departments choosing to ban the vaccine.[61]

1959: Thalidomide—Dr. Frances Kelsey, now a newly minted FDA medical reviewer (see "1937: Sulfanilamide" above), courageously resisted pressure to approve the U.S. release of the German-developed tranquilizer thalidomide, but, under false cover of "clinical trials," U.S. pharmaceutical company Richardson-Merrell (which eventually became Sanofi-Aventis) distributed over 2.5 million doses to 20,000 pregnant women anyway.[62] FDA's senior administrators eventually admitted to birth defects in 17 babies but suppressed evidence assembled by thalidomide survivors that the "unauthorized marketing program" caused far more widespread damage.[63] (For more information about the suppression, see the 2023 book *Wonder Drug: The Secret History of Thalidomide in America and Its Hidden Victims* by Jennifer Vanderbes.)[64]

1962: Triparanol—Fresh from the thalidomide disaster, the Merrell company marketed triparanol (trade name MER/29), the nation's first synthetic cholesterol-lowering drug.[65] Many recipients reported experiencing unsettling and toxic adverse effects, including baldness, loss of body hair, development of a "fish-scale" skin texture, impotence, and unusual blindness-causing cataracts. It later came to light that Merrell had "improperly withheld information already in its files that triparanol had caused cataracts in animals."[66] Merrell paid $50 million to settle civil lawsuits.[67]

These and many other such incidents prompted Mendelsohn to observe, "One of the unwritten rules in Modern Medicine is always to write a prescription for a new drug quickly, before all its side effects have come to the surface." He also commented that for many drugs, "the side effects are the same as the indications" for taking the drug to begin with!

As a pediatrician, Mendelsohn found the drugging of children particularly offensive, dissenting from the runaway prescribing of behavior-modifying concoctions like the central nervous system stimulant methylphenidate (e.g., Ritalin and Concerta) and observing that there was no defining biological test for the nebulous condition called "hyperactivity."[68] In *Medical Heretic*, Mendelsohn frankly warned that many of the drugs given to children were transforming them "into 'brave new world' type zombies."

Because he passed away in 1988, Mendelsohn was not around to witness (but likely would have been horrified by) the tripling of stimulant prescriptions in the U.S. in the 1990s, followed by a further doubling between 2006 and 2016, primarily in children given a diagnosis of "attention-deficit/hyperactivity disorder" (ADHD).[69,70] The "bible" of psychiatry, the *Diagnostic and Statistical Manual of Mental Disorders* (DSM), formalized the subjective ADHD diagnosis in 1980, but again—as with its predecessor "hyperactivity" and all other DSM diagnoses—with no biologically verifiable basis.[71] By 2020, according to the mental health industry watchdog group Citizens Commission on Human Rights International (CCHR), over 3.1 million young people (from birth through age 17) in the U.S. were taking ADHD drugs,[72] including even stronger and longer-acting amphetamine-based stimulants like Adderall and Vyvanse that companies began heavily marketing in the early 2000s. Both methylphenidate and the amphetamine drugs have pharmacologic effects similar to cocaine.[73]

Mendelsohn also surely would have been distressed by the penetration of ADHD drugs into the vulnerable two- to five-year-old market, despite the fact that few such drugs are FDA-approved for children under age six. Between 1991 and 1995,[74] preschoolers' use of stimulant medications increased threefold; by 2011–2012, researchers found that 44% of ADHD-diagnosed children in the two- to five-year age group were taking stimulants or other powerful medications.[75] Cautioning parents, in *How to Raise a Healthy Child*, that drugging kids at any age "obviate[s] the need and the incentive to discover what is really troubling [a] child," Mendelsohn instead recommended common-sense measures such as searching for environmental factors, eliminating food additives from the diet, or checking out other teachers or schools. In his view, these practical pathways were far preferable to letting doctors intervene with a child's delicate biochemistry (see **Ritalin Roulette**).

Ritalin Roulette

The Swiss company Ciba (later Ciba-Geigy, and, since 1996, a division of Novartis) patented methylphenidate as Ritalin in 1950. Currently, ten other companies in addition to Novartis (Ironshore, Janssen, Lannett, Neos Therapeutics, NextWave, Noven, Purdue Pharma, Rhodes, SpecGx, and Vertical) manufacture branded methylphenidate products.[76]

After its FDA approval in 1955, Ritalin's initial use in the U.S. was in psychiatric facilities, where doctors prescribed it, according to health history professor Matthew Smith, for "chronically depressed, schizophrenic and psychotic patients, the 'mentally retarded' and patients recovering from lobotomies."[77] From there, says Smith, its marketing broadened to "troublesome, miserable old people," the "tired housewife," and postpartum mothers, as well as "exhausted businessmen, narcoleptics, convalescents and 'oversedated' patients."

However, it was in 1961, when FDA approved Ritalin for "hyperactive" children, that the drug achieved bestseller status. In the first major behavioral trial of Ritalin—in hospitalized children—in the early 1960s, 70% of the pediatric participants experienced serious side effects; both principal investigators later averred that "ADHD was overdiagnosed and Ritalin was overprescribed."[78]

In *How to Raise a Healthy Child*, Mendelsohn noted that while drug manufacturers, by law, must share information about a drug's side effects with doctors via the *Physician's Desk Reference* (PDR), prescribing doctors do not necessarily share information about "potentially damaging or fatal effects" with their patients. Mendelsohn, therefore, took it upon himself to disseminate the PDR information about Ritalin, including Ciba-Geigy's acknowledgment that the company did not understand how Ritalin worked or know anything about its long-term safety. He noted that the wide range of potential Ritalin side effects admitted to by the company in the PDR included anorexia, blood disorders, depression, insomnia, nervousness, palpitations (and other cardiac and blood pressure irregularities), skin conditions, stunted growth, and toxic psychosis, among others.

Unfortunately, Mendelsohn's warnings about ADHD drugs went largely unheeded, and the drugs remain a source of widespread harm. In a *JAMA* study published in April 2023, for example, anywhere from zero to 25% or more of middle and high school students admitted to "nonmedical" use (that is, abuse) of ADHD stimulants in the past year.[79] Misuse of ADHD drugs was more prevalent than for drugs like opioids or benzodiazepines and, unsurprisingly, was especially likely to occur in schools that had a "large population of students with stimulant medication prescriptions."[80] A 2014 study likewise estimated the "misuse and diversion" of ADHD medications to be common in both high school (5%–10%) and college (5%–35%) students.[81]

Federal agencies have known about ADHD drugs' potential for "abuse and dependence" for decades, but it was not until May 2023 that the FDA belatedly required that an update to the black box warnings for all stimulant drugs communicate the potential for "abuse, misuse, and addiction," *even when taken as prescribed*.[82] Between 2017 and 2022, across all age groups, overdose deaths from stimulants (both prescription and illicit) tripled.[83]

Comparable to Illich, Mendelsohn's take on hospitals was that they were "one of the most dangerous places on earth." In comments that seem particularly foresighted since COVID, he argued that in the hospital setting, "*murder* is . . . a clear and present danger" and "hospitals are *already* getting away with murder" [italics in original]. In *Medical Heretic*, he wrote, "A hospital is like a war. You should try your best to stay out of it. And if you get into it you should take along as many allies as possible and get out as soon as you can." Again, well before the appalling isolation of hospital patients from their loved ones during COVID, Mendelsohn described how hospitals force the surrender of all the "personal factors"—such as family, friends, and home-cooked meals—that ordinarily promote healing and help a person remain upbeat and hopeful. Mendelsohn also opined that nine in ten surgeries were "a waste of time, energy, money, and life," and he lamented that many needless surgeries were being done on children.

At the time of Mendelsohn's writings, his assessment was that patients were receiving 12 different drugs during an average hospital stay. However, as the good doctor elaborated, hospitals also pose chemical dangers to undrugged patients:

> [E]ven if you're not drugged to death or disability, there are other chemicals floating around that can make your [hospital] stay less than healthy. . . . Poisonous solvents used in laboratories and cleaning facilities, flammable chemicals, and radioactive wastes all threaten you with contamination.

Ultimately, Mendelsohn advised would-be patients to exercise responsibility for their own health, suggesting that they simply avoid doctors as much as possible, and especially if not sick. For those determined to head to the doctor's office anyway, he recommended doing proper due diligence beforehand to investigate "possibilities, alternatives, and consequences." He also had other down-to-earth suggestions:

- Ignore badgering to get annual **checkups**. According to Mendelsohn, "not a shred of evidence has emerged to show that those who faithfully submit live any longer or are any healthier than those who avoid doctors." Dr. Nortin M. Hadler, an emeritus professor at the University of North Carolina–Chapel Hill, has sounded similar themes in books like *The Last Well Person: How to Stay Well Despite the Health-Care System* (2004)[84] and *Worried Sick: A Prescription for Health in an Overtreated America* (2008).[85]
- Understand that **lab tests** often "do more harm than good," in part due to their notorious inaccuracy and also due to medical practitioners' inability to interpret accurate results properly.
- Become aware of the inherent dangers of **polypharmacy**.
- When visiting a doctor, **ask questions**. As Mendelsohn put it, "[Y]ou always should be on your guard. Not passively, either. Your job is to make trouble. . . . Subvert the system that will steal your dignity and maybe your life if you let it."

Dr. Barbara Starfield: Implicating the Medical System
"The American public appears to have been hoodwinked."
—Dr. Barbara Starfield

Dr. Barbara Starfield (1932–2011) was an eminent American pediatrician who spent nearly all of her 50-year career at Johns Hopkins University, including a stint as head of the Division of Health Policy. In her obituary in *The Lancet*, a colleague described her as having an intense "fire in her belly to right wrongs and improve the health of people around the world."[86] However, while making much of Starfield's passion for primary care, the obituary said nary a word about the publication that may be one of her most impactful legacies—a pithy three-page commentary she published in *JAMA* in 2000, titled "Is US Health Really the Best in the World?"[87]

In the commentary, Starfield cogently highlighted factors responsible for the United States' abysmal health rankings compared to other highly industrialized (and also less industrialized) nations[88]—a poor standing that persists to this day and is "robustly" consistent across a wide variety of health measures. Pointedly, she called attention to analysts' failure to implicate the medical system and the iatrogenic damage it causes. Using available studies (mostly hospital-based), Starfield judged that medicine was causing 225,000 deaths annually, making it, in her estimation, America's third leading cause of death after heart disease and cancer. This appraisal was 25% higher than that produced six years earlier by Harvard health policy expert Lucian Leape,[89] who speculated that around 180,000 people perished annually from iatrogenic causes.[90]

In all likelihood, Starfield was overly generous in giving iatrogenesis a third-place ranking (versus first or second place), for several reasons. First, medical interventions for heart disease and cancer also cause many serious iatrogenic outcomes, not always recognized as such.[91] French authors reported in 2019 that 40% to 70% of pediatric cancer patients go on to develop another health problem—such as a second cancer—"related to the disease or the treatment," particularly if "treatment-related risk factors" interact with other toxic exposures.[92]

Second, the medical profession's notoriously bad nutritional and other preventive advice surely bears some responsibility for the emergence or exacerbation of heart disease and cancer in individuals who slavishly follow such advice.[93] Where heart disease is concerned, Harvard dissenter Dr. George V. Mann was one of many to characterize the doctor-promoted theory that saturated fat and cholesterol are to blame for heart problems as "propaganda," "dogma," and a "scam,"[94] arguing that the theory was so unscientific it would have made Galileo "flinch."[95] British writer James Delingpole likewise has called the flawed diet-heart theory a "big fat lie," one that dangerously heightens rather than lessens heart risks; he scathingly criticizes the "vast but entirely pointless, corrupt and worthless global industry built over decades on a foundation of junk science, public hysteria and woefully misguided government regulation."[96] Other well-researched critics of conventional heart disease explanations include Danish physician Uffe Ravnskov (author of *The Cholesterol Myths: Exposing the Fallacy that Saturated Fat and Cholesterol Cause Heart Disease*),[97] British author and researcher Zoë Harcombe,[98] and the Weston A. Price Foundation.[99]

Vaccines represent a third iatrogenic pathway for both heart disease and cancer. Package inserts for the vaccines on the childhood schedule reveal an association with a wide range of cardiac adverse reactions.[100] The insert for the Gardasil human papillomavirus (HPV) jab links it to breast, nasopharyngeal, and pancreatic cancers, and an April 2023 lawsuit alleges, ironically, that Gardasil causes the very cancer it is theorized to prevent— cervical cancer—citing "rapidly climbing cervical cancer rates among young women in countries where Gardasil has seen a high uptake."[101] The COVID shots—the most far-reaching medical intervention in living memory—are also plausibly causing both heart disease and cancer.[102,103]

Although studies of iatrogenesis outside of hospital walls were and are scarce, Starfield's *JAMA* commentary directed readers' attention to a 1997 study by pharmacy professors who examined drug-related morbidity and mortality in ambulatory care settings; incorporating their analysis of fatal adverse events nearly doubled Starfield's mortality estimate, adding another 199,000 fatalities to the annual iatrogenic death toll.[104] Possibly even more sobering was the pharmacy experts' calculation that every year, complications from drugs are the catalyst for tens of millions of extra physician visits, prescriptions, emergency department visits, hospital admissions, and admissions to long-term care facilities.[105]

Writer Jon Rappoport has repeatedly reminded his readership of Starfield's seminal article, which he characterizes as "the most explosive revelation about modern healthcare in America ever published," deeming it particularly noteworthy in light of Starfield's "impeccable" insider credentials within medical circles.[106] However, despite the consequential implications of her 2000 publication, Starfield told Rappoport in 2009 (by email) that whereas her work on primary care had been widely disseminated, "including in Congressional testimony and reports," her comments about iatrogenesis and Americans' poor health standing had attracted "almost no attention."[107] As Rappoport summarized, "No major newspaper or television network [ever] mounted an ongoing 'Medicalgate' investigation" after Starfield published her analysis, nor did any federal agency tackle "remedial action." Meanwhile, Starfield wrote to Rappoport, "The American public appears to have been hoodwinked into believing that more [medical] interventions lead to better health."

In further remarks shared with Rappoport, Starfield emphasized the problem of "vested interests," situating health care corruption in the context of wider societal rot:

> [My findings] are an indictment of the US health care industry: insurance companies, specialty and disease-oriented medical academia, the pharmaceutical and device manufacturing industries, all of which contribute heavily to re-election campaigns of members of Congress. The problem is that we do not have a government that is free of influence of vested interests. Alas, [it] is a general problem of our society—which clearly unbalances democracy.

Starfield also told Rappoport in 2009 that studies conducted after publication of her *JAMA* analysis generally had come up with death rates exceeding her calculations. A 2009 update on "errors in medicine" authored by Leape, for example, calmly suggested that medical injury was "causing hundreds of thousands of preventable deaths each year."[108] In 2013 (two years after Starfield's own death), a meta-analysis estimated that there might be two to four million serious, "preventable adverse events" annually in hospitals, with up to 440,000 proving fatal.[109] However, those findings and others published in later years were "quickly 'disappeared' from view," shrouded in a "Wall of Silence" and "intentional amnesia."[110] As another Johns Hopkins-based physician, Dr. Martin Makary, stated in 2016, "We all know how common [medical error] is" and "We also know how infrequently it's openly discussed."[111]

Gary Null: Gruesome Statistics

"The American medical system is the leading cause of death
and injury in the U.S."
—Gary Null

In 2010, Gary Null and physicians Martin Feldman, Debora Rasio, and Carolyn Dean took up the medical muckraking baton, publishing *Death by Medicine* and, in 2011, putting out a documentary by the same title. Null, an "environmentalist, consumer advocate, investigative reporter and nutrition educator," hosts the nation's longest-running nationally syndicated health radio talk show and has directed over 100 full-feature documentary films and written over 70 books.[112]

Right up front, *Death by Medicine's* authors argued that the American medical system "is broken, utterly corrupted by money, and no longer founded on scientific fact." They declared their purpose to be to present "in painstaking detail" the "gruesome statistics" showing that American medicine is not just "a" cause of death and injury in the U.S. but *the* leading cause. Although their book largely sticks with the language of "medical error," they point out that "healthcare is the only business where you keep paying whether you get good results or not. . . . The physician is rewarded for his *efforts*, not for his results" [italics in original].

Null calculated that medicine was killing almost 800,000 Americans every year—255% more than Starfield's figure and 343% more than Leape's. Leape had attempted to contextualize his 1994 calculation by likening the pace of iatrogenic fatalities to "2 unsafe plane landings per day at O'Hare, 16,000 pieces of lost mail every hour, [or] 32,000 bank checks deducted from the wrong bank account every hour."[113] In comparison, Null's 800,000 annual deaths would be "equivalent to six jumbo jets falling out of the sky each day." Extrapolating the annual death toll to a 10-year total would add up to more deaths (close to eight million) "than all the casualties from all the wars fought by the US throughout its entire history."

Null and coauthors emphasized that even though their numbers were higher than previous estimates, the figures probably were still conservative due to the vast underreporting of iatrogenic outcomes, including in children and in outpatient settings. A government official quoted in *Death by Medicine* admitted as much, stating, "the full magnitude of [the medical errors] threat to the American public is unknown." According to Null's review of the literature:

- At best, **1.5%** of all adverse events make it into an incident report.
- Surgical incident reports capture as few as **5%** of adverse events.
- Up to **94%** of adverse drug events may never be identified at all.

A 2013 study in the *Journal of Patient Safety* reiterated the underreporting of iatrogenic harms.[114] The author frankly observed that physicians—with cardiologists alleged to be the worst offenders—"often refuse to report a serious adverse event to anyone in authority"; as a result, "unreported medical errors often [do] not find their way into the medical records of the patients who were harmed." Calculations of iatrogenic injuries and deaths depend on medical records as the primary data source.

According to Null, "business as usual" pressures, the drive to preserve reputations, and motivation to keep lawsuits at bay are all factors that strongly disincentivize the documenting of iatrogenic outcomes by institutions and providers. As Leape not very flatteringly explained in 1994,

> Physicians typically feel . . . that admission of error will lead to censure or increased surveillance or . . . that their colleagues will regard them as incompetent or careless. **Far better to conceal a mistake** or, if that is impossible, to try to shift the blame to another, even the patient" [bold added].[115]

As *Death by Medicine* explains, even well-intentioned providers and researchers face potent disincentives to report problems:

> When honest scientists do exist, they have no power to override the corruption. The price they would pay for writing or speaking the truth about the drug company invasion into modern medicine, or for censuring a colleague for cause, is that the doctor or researcher would be alienated, unable to get grants, unable to publish, possibly even unable to work.

The onus to report problems, therefore, may be on the injured party or their surviving family—but this will happen only if the victims recognize the iatrogenic nature of the harm. Even for those who connect the dots, reporting is no guarantee of follow-up; in the aftermath of the COVID injection rollout, for example, it has not been unusual for patients who managed to file a report with the Vaccine Adverse Event Reporting

System (VAERS) to find their reports mysteriously "disappeared from the system."[116]

Like Illich and Mendelsohn before them, Null and coauthors had no kind words for the pharmaceutical industry or its detrimental influence on Americans' well-being. They noted sobering statistics, such as the fact that (as of 2004) the U.S. pharmaceutical market accounted for almost half (48%) of the global market; that same year, Americans spent four and a half times more on prescription medications than they did in 1990. In 2002, *Death by Medicine* explained, American pharmacies filled 3.34 billion drug prescriptions—and at least one in four recipients experienced side effects, with antibiotics, anti-inflammatory drugs, antidepressants, heart drugs, and chemotherapy being some of the worst offenders.

In comments about vaccines that eerily foreshadowed what was to come with the COVID shots, Null and coauthors described how mandates, liability carve-outs, propaganda, and toxic-drug "solutions" perpetuate a profitable business model:

> Vaccinate infants, children, teens, adults, elders, each one a potentially lucrative marketing niche, even an opportunity to sell drugs to otherwise healthy people. Why not make these vaccinations mandatory? Force us to pay for possible side effects, 'for our own good.' Fright tactics are used to petrify the public into rushing to pay for vaccines that may prove debilitating or worse. All of this is done with a wink and a nod. Not a cent is spent on prevention (except pseudo-prevention through toxic inoculations that do not really prevent disease, and may cause harm); instead, every dollar goes for treatment.

Death by Medicine also showed that for many drugs—52% in a study conducted by the federal government's own Government Accountability Office (GAO) (formerly called the General Accounting Office)—the substance's risks only become apparent *after* FDA approval.[117] On the GAO's list of serious drug risks identified post-approval, as summarized by Null, were "heart failure, myocardial infarction, anaphylaxis, respiratory depression and arrest, seizures, kidney and liver failure, severe blood disorders, birth defects and fetal toxicity, and blindness."

In the book's Table 2, *Death by Medicine* lists a variety of pathways to iatrogenic mortality, ranging from adverse drug reactions to negligent care (e.g., hospital bedsores, nursing home malnutrition) to unnecessary procedures and more. However, some of the book's most ghastly information addresses the problem of unnecessary surgeries and their sometimes fatal outcomes. *Death by Medicine* reported a tripling of unnecessary surgeries between 1974 and 2001, from an estimated 2.4 million (fatal for almost 12,000 individuals) to 7.5 million (fatal for over 37,000). While not going so far as to deem 90% of surgeries pointless, as Mendelsohn did, Null estimated that "the proportion of unwarranted surgeries could be as high as 30%." Whether rated as "necessary" or not, the most common surgical procedures—which include cataract surgery, Cesarean delivery, inguinal hernia operations, knee arthroscopy, back surgery, and removal of anatomic structures (for example, the tonsils, appendix, or uterus)—all are rife with iatrogenic risks (see **Iatrogenic Complications from Surgery**).

Iatrogenic Complications from Surgery

According to *Death by Medicine*, the list of iatrogenic complications from surgery "is as long as the list of procedures themselves." Among the hair-raising possibilities mentioned are paralysis associated with catheterization[118] and surgical equipment left inside patients. The latter is "one of the more common acts of negligence," according to one law firm, which notes that the range of items left inside patients' bodies includes "sponges, scalpels, scissors, drain tips, needles, clamps, forceps, scopes, surgical masks and gloves, tubes, and measuring devices."[119] Patients generally need to undergo another potentially risky procedure to remove the abandoned item(s)—that is, if the items are discovered!

Death by Medicine also mentions "wrong-site" surgery, a subset of "wrong-site, wrong-procedure, wrong-patient errors" (WSPEs)—defined by the Agency for Healthcare Research and Quality (AHRQ) as "patients who have undergone surgery on the wrong body part, undergone the incorrect procedure, or had a procedure intended for another patient."[120] The AHRQ admitted in 2019 that the official rate of WSPEs would probably be "significantly higher" if the agency had access to data from ambulatory as well as hospital settings. In August 2023, *Medpage Today* reported in its "weekly roundup of healthcare's encounters with the courts" that a man's estate was suing several health care organizations for operating on the wrong eye and leaving him "essentially blind."[121]

Also in 2019, researchers commented on the growing incidence of "iatrogenic wounds"—surgical interventions that compromise skin integrity or damage the subcutaneous soft tissue or deep tissue.[122] The authors hypothesized that such wounds are on the rise due to "the continued expansion of surgical indications" and "an increase in difficult surgeries" as well as "the constant emergence and application of new implantable biomaterials." They observed that improperly handled iatrogenic wounds "have a very poor prognosis and will cause serious physical and psychological harm to patients."

In 2023, ProPublica shone a light on the rampant performance of unnecessary surgery—often with government and institutional complicity—in its investigation of "profit-driven procedure mills, in which doctors can deploy any number of devices in the time it takes to drill a tooth and then bill for the price of a new car."[123] The report's title—"In the 'Wild West' of outpatient vascular care, doctors can reap huge payments as patients risk life and limb"—aptly summarized the rogue health system behavior and corresponding risks to patients.

ProPublica focused on atherectomies, an outpatient endovascular procedure that involves the use of a blade or laser to remove arterial plaque buildup and repair blood flow, often in the legs. Some researchers dismiss atherectomy as a "niche" intervention of dubious effectiveness, noting the failure to conduct rigorous studies.[124] ProPublica discovered that in 2008, the Centers for Medicare and Medicaid Services (CMS) essentially launched an atherectomy boom by "turbocharging" its insurance payments to doctors, opening the floodgates for "unfettered profiteering"; between 2013 and 2021, atherectomies doubled and CMS payments to doctors tripled. (CMS ignored ProPublica's request for an interview or written response.)

Concerned about the "disproportionately higher" use of atherectomy compared to interventions such as angioplasty or stenting, the authors of a 2019 study found that atherectomy patients were much more likely to experience subsequent amputation or a "major adverse limb event" than recipients of the other procedures.[125] In their toned-down conclusion, they wrote, "[A]lthough emerging endovascular technologies may be popular in contemporary practice, the related increased risk of long-term adverse outcomes may caution against widespread use."

To further illustrate atherectomy's perils, ProPublica profiled a suburban Maryland doctor who earned over $30 million from CMS for a decade's worth of "medically unnecessary and invasive vascular procedures" that allegedly killed at least one woman and resulted in a man having a leg amputated. Despite years of complaints and various lawsuits, it was not until October 2022 that the state's medical board belatedly suspended him and fined him a token $10,000, ordering him to "enroll in an ethics course." The daughter of the deceased woman told ProPublica, "I trusted doctors, but now I'm starting to think that maybe they shouldn't be as fully trusted."

In 2022 and 2023, news outlets reported on a New Hampshire hospital's complicity in the "troubling record of medical errors" (to put it mildly) of a top heart surgeon who racked up 21 malpractice settlements—with 14 linked to patient deaths—before his "abrupt" retirement in 2019 at age 63.[126] According to a "chilling" investigation by the *Boston Globe*—which pointed out that the surgeon in question had "one of the worst surgical malpractice records among all physicians in the United States," contrasting sharply with the one to two malpractice claims that mark most surgical specialists' careers—senior hospital officials "knew the truth and its consequences" but persisted in celebrating the doctor "as a star" and promoting his services "in glowing terms."[127] Hospital management had heard about the problems as far back as 1997 but "resisted reining in one of their leading rainmakers," ignoring repeated complaints from the surgeon's horrified colleagues.

To its credit, the *Globe* went beyond merely blaming one treacherous provider to emphasize the wider iatrogenic lessons:

> This case out of a little-known community hospital . . . reveals painful truths that apply far beyond its halls and operating rooms and **point to some common realities in today's health care world**: Medical consumers . . . are often kept in the dark about the performance history of their physicians, even when that history is grim [bold added]. And hospital officials can in some cases evade accountability for years, even when confronted repeatedly by alarmed medical staff, as happened in this case.

A 2023 Reuters investigation shows that doctors would not be able to "buy their way out of trouble" without the say-so of an unperturbed and wily government.[128] Reuters describes situations—quite common—in which criminally inclined physicians pay negotiated settlements and then "walk free," only to continue engaging in the same harmful practices. In the case of a surgeon who paid a civil settlement after federal prosecutors alleged he "had performed scores of medically unnecessary cardiac procedures,"

> [He] faced no judge or jury. He did not admit to wrongdoing. He maintained his license to practice. What's more, neither [the doctor] nor government officials were required to notify patients who purportedly were subjected to vascular surgical procedures they didn't need.

After a nurse-whistleblower subsequently reported the same doctor again, stating that he had resumed his unethical behavior, prosecutors mounted—but then dropped—charges of felony fraud in favor of another civil settlement.

Nor do such settlements come solely from impenitent providers. Reuters found that between 2013 and 2022, "more than 2,200 hospitals and healthcare companies . . . negotiated civil deals to sidestep prosecution for alleged offenses that included paying bribes, falsifying patients [sic] records and billing the government for unnecessary patient care." In all, Reuters discovered, the U.S. government collected over $26.8 billion in civil settlements over the decade, making health care settlements a lucrative revenue stream for the Treasury Department. According to Reuters, the government justifies its bias in favor of civil settlements (versus criminal prosecution) by stating that its aim is for "accused wrongdoers [to] stay in business, thus preserving public access to vital healthcare services"!

Death by Medicine, too, addressed the issue of government complicity with medical-pharmaceutical crimes, zeroing in especially on the role of a negligent and deceitful FDA. At the same time, the book's authors pointed out, it is not just the FDA and drug companies who have a "cozy" symbiotic relationship with one another; numerous other players—"media, scientists, professors, universities, hospitals, governmental agencies, such as the . . . EPA [Environmental Protection Agency] and the CDC [Centers for Disease Control and Prevention]"—all "banquet at the pharmaceutical table."

Although the FDA's subservience to industry has attracted criticism for decades—including from U.S. senators and the FDA's own staff[129]—the regulatory capture has steadily worsened and become more entrenched over time.[130] "User fees" are one of the principal factors responsible for FDA laxness. The somewhat euphemistic term describes payments to FDA by the companies that "make and market FDA-regulated products"—including drugs, certain biologics (called "biosimilars"), and medical devices.[131] Commenting on the user fee arrangement, *Death by Medicine* quoted a 20-year veteran of the FDA's own Drug Safety Office, who pointedly told a Senate committee, "He who pays the piper calls the tune."

The first of multiple Congressionally-enacted User Fee Acts came on the scene in 1992. As pharmacy professor C. Michael White explained in 2021:

In 1987 [before the advent of user fees], it took 29 months from the time a new drug application was filed by the manufacturer for the FDA to decide whether to approve a medication in the U.S. In 2014, it only took 13 months and by 2018, it was down to 10 months.[132]

At present, nearly half (46% or $3.3 billion) of the FDA's total budget comes from user fees,[133] but even that percentage understates the true magnitude of the pharmaceutical industry's quid pro quo influence, with user fees financing fully "two-thirds of the drug regulation budget, and the work of at least 40% of the FDA's 18,000 employees."[134]

Another factor favoring manufacturers over meaningful regulation is the patchwork of federal legislation that encourages accelerated approvals of new drugs. Relevant legislation includes the Orphan Drug Act (1983), the Accelerated Approval Program (1992), Priority Review (1992), the Fast Track process (1997), and the Breakthrough Therapy designation (2012). For some fast-tracked applications, the FDA churns out approvals in under two months.[135] A 2022 *Kaiser Health News* report showed that while these high-speed programs have clear advantages for companies, they tend to result in drugs that are both expensive and dangerous.[136]

Death by Medicine concludes by lamenting the closure of the Office of Technology Assessment (OTA)—in the authors' view, "the US government's last honest agency." The OTA once had the duty to provide Congress with "objective and authoritative analysis of complex scientific and technical issues," but Congress killed the agency in 1995,[137] "[s]hortly after the OTA released a report that exposed how entrenched financial interests manipulate healthcare practice."

Dr. Peter Breggin: The Psychopharmaceutical Complex

*"Psychiatric medications are breaking the spirits and the brains
of our children."*
—Dr. Peter Breggin

Dr. Peter Breggin is a psychiatrist, bestselling author, and expert in clinical psychopharmacology (the study of the effects of drugs on the mind and behavior) who has served as a medical expert in criminal and civil cases involving psychiatric drugs. Celebrated by many as the "conscience" and the "Ralph Nader" of psychiatry, Breggin's decades-long career has been marked by "principled, courageous confrontations with organized psychiatry, drug companies, and government agencies"—the entities that make up the "psychopharmaceutical complex."[138] His numerous books include *Toxic Psychiatry*

(1991), two editions of *Brain-Disabling Treatments in Psychiatry* (1991 and 2007), *The Antidepressant Fact Book* (2001), *The Ritalin Fact Book* (2002), and *Talking Back to Prozac* (2014), among others.[139] In 2021, Breggin and his wife Ginger Breggin published *Covid-19 and the Global Predators: We Are the Prey*, a deep dive into the years of planning carried out by "predatory globalists" to "reorganize the world in the name of public health."[140]

In 2008, Breggin's book *Medication Madness: The Role of Psychiatric Drugs in Cases of Violence, Suicide, and Crime* summarized shocking evidence that psychiatric drugs—with hundreds of millions of prescriptions written annually—do more harm than good. One of the book's key take-home messages is that all such drugs are "brain pollutants" that impair brain function (see **Types of Psychiatric Drugs**). *Medication Madness* makes a strong case that psych drugs are both dangerous and useless, and highlights the risks of iatrogenic (medically induced) addiction.[141] Drawing attention to the relationship of drug addiction to suicide, Breggin has noted that such addiction "is the most frequent identifiable factor in the suicide of young people"[142]—and suicide is the second leading cause of death in preteens, teens, and young adults ages 10 to 24 years.[143]

Types of Psychiatric Drugs

Antidepressants: Drugs billed as "antidepressants" have been in the pharma mix since the 1950s and currently include monoamine oxidase inhibitors, trycyclics, selective serotonin reuptake inhibitors (SSRIs), serotonin-norepinephrine reuptake inhibitors (SNRIs), and "atypical" antidepressants. SSRIs, the most commonly prescribed category, include Celexa, Lexapro, Paxil, Prozac, and Zoloft. Doctors also prescribe the antidepressant Strattera (an SNRI) for ADHD.

Antipsychotic agents (neuroleptics): Doctors deploy antipsychotic drugs after they have conjured up diagnoses such as "schizophrenia," "bipolar disorder," "mania," and "delusional disorder"—never considering those conditions' possible iatrogenic causes. Commonly prescribed antipsychotics include Abilify, Haldol, Risperdal, and Zyprexa. The drugs' "primary" or "therapeutic" effect, says Breggin, is "severe apathy and indifference" as well as "a general crushing of spontaneity and autonomy"; stated another way, they are the equivalent of a "pharmacological lobotomy."

Mood stabilizers: Doctors use this class of drugs (which includes lithium and several antiepileptics) for, as Breggin puts it, "the long-term control of so-called bipolar disorder." He observes that physicians also dispense mood stabilizers to those "who have but the faintest signs of a manic-like problem, such as irritability and mild mood swings in adults or temper tantrums in children."

Stimulants: Handed out like candy for ADHD, stimulants include amphetamine- and methamphetamine-based and methylphenidate drugs such as Adderall, Concerta, Ritalin, and Vyvanse. (See the earlier section about Dr. Robert Mendelsohn and his condemnation of this drug category.) According to Breggin, "All the classic stimulants can cause addiction, violence toward self and others, depression, mania, and a broad array of bizarre mental reactions and behaviors." He sadly adds, "Millions of children are having the spiritual stuffing knocked out of them by stimulant drugs." Nearly all are Schedule II narcotics—drugs "with high potential for abuse and/or addiction."

Tranquilizers/sleeping pills: This category includes benzodiazepine tranquilizers and sleeping pills (for example, Halcion, Valium, and Xanax) as well as non-benzodiazepine sleeping pills (e.g., Ambien, Sonata) and barbiturates. Long-term exposure to Xanax and other "benzos" can lead to "permanent difficulties with memory function, learning, and the clarity of . . . thinking." In class-action lawsuits related to the non-benzo sleeping pills, attorneys have coined the term "Ambien zombies" to describe some of the "bizarre" blackout behaviors induced by the medication.[144]

According to a 2021 report in *Insurance Journal*,[145] one in five Americans take prescription medication for "stress, anxiety or depression," and the number of people taking such drugs has been rising, especially since COVID. (Eighteen U.S. states clocked 10% to 20% increases between

August 2020 and March 2021.)[146] Another factor responsible for the proliferation of psych drugs is society's growing "psychiatrization," a multifaceted
concept referring to the "expansion in the number and inclusiveness of psychiatric diagnoses" that convert socially undesirable behaviors into drugga-
ble "symptoms."[147] Critics of psychiatrization suggest that we need to "save
normality from the relentless encroachment of diagnosable pathology."[148]

Citing the "ADHD" diagnosis in the DSM as an example of "the global
spread of psychiatric ways of being a person," European researchers argued
in 2022 that ADHD in children is an arbitrary "scientific conceit" that
strongly "reflects the DSM's political, cultural, and financial role in the
psychiatrization of children's everyday lives."[149] Their perspective, consistent with that of Mendelsohn and Breggin, is that we can best understand
ADHD as a "social category"—one that "eliminates human diversity and
enforces the standard model of what an individual should behave and be
like in order to navigate within the cultural boundaries of normalcy and
be a productive citizen."

Breggin has long criticized the "theoretical edifice" that psychiatrists
use when they prescribe antidepressants and other psychiatric medications—namely, the unproven premise, which he calls "hocus-pocus," that
individuals suffer from "biochemical imbalances" in the brain and that
the drugs correct the "imbalances." A systematic review published in 2022
conclusively laid this theory to rest;[150] as a clinical psychologist summed
the matter up in *Psychology Today*, "if you hear a medical professional using
the term 'chemical imbalance' to explain depression, you are hearing a fic-
tional narrative (or a sales pitch), not scientific fact."[151] After the review's
publication, a writer observed in *Slate* that the unsoundness of the chemical imbalance theory had been "an open secret within mental health circles
for at least a decade."[152] Breggin points out that "no biochemical imbalances have been identified in the brains of patients with diagnoses such as
anxiety disorders, depressive disorders, bipolar disorder, or schizophrenia,"
emphasizing that psychiatric disorders "have no biological markers, no
known physical causes, and no rational physical treatments."

With a hint of dark humor, Breggin's take is that "If you have a biochemical imbalance in your brain, the odds are overwhelming that your
doctor put it there with a psychiatric drug." Along those lines, animal

models suggest the possibility that doctors who prescribe antidepressants for pregnant women may even be setting multigenerational effects in motion;[153] in 2020, researchers reported concerns that prenatal SSRI use might affect fetal brain development and increase the risk of a subsequent diagnosis of "autism spectrum disorder" (ASD) in offspring—with "ASD" itself being an "artifact" and cover term designed to distance pharmaceutical, chemical, or other toxins from responsibility for the damage they cause.[154]

In Breggin's professional assessment, psychiatric drugs' harmful effects insidiously compromise a person's mind and emotions "before [they] realize what is happening." He characterizes as "medical spellbinding" people's utter inability to connect the drug(s) they are taking with the feelings and behaviors that follow—which can range from apathy to suicide to mania to homicide.[155,156,157,158] In his view, medical spellbinding plays out in a variety of troubling ways, and "no two cases of spellbinding are alike." He describes a horrifying list of potential symptoms and sensations triggered by psychiatric drugs (see Table 2), encompassing reactions so "bizarre," they can "drive people to suicide and violence, and to madness" and permanent disability.

Table 2. A Partial List of Psychiatric Drug Reactions
Blunting of emotions (becoming "zombie-like")
Dulling of self-awareness
The feeling of a "mental straitjacket" (or "cement in the brain")
Cognitive toxicity ("deterioration of mental functions related to thinking such as learning, attention, and memory"), including dementia
Disinhibition and loss of emotional control
Impulses that emerge "out of the blue"
Obsessive-compulsive behavior
Manic-like episodes
Irritability, aggression, and hostility
Akathisia (inability to sit still) and other movement disorders[159]
Inability to appreciate right and wrong
"Indescribable mental and physical pain" inside the head
Derealization (remoteness from reality)
Depersonalization (remoteness from oneself)

In 2017, veteran Danish researcher Peter Gøtzsche described SSRI-associated risks of suicide and homicide in *The BMJ*, asserting, "It can no longer be doubted that antidepressants are dangerous."[160] The preceding year, Gøtzsche had estimated that "psychiatric drugs alone" might be "the third major killer."[161] Noting the unpredictability of SSRI reactions, Gøtzsche observed that "many have committed SSRI-induced suicide or homicide within a few hours after everyone thought they were perfectly okay." Gøtzsche co-founded the Cochrane Collaboration, an organization that systematically reviews health research on specific topics. In 2018, Cochrane chased him out for being too critical of pharma[162] and for daring to suggest that "A life without drugs is possible for most of us most of the time."[163]

Emphasizing psychiatric drugs' brain-disabling effects, Breggin explained in *Medication Madness*:

> As illustrations of the overall brain-disabling principle, the apathy or euphoria created by antidepressants is misinterpreted as an improvement in depression—the blunting of all emotions and self-awareness caused by antipsychotic drugs is seen as an improvement in the psychosis; and the generalized sedation and suppression of brain function caused by antianxiety drugs is viewed as a treatment for anxiety. In reality, no specific improvements have occurred in the underlying depression, psychosis, or anxiety. **Instead, the brain has been partially disabled**, artificially changing the individual's mood and rendering the patients less able to feel, to perceive, or to express their underlying mental condition or outlook [bold added].

Two studies published in 2019 underscore the brain-disabling effects of drugs given to youth diagnosed with "ADHD." One study with teens and young adults found that new users of amphetamine-based Adderall-type drugs had more than double the risk of experiencing psychosis as those who started out with a methylphenidate drug like Ritalin or Concerta—1 in 486 versus 1 in 1,046—with adverse reactions answered with antipsychotic medications that further perpetuate the drugging cycle.[164,165] Lest unwary parents conclude that methylphenidate drugs are a "safer" option, however, the second study reported MRI-observable brain changes in ADHD-diagnosed 10- to 12-year-old boys within four months of methylphenidate use.[166]

Breggin classifies benzodiazepine tranquilizers as "the most spellbinding drugs of all," citing the substances' ability—documented since at least the late 1970s—to generate "hallucinations, delusions, paranoia, amnesia, delirium, [and] hypomania." In 1990, Cornell authors warned that "current evaluations of benzodiazepines . . . demonstrate clearly that they produce tolerance and dependence in short and long-term administration," as well as abuse and addiction.[167] Doctors have largely ignored the warnings. By 2023, concerned researchers at Yale and elsewhere were describing the U.S. as "a nation on benzodiazepines"—suggesting that benzos had become "deeply entrenched" in modern medicine, "to the tune of an estimated 65.9 million office visits per year."[168]

The "callous" and "cavalier" overprescribing of drugs such as benzos, according to Breggin, is one of many factors showcasing psychiatrists' and other doctors' dereliction of duty and failure to recognize or admit to the havoc wrought by mind-altering drugs—despite the decades-long presence of "substance-induced mood disorder" as a formal DSM diagnosis that actually makes sense.[169] Breggin states, "While psychiatrists are eager to make believe they are treating real diseases, they rarely admit that they are causing them"; nor do they acknowledge that medication madness "is a real neurological disease." Discussing humans' aversion to "being found wrong," Breggin unhappily wrote in 2008, "I cannot exaggerate how reflexively my colleagues reject any suggestion that their drugs could be making their patients worse, let alone crazy."

This key fact pattern is apparent across the 50-plus shocking but riveting real-life cases described in *Medication Madness*; the stories reveal that doctors, more often than not, increase a person's dose at the first sign of trouble—continuing patients on the very medications that "have driven them over the edge." They frequently pile on more drugs, as in a case profiled in 2022 in the *New York Times* of a high school girl prescribed a whopping 10 psychotropic medications.[170] In only one of the dozens of cases recounted in *Medication Madness* did a psychiatrist concede that a medication was problematic and "prevent a tragedy by stopping it"; in that instance, a high school girl had become "obsessed with the idea of stabbing her mother in the back" during her second week on Prozac. Echoing Illich, Breggin characterizes the family doctors, pediatricians, and psychiatrists who "clumsily" work with distressed children as "technologists" who dispense drugs "with much less acumen and even less success than an auto mechanic addressing a knocking sound in the motor." Distressingly, when things go wrong, doctors often resort to gaslighting the patients (or, in the case of youngsters, their parents).

The stories in *Medication Madness* convey a number of other key points:

1. Most negative reactions to psychiatric medications affect people who are taking **"routine dosages"** or even **"unusually tiny"** doses. Toxic symptoms can occur within a day or two of starting a drug— that is, after just one to two doses. As Breggin summarizes, "if the dose is sufficient to have any effect on the brain and mind, then it is sufficient to cause mental disturbances." He also notes that even short-term and low-dose use can produce "lasting and probably permanent [brain] abnormalities."

2. A **change in dosing** often precedes and triggers bad reactions— importantly, the change can be "either up or down," which is to say that reactions can arise when individuals attempt to stop taking the drug. Breggin cautions that drug withdrawal is best undertaken under experienced clinical supervision.

3. Clinical trials underestimate psych drugs' toxicity and adverse effects because the **duration of exposure** in a study is much shorter than people's real-world use, which often continues for many months or years.

4. Adverse reactions tend to be marked by a "relatively sudden onset and rapid escalation of abnormal thoughts and behavior" that are completely **out of character** for the affected individuals, often causing them to act "in ways that would ordinarily terrify and appall them."

5. Discussing children, Breggin says that a child's desperate brain attempts to "compensate" for toxic drug effects, "producing a **complex, unstable condition**."

6. Even on the "mild" or more subtle end of the medical spellbinding continuum, psychiatric medications can "impair or ruin the person's

quality of life," making people "more irritable, less optimistic, or more emotionally shallow, without realizing that they have changed."

7. Individuals who take more than one medication at a time are vulnerable to scary **synergistic effects**, leading to the tame recommendation by some researchers for "psychotropic monotherapy" (the use of one drug at a time).[171] As Breggin puts it, "When combined with other drugs, a small dose can become a large one, and a large dose can become a mammoth one."

8. Medical madness "can long outlast the last dose," with some individuals remaining "erratic" for months or even years **after withdrawal**.

9. **Nonpsychiatric drugs**—including "corticosteroids, anti-Parkinsonian drugs, anti-epileptics, antiretrovirals, antibiotics, anticancer drugs, analgesics, drugs targeting endocrine and cardiovascular disorders, immunosuppressants, skeletal muscle relaxants and bronchodilators" according to a study in the journal *BJPsych Advances*— are also capable of causing "medication madness."[172]

10. The often dire consequences of medical spellbinding affect not just the individual taking the drug(s) but also "families, innocent strangers, and **whole communities**."

From Breggin's vantage point, the "combined powers" of the psychopharmaceutical complex—consisting of drug companies, the insurance industry, psychiatry, organized medicine, and the FDA and other federal agencies—generate "an enormous amount of propaganda to convince people to overcome their natural and healthy skepticism about taking psychiatric medications." Like the authors of *Death by Medicine*, Breggin has long criticized the FDA as a captured "drug-promoting agency"[173] that is "grossly fail[ing] America's children"—prone to taking "too little, too late" action, if any, to warn the public about psych drugs' risks. Jon Rappoport characterized the FDA's "sordid" behavior in 2019 as follows:

That agency went rogue a long, long time ago. It takes no responsibility for launching killer chemicals on the population. It operates as a colluding partner with the pharmaceutical industry. Trusting the FDA to protect people from drugs such as Prozac is like trusting a PR company, hired to promote war, to maintain the peace.[174]

In connection with a Prozac legal case, Breggin at one point learned through Freedom of Information Act (FOIA) requests that Eli Lilly and FDA had jointly agreed to "expurgat[e] some of the most damning information from the [Prozac] label shortly before its publication." After the 1989 release of Prozac, the FDA dragged its feet for more than 15 years—ignoring "mountains of evidence" and hundreds of reports of serious adverse reactions—before public pressure forced it to add a black-box warning for children and adolescents about the risk of drug-induced suicidality (also called "treatment-emergent suicidal events") from Prozac and other antidepressants.[175] In 2007, FDA grudgingly expanded the suicidality warning to 18- to 24-year-olds.[176] Responding to subsequent (and unconvincing) pushback from the drugs' apologists, American and Australian researchers reexamined the evidence in 2020 and affirmed that the black-box warning "is firmly rooted in solid data."[177]

Interestingly, the warning's initial wording made the "causal role" of antidepressants in inducing suicidality explicit, but, according to Breggin, the FDA chose to delete the phrase about causality in the final version. Moreover, the "FDA's parsing of a warning into various age brackets" was "quite unprecedented," representing a form of "pander[ing] to the drug companies' needs to obscure the reality that antidepressants cause suicide in [both] children and adults"—in other words, in people of all ages.

In 2020, in a frankly worded editorial in *Frontiers in Psychiatry*, Swiss-Canadian professor Dr. Michael Hengartner outlined pharmaceutical industry flimflam related to psychiatric drug research and emphasized how little rationale there is for pediatric prescribing of antidepressants.[178] He wrote:

The use of antidepressants in children and adolescents has a troubled history. . . . It is a history characterized by **systematically biased research, financial conflicts of interest, and professional reck-lessness** [bold added]. . . . It is now well-established that most pediatric antidepressant trials were industry-sponsored and had serious methodological limitations; many trials remained unpublished due to unfavorable results, and those published were mostly ghost-written, selectively reported efficacy outcomes and misrepresented the true rate of treatment-emergent suicidal events.

Noting how the FDA looks the other way when manufacturers engage in research shenanigans,[179] Breggin observes that in the clinical trials for Prozac, the antidepressant "overstimulated so many patients . . . that Eli Lilly decided to break the rules of the trials by giving tranquilizers and sleeping medications to many of the subjects." When Eli Lilly later informed FDA about its maneuver, FDA "retrospectively permitted this breach of its own rules. In effect, instead of approving Prozac . . . the FDA approved Prozac in combination with addictive tranquilizers, without ever informing the medical profession or the public about this ruse."

One might expect black-box warnings and the occasional news stories that leak out about psychiatric drugs' destructive, slippery-slope effects to dampen Americans' enthusiasm for the drugs, but the public's misgivings do not yet appear to have reached critical mass. As Breggin observes, "Especially in regard to psychiatric drugs, patients take them because they have faith in 'science' and faith in their doctor. . . . The mood-elevating effects are almost always short-lived but they encourage the individual to keep hoping that one or another drug will finally provide sustained relief from suffering." Where children are concerned, Breggin does not blame "bamboozled" parents but chastises his colleagues "for promoting the use of medication to subdue and control children . . . a form of technological or pharmacological child abuse."

Drug industry and government propaganda is now convincing many parents to embrace aggressive mental health screening, which Breggin characterizes as a "psychopharmaceutical steamroller" and Jim Gottstein (of the PsychRights public interest law firm) as a "drugging dragnet."[180,181] A 19-country survey published in *JAMA Network Open* in 2023 reported that fully 93% of parents expressed interest in regular mental health screening for their children, with 65% endorsing annual screening and another one in four (23%) favoring *quarterly* screening![182] Researchers at the University of Massachusetts argued in 2020 that there is no evidence

to support mental health screening in youth; those authors blamed commercial interests and institutional corruption for the global drive to "'scale
up' the diagnosis and treatment of mental illness"—a push that not only
presupposes a questionable "global mental health care crisis" but "positions Western psychiatric treatment as the main solution."[183]

In fact, promotion of the drugging of young people by the World
Health Organization (WHO) and sister organizations has been wildly successful already. Worldwide, use of psych drugs has increased substantially
among children and adolescents since the mid-2000s.[184] In England, antidepressant prescribing for children ages five through 12 years increased
by 41% between 2015 and 2021, including a spike during the COVID
lockdowns; commenting on pediatric overprescribing, a UK psychiatrist
warned, "there is a very high risk that children who start taking these
drugs when very young will continue taking them for many years and into
adulthood."[185]

In the U.S., the quarter-century between 1987 and 2014 saw a 14-fold
increase in antidepressant use in Medicaid-insured youth under age 20.[186]
A 2021 study in *JAMA Pediatrics* traced an even more worrisome trend
for the period between 1999 and 2015—a 188% increase in "psychotropic polypharmacy" in youth under age 18.[187] Over 85% of the kids taking multiple drugs had an ADHD diagnosis, and stimulants were the top
class of drug prescribed, but disturbingly, antipsychotic use (prescribed for
"bipolar disorder") doubled over the time period. The organization Mad
in America points out that "the increasing prevalence of bipolar disorder
diagnoses in children" could be an outcome of the rampant prescribing
of stimulants and antidepressants, both of which can produce manic-like
effects that lead to unwarranted diagnoses and further medication.[188]

Breggin argues that "depression," "anxiety," "schizophrenia," and other
such diagnoses must be understood by looking at "an individual's psychological and spiritual life in the context of family, culture, and society." Many other professionals agree that the narrow, pharma-dominated
view of "mental illness" and treatments is seriously flawed. One critic asks,
"What if mental illness is a complex social, economic and trauma-related
issue that the very limited biomedical model can't answer?"[189] Another
writer calls attention to the wide range of factors that can contribute to

challenges conveniently dumped in the "mental illness" basket—ranging from the gut microbiome to environmental and socioeconomic influences to "interpersonal stressors."[190] As these individuals lament, when they express doubts about the appropriateness of psych medications, they are treated as "heretics."

Some of the most poignant and heartbreaking observations in *Medical Madness* have to do with the ways in which psychiatric drugs "shred the human spirit," with medicated individuals—and especially the children who are our "best and brightest"—having "the zest, spark, or spirit rubbed out of their lives without appreciating what is happening to them." Breggin argues that with the widespread use of psychiatric drugs,

> [W]e are creating generations of drug consumers who have no idea how to live with a clear brain and mind, and how to improve their lives through self-understanding, personal responsibility, principled living, and higher ideals.

Since 2020, Breggin and others such as the Solari Report's Catherine Austin Fitts have issued strong warnings about the weaponization of mental health—a tactic used throughout history by authoritarian and totalitarian regimes to neutralize dissenters and impose other forms of social control.[191] In a December 2022 interview, Sherry Strong of Children's Health Defense Canada and Breggin discussed psychiatry's history as a tool of oppression used by "strong-arm governments," including to support mass murder, exemplified most notably in Nazi Germany.[192,193] Strong cited several Canadian examples of forced institutionalization and drugging of individuals who refused COVID vaccines, questioned the COVID shots' appropriateness for children, or spoke out about stillbirths related to the COVID injections. Canada's Centre for Addiction and Mental Health (CAMH) assessed COVID vaccine hesitancy in a 2021 mental health survey and pondered, somewhat ominously, "how hospitals, including CAMH, [could] best respond."[194] Unlike other hospitals in the Toronto region, CAMH included "people with a mental illness or substance-use diagnosis," as well as individuals with dementia, on its vaccine "eligibility" list.

Although not overtly related to the weaponization of mental health, France's National Assembly in early 2024 approved a sinister piece of legislation that would criminalize "any individual criticizing or encouraging others to avoid mainstream medical practices or health policies."[195] Medical dissent could be punishable by up to three years in prison and a fine of up to 45,000 euros (almost $49,000).[196,197]

Jon Rappoport and Celia Farber: The Spin Machine

"Build a false narrative and people will come in droves."
—Jon Rappoport

"On any given story, a dominant narrative takes shape, and journalists form a herd around it."
—Celia Farber

Widespread harm could not remain an endemic feature of Western medicine, decade after decade, without the back-up of a mighty propaganda apparatus tasked with contriving obfuscatory and fraudulent narratives. Media subservient to U.S. intelligence agencies and funded by agenda-driven billionaires serve as the public face of a spin machine that churns out carefully crafted soundbites.[198,199]

Prompted by what he learned in the 1980s diving into the underbelly of AIDS—resulting in his 1988 book titled *AIDS Inc: Scandal of the Century*[200]—writer Jon Rappoport has spent decades reporting on the medical cartel and its media lackeys at his blog (and, more recently, at Substack).[201,202] He often mentions the role of a craven and self-censoring press that, without any "prodding," toes the party line. In a 2018 article titled "Media won't investigate medically caused death numbers,"[203] Rappoport described a press corps fully willing to "defend the Crown" (a.k.a. the cartel):

Over the years, I've talked to reporters who are solidly addicted to obfuscations. Like any addict, they have an army of excuses to rationalize their behavior. The medical experts are worse. . . . When you peel the veneer away, they are enablers, persons of interest, co-conspirators. There is nothing quite like a high-minded, socially-positioned, card-carrying member of the King's circle of protectors. The arrogance is titanic. Because what is being hidden is so explosive.

One of Rappoport's most insistent insights—one that he has reiterated for every declared "epidemic" since AIDS, including COVID—has to do with the promotion of a "one disease, one cause" (or often, "one disease, one germ") narrative, rigid stories marketed with elaborate fear campaigns that often serve to deflect attention away from man-made causes of poisoning and toxicity (see **The "One Disease with One Cause" Narrative**).[204] Rappoport calls this "the motto engraved on the gate of the medical cartel," a "con job" that he equates with stage magic.[205] He summarizes the pharma-friendly tactic in a blog titled "The history of Big Pharma any idiot can understand":

> We SAY we've discovered a new disease. We SAY we know the cause. We SAY we have a drug to kill the germ and a vaccine to prevent the germ from taking hold. That isn't research. That's marketing.[206]

At the same time, marketing a "disease *without* a cause" is an equally effective business model, according to Rappoport:

> You make a list of symptoms. You say many people are experiencing this cluster of symptoms. You give a label to this list of symptoms. A name. The name of a disease or disorder or a syndrome. Over time, through promotion, the name sticks. You fund research to find the cause of the disease. This research can stretch out for a long time. Possibly forever. Meanwhile, you develop and sell drugs to treat the disease.

Illustrating this circular logic using the example of Parkinson's disease (PD), Rappoport points to the flimsy PD definition provided by the Parkinson's Foundation: "an extremely diverse disorder" caused by an unknown "combination of genetic and environmental factors."[207] The Foundation's list of 31 "PD" symptoms includes a wide range (such as apathy, fatigue, dizziness, and slowness of movement) that are not unique to PD, and the Foundation admits that "no two people experience Parkinson's the same way." Nevertheless, remarks Rappoport, "We know that [it's a disease] because it has a label. PD. And the label is the proof."[208]

The "One Disease with One Cause" Narrative

As Jon Rappoport has many times observed, the AIDS era taught him that "frightening diseases" sell—and especially narratives about scary viruses.[209] He comments, "[S]elling fear of THE VIRUS is big business. . . . [V]irus-stories are shaped and managed and written and managed and broadcast according to a plan that has nothing to do with actual disease."[210] In a blog titled "For alert minds: the art of the covert narrative,"[211] Rappoport outlines a few salient but neglected facts about past "outbreaks":

- **West Nile (U.S.)**: "Cases" correlated with "centers of spewing industrial pollution."
- **SARS (Hong Kong)**: A sizable proportion of "cases" came from an apartment complex "where feces were leaking into the internal water supply."
- **Swine flu (La Gloria, Mexico)**: Afflicted individuals worked at "a large commercial pig farm, where lagoons of pig feces were baking festering in the sun," a situation worsened by the subsequent spraying of toxic chemicals over the lagoons.
- **Ebola (West Africa)**: Many inhabitants already suffered from "horrific conditions" that are "endemic, chronic, and long-term."

As an interesting footnote to the Ebola story, Liberia's lead newspaper published reports at the height of the 2014 Ebola scare alleging that "agents" across the country were pouring formaldehyde into community water supplies to induce "Ebola-like symptoms," and also cited accounts of individuals "dressed as nurses" injecting "Ebola vaccines" thought to be "formaldehyde-water mixtures."[212]

The *Washington Post* vociferously dismissed these reports as "rumor-laden" conspiracy.[213]

Describing the crafting of these disease narratives, Rappoport asks, "[I]s it possible that jungles and Africa and China are typically chosen for virus fairy tales because, in the minds of many Westerners, they satisfy a requirement of 'strange,' 'different,' 'primitive,' and so on? We're talking theater here—and when you stage a propaganda play (fiction), you want to tap into the reflex instincts of the audience."[214]

Rappoport's insights about disease narratives developed in the 1980s when he looked into the framing of AIDS as a "syndrome" defined by one or more of over two dozen disparate "AIDS-defining illnesses" for which an obscure microbe called HIV reportedly held sole devious responsibility.[215] It quickly became evident to him that there were many other variables (including drug abuse, prison and military medical experimentation, pesticide exposure, and other lifestyle factors, to name just a few) that could explain the 27 symptoms called "AIDS." He also studied the "unreliable, deceptive, and useless" tests used to diagnose "HIV."

During the height of AIDS, Rappoport asked a leading virologist about the HIV theory of causation, and the scientist admitted—off the record—"[T]hey really haven't proved HIV causes AIDS." Many years later, in 2014, researcher Patricia Goodson published "Questioning the HIV-AIDS Hypothesis: 30 Years of Dissent" in *Frontiers in Public Health*—wherein she reasonably and methodically reminded readers of the plethora of non-HIV explanations for AIDS, including "pharmacological (drug) factors, immune disbalance factors, latent infection overload, and malnutrition."[216] She also noted that the question of causation was not trivial but had "life-and-death" ramifications, and she pointed out that the scientific orthodoxy had engaged in multiple "ethically questionable actions" to defend its tenuous position. It turned out, however, that even at the remove of several decades, the HIV-causes-AIDS model was still sacrosanct. After downgrading Goodson's article to an "opinion" piece in 2015,[217] the journal fully retracted it in 2019—after almost 92,000 views—claiming that because Goodson's article had attracted five times more attention than articles promoting the official position, it posed a "potential public health risk"![218]

In 2020, when officials began using the same questionable diagnostic tools to drum up hysteria about a microbe they called "SARS-CoV-2"—including in people who had no symptoms whatsoever—Rappoport recognized the pattern. Without denying that some people were experiencing serious flu-like illnesses and deaths, he suggested (on April 1, 2020) that readers might want to be careful about jumping to conclusions about unitary causes:

> There are people in Wuhan who have pneumonia because of the horrendous air quality in the city. There are people in New York who have ordinary flu-like illness. There are people in Italy who have histories of multiple, long-term, serious health conditions . . . made far worse through treatment with toxic drugs. There are people in hospitals around the world who, after being diagnosed with COVID, are dosed with powerful toxic antiviral drugs. There are people on breathing ventilators who are being given too much oxygen and too much pressure—and their lungs collapse. There are perfectly healthy people who are testing positive for the virus because the test is irreparably flawed. All these people are called "COVID cases."[219]

In a March 2023 article in Britain's *Off-Guardian*, journalist Kit Knightly reiterated many of Rappoport's cautions, summing up 40 facts about "Covid" [quotation marks in original],[220] 12 of which (pertaining to symptoms, diagnosis, and cases) are listed in Table 3. Again, none of Knightly's facts and explanations negate people's lived experience of illness; instead, they point to holes in the official one-cause-one-disease narrative.

Table 3. 12 "Covid" Facts about Symptoms, Diagnosis, and Cases
Fact: "Covid19" and the flu have identical symptoms. After pointing out the CDC's stance ("You cannot tell the difference between flu and COVID-19 just by looking at the symptoms alone because they have some of the same symptoms"),[221] Knightly noted that "coincidentally" in 2020, "cases of flu reportedly dropped to almost zero."
Fact: "Ground glass opacities" [lung findings on chest x-rays/imaging] **are not unique to "Covid."** According to Health.com, noted Knightly, "GGOs are not specific to COVID. They can show up due to other health conditions and infections that affect the lungs."[222]
Fact: A loss of smell and taste is not unique to "covid." Johns Hopkins Medicine lists 10 causes of smell and taste disorders (and does not even mention COVID): illness, head injury, hormone changes, dental or mouth problems, nasal polyps, exposure to chemicals, medicines, radiation therapy for head or neck cancer, snorted cocaine, and cigarette smoking.[223]
Fact: It is not possible to clinically diagnose "Covid19." Knightly commented, "Since 'Covid19' has no unique symptomatic profile, and since ALL major symptoms of 'Covid' can potentially apply to literally every common respiratory infection, it is impossible to diagnose 'Covid19' based on symptoms."
Fact: PCR tests were not designed to diagnose illness. Knightly quoted PCR inventor Kary Mullis as saying, "PCR is just a process that allows you to make a whole lot of something out of something. It doesn't tell you that you are sick."[224]
Fact: PCR tests have a history of being inaccurate and unreliable. PCR tests react "to DNA material that is not specific to Sars-Cov-2."
Fact: The CT [cycle threshold] **values of the PCR tests are too high.** "Based on what we know about the CT values, the majority of PCR test results are at best questionable."
Fact: The WHO (twice) admitted PCR tests produced false positives. *Off-Guardian* discussion of WHO memos here and here.[225,226]
Fact: Lateral flow tests [rapid do-it-yourself home test kits] **are unreliable.** "LFT and PCR results will often contradict one another. Meaning you can test positive on one, but not the other."
Fact: The scientific basis for all "Covid" tests is questionable. "If the [Corman-Drosten] paper [upon which the PCR assays are based] is questionable, every PCR test is also questionable."

> **Fact: "Covid case" numbers are inherently *meaningless*** [italics in original]. "If you cannot reliably test for the disease in a lab, and cannot identify it via a unique symptom profile, and many 'cases' are recognised as 'asymptomatic,' then 'Covid19' becomes a label with no meaning."

> **Fact: "Covid deaths" were created by statistical manipulation.** "Removing any distinction between dying *of* 'Covid,' and dying of something else *after testing positive for Covid* will naturally lead to completely meaningless numbers of 'Covid deaths'" [italics in original]. (Note: In 2021, two Oregon senators filed a grand jury petition alleging CDC and FDA inflation of COVID death data.)[227]

In his conclusions, Knightly also pointed to the simultaneous engineering of other causes of death—for example, deaths linked to the increased "poverty, malnutrition, drug and alcohol abuse and mental health problems" produced by the lockdowns, as well as hospital "guidelines" that killed patients with remdesivir, do-not-resuscitate (DNR) orders, and ventilators; these helped to create "increases in excess mortality which could be officially blamed on 'Covid.'" A real pandemic, Knightly added, "would not require corrupt testing practices and statistical sleight-of-hand to spread."[228]

Writer Celia Farber—who, like Rappoport, cut her journalistic teeth investigating AIDS in the eighties—experienced first-hand what happens to reporters who try to steer an independent course of inquiry. As she recounts in the 2023 Afterword to her re-released 2006 book, *Serious Adverse Events: An Uncensored History of AIDS*,[229] she was a young journalist starting out at *SPIN* magazine when, innocently enough, she began chronicling the putative new disease. Curious as to how the government could declare HIV the cause of AIDS by "fiat"—a historical moment Farber later characterized as both "scientific disgrace" and "theater of the absurd"—she also was baffled by the herdlike mentality that caused most journalists to behave as though "the only story about AIDS worth reporting was the government's version."

Several months prior to her book's 2006 publication, Farber published an AIDS exposé in *Harper's* that ran counter to the "government's version." The upholders of the official narrative used the article as a pretext to orchestrate a "firestorm" that ended up sending her into years of

professional exile. (Fans of Farber's fearless and distinctive writing currently can find her at her Substack blog, *The Truth Barrier*.)[230]

Farber's account in her book of the post-*Harper's* blitzkrieg illustrates
some of the tactics the medical establishment uses to keep would-be ques-
tioners tethered to official orthodoxies:

> The attack came in the form of a thirty-seven-page "accusation doc-
> ument" with eight authors, including Dr. Robert Gallo [propagator
> of the theory that HIV causes AIDS]. . . . None of the "errors" that
> the accusation document cited were genuine errors. Instead, they
> described *thought crimes* [italics added]. I had deviated from the
> core Weltanschauung [worldview] that AIDS research was sacred
> and must always continue unfettered, unquestioned, and massively
> funded—and that members of the AIDS Inc. research apparatus are
> saints, tirelessly saving lives. Even when they kill people.

The engineered uproar not only ensured a "swift death" for Farber's
book but guaranteed her "blacklisting" from the high-profile magazines
(*Esquire, Rolling Stone, Salon*, and many more) that previously had published her to great acclaim. As she sadly observed in her 2023 Afterword,
"there was only one thing in store for those who spoke out against the prevailing narrative: reputation destruction, targeting, and mobbing." Farber
and Rappoport both interviewed New-York-based physician Dr. Joseph
Sonnabend during the early AIDS years, and Sonnabend told Farber, "The
real villains are . . . [t]he journalists. We have traditionally depended on
the press to protect us from nonsense like this—not anymore." And as the
fate of Goodson's article (and many others) illustrates, scientific journals
also play a major role in propagating and enforcing monolithic official
narratives, through censorship as well as publication of what *Lancet* editor-in-chief Richard Horton, in 2004, dubbed "McScience."[231,232]

Like Rappoport, Farber recognized AIDS as a seismically important
chapter in the playbook for death by medicine—and in hindsight, as
the "dress rehearsal" for tactics deployed even more widely and skillfully
during COVID.[233] As professor and propaganda historian Mark Crispin
Miller wrote in his 2023 preface to Farber's re-released book, "COVID-19

was a movie that we'd seen before." The group Rethinking AIDS (now transitioned to "Rethinking AIDS/Unmasking Covid")—an alliance of scientists, physicians, and other researchers who first came together in the 1980s to examine "the extremely questionable claims being made about HIV and AIDS"—rapidly reached the same conclusion, describing the new "movie" as using:

> The same scare tactics, the same contradictions, the same wrong predictions, the same bogus tests, the same deadly treatment approach, all this but more global, more coordinated, and even more socially destructive. . . . The goal of science is the truth. The goal of marketing is the sale. "HIV/AIDS" and "Corona/Covid" were not discovered; they were *sold*, primarily through fear and lies of omission" [italics in original].[234]

Running the AIDS gauntlet taught the uncompromising Rappoport and Farber not only how to deconstruct false epidemics but how to zero in on the phony political and cultural narratives that buttress engineered health crises.[235] This positioned them to publish deft critiques of the "coronavirus" and "COVID" stories as soon as the official account began to unfold—and to doubt the good faith of the agencies and individuals promoting weirdly uniform talking points. In fact, as soon as the ready-made stock images of a "coronavirus" made their much-ballyhooed debut in January 2020, Rappoport began writing—producing a canon of 500 articles over the next two and a half years that itemizes a "massive scientific and medical fraud."[236]

Farber found officials' mobilization of COVID fears and their "death-sentence" rhetoric—what Mark Crispin Miller refers to as "terroristic fakery" and "wizardry"—eerily familiar. As her book explained about AIDS, officials fed the public a "terrifying message: Those with HIV would, without a doubt, die. They were entrapped by a scientific hypothesis and left no method of escape." That drumbeat of terror sent hundreds of thousands of healthy individuals running for the fatally toxic, fast-tracked drug AZT,[237] which eventually killed an estimated 300,000 gay men. AZT even had "a 'Skull and Crossbones' on the package insert, warning . . . pharmacists not to open the bottles nor touch the capsules for fear of acute poisoning."[238] Author John Lauritsen (deceased in 2022), who wrote the 1990 book *Poison by Prescription: The AZT Story*, concluded, "I don't think 'murder' is too strong a word to use when you have a drug like AZT, approved on the basis of fraudulent research."[239]

In a July 2023 blog,[240] Farber pointed out that AZT was not the end of the story where toxic AIDS drugs are concerned. Officials later used fear and the false and mechanistic "HIV causes AIDS" story to build the scaffolding for the creepy medical recommendation to take "pre-exposure prophylaxis" (PrEP)—toxic drugs claimed to "prevent HIV" that, in practice, have kept the "HIV" iatrogenic death toll going. Despite escalating lawsuits and PrEP drugs' association with bone loss, kidney failure, and death,[241] the manufacturer Gilead raked in $2 billion from its oral brands Truvada and Descovy in 2022,[242] and the FDA approved a third PrEP option, the injectable drug Apretude, in December 2021.[243] Lamenting the distressing mass medication of perfectly healthy people, Farber concluded: "I know [of] no cult more asphyxiating, dishonest, or deadly than the HIV/AIDS cult. The cult's core faith is that there is no 'treatment' protocol, no drug regimen, too toxic to be worshipped and violently defended."

Having observed the weaponization of fear during and after AIDS, Farber was ready, in March 2020, to ask, "At what point do you become unable, or unwilling to tolerate any more fear?"[244] Greatly concerned by the snow job underway, she described the mass media's "relentless terror pollution":

> [The media] get to "own" your mind and soul, inject and project right into it, mediate, frighten, confuse you, cause you to abandon your hopes, as learned helplessness finally sets in. . . . You are but a receptor site for para-governmental transnational corporate demoralization campaigns.

In Farber's view, AIDS fundamentally amounted to an "informational war . . . about how and why we think we know what we think we know." In 2021, she spelled out the essence of the war in comments included in Chapter 3 ("The HIV Pandemic Template for Pharma Profiteering") of Robert F. Kennedy Jr.'s bestselling book, *The Real Anthony Fauci*.[245] She subsequently deemed her insights about what happened after Fauci took the helm at the National Institute of Allergy and Infectious Diseases (NIAID) to be her "most important realization" about the propaganda machinery forged and refined during AIDS, and about its impact on American science and reverberations during COVID:[246]

> What Fauci did was he made political correctness the new currency, of his funding empire. . . . Fauci had, by 1987 . . . an apparatus that included mass media, psychological operations, public health—this octopus that just straight-up throttled the entire scientific tradition of Western civilization. Evidence based science and the discourse culture that goes with it—gone. That's what he did. It's no small feat. He destroyed American science by snuffing out . . . the spirit of open inquiry, proof and *standards* [italics in original].

The "throttling" Farber describes was also one of the threads that Rappoport explored in *AIDS Inc.*, where he noted that many scientists—protective of their jobs or grants—refused to talk to him on the record despite having "serious grievances about the way AIDS [was] being researched and explained to Americans." Discussing government-backed "totalitarian science" back in 2017, Rappoport suggested that knowledge is "irrelevant" to the government's fleet of "reality builders" (e.g., the press, corporations, foundations, academic institutions, and so-called "humanitarian" organizations), who all collude in "fabricat[ing] something that *looks like* the truth" [italics added].[247] Rappoport continued:

> The most useful politicians—as far as official science is concerned— are those who automatically promote its findings. Such politicians are lifted into prominence. They are champions of the Science Matrix. They never ask questions. They never doubt. They never make waves. They blithely travel their merry way into new positions of power, knowing they have enormous elite support behind them. When they need to lie, they lie.[248]

In short, Rappoport presciently argued in 2017, "official" science functions as a tool to "force an agenda of control over the population"; his 500-plus articles on COVID provide a methodical case study, illustrating the progression from lockstep to lockdown and tyranny.[249] In January 2021, he warned readers that while China's overnight lockdown of 50 million had provided a precedent that the rest of the world's leaders could then cite and copy, those initial lockdowns were only "phase one"—a preparatory step to "prepare" citizens for "phase two"—a "Lockdown Civilization" amounting to a "technocratic revolution."[250] (According to some analysts, the current push for "15-minute cities" is setting the stage for "climate lockdowns.")[251]

Rappoport's use of the word "technocratic" is no accident. Technocracy historian Patrick Wood describes technocracy as a socially engineered "resource-based economic system" in which "unelected and unaccountable technocrats" control both production and consumption.[252] In a 2005 "wake-up call" flagged by Wood, Twila Brase of the Citizens' Council on Health Care warned that technocrats were "taking over the practice of medicine" with the help of tactics such as "health data collection, guideline creation, intrusive clinical surveillance, pay-for-performance strategies, and centralized medical decision-making"[253] (see **Chapter Five** for a discussion of "evidence-based medicine"). In the current era of "surveillance capitalism," in which "data is the new gold,"[254,255] another writer has argued that electronic medical records (EMRs), "ensconced in a technocratic paradigm," are "facilitat[ing] and contribut[ing] to technocratic operations."[256] Wood's curated *Technocracy News & Trends* news service documents many other examples of the technocratic domination of medicine and public health.[257]

Emphasizing that technocratic medicine is "not patient-friendly," Brase warned that "the public should be alarmed" at the loss of "compassionate, first-do-no-harm, to-my-own-patient-be-true ethics of medicine." Catholic physician John Travaline of Temple University's school of medicine concurred in 2019, warning of the ethical "perils inherent in a technocratic model of medicine."[258] The features of technocratic medicine he outlined would not surprise anyone who has recently interfaced with an increasingly impersonal and unaccountable health care system (see **Features of Technocratic Medicine**). During COVID, stories about the health care system's willingness to accept federal incentives to commit homicide have become too numerous to count.[259,260]

Features of Technocratic Medicine

According to Temple University's John Travaline, medical technocracy "compromises good, compassionate care" and "threatens the practice of medicine and health care in general."[261] In his view, technocratic medicine:

- **Subordinates** patients to institutions and institutional goals
- Is "primarily **profit driven** and not necessarily value driven"
- Has no room for **moral reasoning** or analysis
- Overvalues **technology**
- Seeks to control **nature**
- Medicalizes **death**
- Is "**intolerant** of alternate ways of doing things"
- "Results in **domination and alienation** of individuals"

Travaline bravely urges his fellow physicians to "push back" and "[r]espectfully decline to participate in certain aspects of the [technocratic] system, though doing so might entail professional risks to the physician such as income loss or loss of practice privileges with an institution."

Matthew Cole, a political scientist who writes for entities that include The Orwell Foundation (the focus of which is to "celebrate honest writing and reporting" and "confront uncomfortable truths"), published an extensive post-COVID analysis of modern technocracy in *Boston Review* in August 2022.[262] For Cole, technocracy involves "ensembles of actors and institutions, typically but not always national or supra-national, that concentrate power among unelected experts and make binding decisions on the basis of expertise, as opposed to offering merely advisory input." Cole noted that "technocratic neutrality" often helps disguise "policies that benefit economic elites at the expense of ordinary citizens . . . as what simply must be done, rather than what a certain class or interest group insists on having done." The consolidation of centralized power and billionaire wealth enabled by the COVID policies that shut down small businesses (putting many out of business permanently) certainly underscores that latter point.[263] For these reasons, Cole maintains that it is vital for "those who care about good policy, . . . deep democracy or public citizenship" to subject technocracy to "serious scrutiny."

Cole remarks that central banks should top the list of "unelected experts"—as they represent the "epitome of technocratic power"—but suggests that public health and regulatory agencies also rank high on the list of unaccountable organizations:

> [B]oth the United States and the EU delegate a profound level of decision-making power to administrative agencies that receive little meaningful oversight from either the public or elected officials—not only central banks but institutions like the CDC, the Environmental Protection Agency, and the Food and Drug Administration.

In March 2023, reflecting on three years of COVID-related medical malfeasance, Karen Hunt speculated in *Off-Guardian* that technocrats may be conditioning the public to accept a "radical" medical/pharmaceutical agenda of forced medication.[264] As she summarized [bold and caps in original], the progression would go something like this:

- 1950s: "[T]hese drugs will make you feel better, **just try them**."
- 1960–2020: "**WE RECOMMEND** these drugs if you don't want to be sick, depressed or dead."
- 2020–2023: "YOU MUST TAKE these drugs or else you endanger your own life and the lives of those around you."
- Forthcoming: "**YOU ARE REQUIRED** to take these drugs by law and if you don't, you will go to prison for endangering the planet."

Forced medication has long been an acknowledged practice in psychiatric institutions,[265] including in persons held against their will who have "no mental health conditions and no prior mental health history."[266] As officials weaponize mental health and states like New York push for draconian "quarantine camp" regulations allowing the Department of Health to "lock you up or lock you down according to their whim for however long they want,"[267] the prospect of forced medication—including forced vaccination—no longer seems so far-fetched.

Rappoport, Farber, and *Off-Guardian's* Hunt have little patience with those who are unwilling to look the medical death machine squarely in the face. Rappoport unapologetically comments:

> Apparently, even many "alternative" journalists and doctors are keeping a piece of their souls in the official prison of fake medicine and fake science. On purpose. They want to hedge their bets. They want to go halfway, but not all the way. They want to admit some things, but not other things.[268]

Where doctors are concerned, he states:

> I'm talking about the armies of doctors who make their living diagnosing and treating one disease after another and giving vaccines left, right, and center. Most of the time, they're grossly dangerous and ineffective. They write toxic drug scripts like there's no tomorrow. They front for pharma.[269]

For those who want to be healthy, Rappoport's counterproposal is simple: "[F]reedom from harmful medical treatment is necessary for vitality to flourish."[270]

Sasha Latypova and Katherine Watt: Legalizing Democide

"We are observing a global crime scene."
—Sasha Latypova

"Observed harms caused by use of biochemical weapons labeled as vaccines . . . are intentional."
—Katherine Watt

Alexandra "Sasha" Latypova and Katherine Watt have been two of the most forthright voices discussing the "medicalized totalitarianism" that has been consolidating its hold on the U.S.—and the world—since 2020. Their independent research, published on Substack at *Due Diligence and Art* and *Bailiwick News*, respectively, has brought the weaponization of COVID-era medicine into sharp focus.[271,272]

Latypova is a former pharmaceutical industry entrepreneur and expert on drug manufacturing practices. Like Toby Rogers (mentioned in the Introduction), Latypova has come to view the injuries and deaths resulting from COVID injections as "a crime of mass murder and attempted mass murder by poisoning."[273] Watt, "a Roman Catholic, American, Gen-X writer, [and] paralegal," likewise believes that a "brutal" program of mass murder is underway,[274] enabled by a "pseudo-legal" framework carefully constructed through the use of "Presidential executive orders, Cabinet declarations, . . . administrative agency regulations..., and changes to the United States Code."[275] Though the public may find it gruesome to examine the notion of democide ("murder by government" or by "officials acting under the authority of government"),[276] the past four years' deadly events suggest that it may be risky to rule out the possibility.

Both individually and in collaboration, the meticulous and game-changing analysis by Latypova and Watt has helped expose the chain of command behind the lethal rollout of COVID "countermeasures" and the declaration of a Public Health Emergency (PHE) in the U.S.,[277] with the Department of Defense (DOD) in the lead.[278] As Latypova, Watt, and others noticed early on, DOD was at the top of the Operation Warp Speed (OWS) organizational chart as "Chief Operating Officer," while HHS was assigned a subordinate role as "Chief Science Advisor" (see **A Glaring Military Role**).[279] Within HHS, the agency's most "militarized" branch, the Biomedical Advanced Research and Development Authority (BARDA), was particularly active.[280] Military insiders have commented that the organizational structure of DOD's Defense Advanced Research Projects Agency (DARPA) "served as a model for OWS."[281] Other OWS partners included CDC, FDA, DOJ, the Department of Homeland Security (DHS), and pharmaceutical manufacturers Pfizer and Moderna.

Summarizing DOD's role, Latypova comments:

All covid countermeasures were ordered by [DOD], typically as "demonstrations" and "prototypes" via Other Transactions Authority [OTA] contracts. [*Note: See explanation of OTAs below.*] DOD partnered with HHS in order to over-ride the OTA restrictions of both, the DOD and HHS. DOD oversaw the development, manufacture, and distribution of the countermeasures.[282]

A military insider celebrates the OTA mechanism as a "flexible contracting vehicle" that makes it possible to "bypass the usual procurement regulations": "OTAs are . . . not subject to many of the statutes and regulations that . . . most federal contracts are subject to . . . and . . . were originally described by statute as 'a more permissible form of contract.'"[283] He adds, "OTAs are thought to attract innovative companies that otherwise would not have sought to do business with the federal government." Conveniently, the OTA contracts for OWS eliminated "all liability for the manufacturers and any contractors along the supply and distribution chain," as revealed when hundreds of partially redacted DOD contracts for COVID countermeasures became publicly available through FOIA requests.[284] As discussed below, even in the absence of contractual provisions, the PREP Act provides a liability shield for virtually any entity or individual along the supply and distribution chain.

The implications of military control are significant.[285] Latypova and Watt consider the COVID PHE[286]—as well as other PHE declarations, past and future—to have "largely the same legal status as war declarations,"[287] and they characterize the DOD-led rollout of the countermeasures as "an act of war."[288] Legal scholars candidly observe that "the Constitution is often reduced at best to a whisper during times of war," despite the fact that the Constitution "contains no 'emergency power' or general 'suspension' clause."[289] Latypova and Watt argue that the declaration of a PHE made it possible for the U.S. leadership to effectively suspend the Constitution, establishing, in Watt's words, a "democidal American public health-police state."[290]

Both writers also emphasize the global scale of the crime, showing that around the world, governments have all been using the same tools ("toxic by design" injections and informational warfare) and implementing the same "malignant" policies—including lies and cover-ups, gaslighting of the injured, suppression of dissent and whistleblowers, collusion with media, and perverse financial incentives that encourage and reward medical killing.

A Glaring Military Role

By September 2020, the military's role in Operation Warp Speed (OWS) was so glaring that it forced even mainstream news outlets such as *STAT* to acknowledge OWS as a "[291]highly structured organization in which military personnel vastly outnumber civilian scientists." With a senior federal health official reportedly "struck by the presence of soldiers in military uniforms walking around the health department's headquarters in downtown Washington," *STAT* provided other colorful details:

- The military's "extensive involvement" in the COVID shots' development and distribution was "a departure from pandemics of the past."
- Many of the "roughly 60 military officials . . . involved in the [OWS] leadership" had "never worked in health care or vaccine development."
- The military helped "prop up . . . vaccine manufacturing facilities—flying in equipment and raw materials from all over the world" and "set up significant cybersecurity and physical security operations," with powwows taking place "in protected rooms used to discuss classified information."
- OWS's "go-to vaccine coordinator" was someone who had "cut his teeth working on high-tech military projects," including human-implantable sensors.
- OWS paid for extensive security cameras around the facility of a materials science company manufacturing vials for the effort.
- Within HHS, BARDA's role in OWS was prominent, while CDC and FDA were "largely absent."

Latypova's previous work as a pharmaceutical and medical device research and development (R&D) executive endowed her with the knowledge and competencies to understand the rules and regulations by which HHS, the FDA, pharmaceutical companies, and "all public health actors" are supposed to abide. In her Substack orientation for new readers, she bluntly suggests that disregard for rules and regulations has become the norm:

> [T]he federal and most of the state governments are gone and captured. Whatever is running the federal gov[ernment] agencies (e.g., HHS) really intends to kill you, or at least substantially injure you, damage your reproductive capacity and repossess your assets in the process.[292]

Watt, for her part, has made enormous contributions by dissecting the legal and military underpinnings of PHEs and the other mechanisms that she believes have helped put a "kill box system" in place.[293] According to her research,[294] those mechanisms include "[a]t least six Congressionally-authorized statutory frameworks and related budget appropriations reinforced through Presidential Executive Orders and related executive branch declarations, and implemented through hundreds of regulatory amendments." Diving into these consequential legal weeds, Watt provides a sobering overview of the six layers:

- **1969:** According to Watt, the **Chemical and Biological Warfare Program** seems to authorize the use of human subjects for weapons R&D and allows executive branch suspension of statutes and regulations under "national emergency" conditions, "including apparent nullification of informed consent rights;" later amendments redefined bioweapons as "medical countermeasures."[295]
- **1983: PHE** legislation, says Watt, represented a "key turning point" that "added a new category of national emergency under which Constitutional and statutory protections for American lives, liberties and property . . . could apparently be suspended unilaterally,"

with subsequent amendments concentrating the power to declare a PHE into the hands of the HHS Secretary.[296]

- **1986:** In Watt's view, the **National Childhood Vaccine Injury Act** and **National Vaccine Injury Compensation Program (VICP)** established "a legal model and precedent providing civil and criminal immunity for producers, 'vaccinators' and others who manufacture and/or use products classified . . . as 'vaccines.'"[297] Children's Health Defense (CHD) concurs that the 1986 Act, which in essence abolished vaccine injury lawsuits, fundamentally altered the vaccine policy landscape in the U.S.[298] Congress then replicated the liability immunity model for "medical counter-measures" in its Countermeasures Injury Compensation Program (CICP).[299]

- **1997 and 2004:** After **expanding access to unapproved thera-pies and diagnostics**, notes Watt, Congress established the 2004 **Emergency Use Authorization (EUA)** program as part of the Project BioShield Act, authorizing "the HHS Secretary, at his or her sole discretion, to knowingly and deliberately suspend ordinary federal drug safety regulation;" Watt considers the EUA program "the key expansion that apparently enabled the Covid-19 'vaccine' bioagent attack on the American people."[300,301]

- **2002: National All-Hazards Preparedness for Public Health Emergencies**, says Watt, "expanded and centralized the managerial . . . chain-of-command, establishing parallel offices or directorates" within HHS, DOD, DHS, DOJ, and other federal agencies.[302]

- **2015: Research projects: transactions other than contracts and grants**: This layer of the legal framework allows DOD to use OTA contracts to carry out so-called "prototype projects"; this reduces Congressional oversight and "suspends most normal financial con-trols on federal spending."[303]

The in-your-face goal of the 1986 National Childhood Vaccine Injury Act was to "insulat[e] manufacturers from product liability suits," and it has achieved this goal in spades.[304] As CHD Chairman-on-leave Robert F. Kennedy, Jr. often points out, the Act enables vaccine manufacturers to sidestep responsibility for their products "no matter how toxic the ingredients, how negligent the manufacturer or how grievous the harm."[305] The taxpayer-funded, no-fault compensation mechanism created by the Act—the VICP—"wrests jurisdiction from traditional courts,"[306] but in the process eliminates the transparency and document discovery ordinarily associated with litigation. (For more information on the 1986 Act and related issues, see the CHD eBook, *Conflicts of Interest Undermine Children's Health*.)[307]

According to HHS, the VICP is supposed to function as an "accessible and efficient [administrative] forum for individuals found to be injured by certain vaccines;"[308] in practice, petitioners more often than not fail to obtain any compensation at all, and those who do must first slog their way through a combative process that imposes an almost impossible burden of proof. In 2015, Stanford law professor Nora Freeman Engstrom pronounced the VICP's performance "discouraging," "bleak," and "gloomy," concluding that "The VICP experience casts significant doubt on health courts' ability to offer [administrative] advantages" and suggesting that it "ought to shake public confidence" in a mechanism that Congress initially presented as a beneficial alternative.[309] In the even franker assessment of CHD attorney Rolf Hazlehurst, "The Vaccine Act has succeeded wildly at protecting the pharmaceutical industry and has failed miserably at compensating vaccine-injured children."[310]

The many VICP features that work against Congress's promises include the following:

- Decades into the VICP's creation, many parents, attorneys, and health care professionals remain unaware of the program's existence.[311]
- Petitioners face a three-year statute of limitations from the time of vaccine injury.[312]
- DOJ lawyers represent and aggressively defend the interests of HHS and, together with the "special masters" who serve as arbiters (appointed by U.S. Court of Claims judges), have created a process that is "inescapably" adversarial and "inquisitorial."[313] Attorneys representing the vaccine-injured have hinted at special master bias,

stating "the biggest factor in winning or losing a case . . . is which special master is assigned your case."[314]

- The challenges of proving vaccine injury causation—particularly in the context of the government-industry assertion that vaccines are almost always "safe and effective"—adds further "drag on the system." Drawing on what has been learned from occupational injury claims, Engstrom suggests that health conditions which are "non-signature, latent . . . and can arise synergistically from the interaction of several substances"—characteristic features of many vaccine injuries—make for "long delays, inconsistent outcomes, high rates of attorney retention, high levels of formal contestation, and a high degree of undercompensation."[315]
- Even for those who do obtain compensation, "adjudications within the VICP often take years and, in fact, take *longer* than litigation, to judgment, within the traditional tort system" [italics in original].[316]

The CICP presents individuals injured by EUA countermeasures with even more barriers, including a far shorter one-year statute of limitations for filing claims.[317] After the CICP awarded less than $4,700 to three individuals seriously injured by COVID vaccines (an average of $1,545 each), leaving 8,130 claims unaddressed as of April 2023, CHD's Laura Bono commented,

A payout of roughly $1,000 for myocarditis when the mortality rate increases to 50% within five years of diagnosis is absolutely insulting. . . . While victims linger with their injuries, paying out-of-pocket for expenses, or at worst die, the industries run to the bank.[318]

Returning to the role of DOD, Watt's painstaking analysis of existing legislation has given her the confidence to allege that "The US military is actively engaged in an organized criminal enterprise to injure and kill large numbers of military personnel and civilians without detection or legal impediment."[319] As regards military personnel, history reveals numerous examples of DOD willingness to experiment on its soldiers. A 2015 segment on National Public Radio reported that between 1922 and 1975, the military tested substances designed "to induce symptoms such as 'fear, panic, hysteria, and hallucinations'" on tens of thousands of troops, while withholding information about potential risks.[320] (Soldiers later filed a class action lawsuit seeking to obtain disclosure and proper medical care.) In the late 1990s, as extensively documented by Dr. Meryl Nass (member of CHD's Scientific Advisory Committee and founder of Door to Freedom),[321] the military mandated an anthrax vaccine even though "science to support the program did not exist;" when service members either refused the vaccine or developed illness following vaccination, they were met with "court-martials, fines, and less-than-honorable discharges."[322] A similar response greeted soldiers who refused the mandated COVID shots, with relentless "bullying, coercion and pressure" to take the jabs; some members of the military report that the policy has led to significant attrition, provoking a "manpower crisis."[323]

Watt summarizes her analysis as follows [excerpts rearranged into a list with bold added]:

- "One of the most useful tools in the arsenal—because it strikes an effective balance between . . . speed and deniability—is the deployment of prohibited biochemical weapons **labeled as** FDA-authorized or FDA-approved '**vaccines**.'" (Elsewhere, Watt maintains that "All FDA activity that appeared to be license-related, pertaining to all biological products manufactured since May 2019, has been fraudulent, performative, charade, pretextual, and any other word or phrase that means not real, not substantive, not legally relevant."[324]

- "The most recent and most visible phase of the program launched in the US in early 2020, under the title Operation Warp Speed, and resulted in global deployment of **psychological fraud and control programs** including terrorizing propaganda; social isolation; mask mandates; diagnostic tests; manipulated data presentations . . . ; prohibition on treatments for symptoms; and financial coercion of hospitals and nursing home death protocols (sedatives, ventilators and toxins)."

- "These components were followed by distribution [in the U.S.] of **three brands of biochemical weapons** (Pfizer-BioNTech, Moderna and Johnson & Johnson) with an unknown number of different batch formulations."
- "Independent researchers have identified some but not all components of some vials . . . including heavy metals, genetic code fragments, and many other contaminants not listed on applications submitted to regulators by manufacturers, who are **working under redacted contracts for the US Department of Defense**."[325]

When the HHS Secretary opens the door to EUA medical countermeasures, the FDA is empowered to authorize "unapproved medical products or unapproved uses of approved medical products," but only if "certain criteria are met"; notably, there must be "no adequate, approved, and available alternatives."[326] Early in 2020, doctors in the U.S. and elsewhere identified at least two inexpensive drugs—hydroxychloroquine and ivermectin—that not only produced promising results but had established safety records. Through misrepresentation of science, censorship, and attacks on the professional reputations and licenses of doctors who promoted the two drugs, FDA was able to make a tenuous case for awarding EUA status to far riskier countermeasures like remdesivir and the COVID shots.[327,328,329]

Following their jury-rigged assertion that the EUA criteria had been met, the FDA and CDC continued "knowingly and willfully maintaining a fraudulent pseudo-'regulatory' presentation to the public," according to Latypova.[330] During and after the rollout of the COVID injections, for example, CDC claimed that the shots had undergone "the most intense safety monitoring in US history" and had met the FDA's "rigorous scientific standards for safety, effectiveness, and manufacturing quality needed to support . . . EUA."[331] In truth, as Latypova reminded *The Defender* in a June 2023 interview, EUA countermeasures are "not subject to the same regulations as typical pharmaceutical manufacturers, distributors or regulatory agencies;" again, DOD and BARDA contracts with the manufacturers of the COVID shots "were structured such that [normal consumer] protections weren't required."[332]

Spelling out the scale of these "wartime" arrangements, Latypova writes:

The underlying FDA authorizations and approvals under EUA statutory authority and Investigational New Drug regulatory frameworks all violated drug safety laws governing clinical trials, product
labeling, product serialization, importation, product distribution,
product quality control testing, dispensing and other parts of the
national drug supply oversight system.[333]

Sticking with military phraseology, Latypova and Watt describe the
mRNA-lipid nanoparticle (LNP) compounds passed off as COVID vaccines as "cellular genetic dirty bombs," where "The nurse-with-needle is
the bomber," the LNP is "the suitcase used to smuggle the bomb into
cells," and the genetically modified RNA is the "bomb."[334] (See **Chapter
Three** for more discussion of the mRNA and LNP technologies.) And if
there were a covert goal to wreak medical carnage, one would have to concede that the "dirty bombs" have been successful, generating a heretofore
unprecedented scale of post-vaccination deaths, injuries, and disabilities,
both in the U.S. and internationally.[335,336,337]

Looking at the historical record, Watt now considers vaccine production facilities to be "indistinguishable from bioweapon production facilities," and vaccines to be "indistinguishable from bioweapons,"[338] and—as
she discusses in a post titled "93 biochemical weapons to decline whenever
a medical mercenary offers them to you or your children" (referring to
the nearly eight dozen FDA-licensed vaccines)—she does not exempt pre-
COVID vaccines from this assessment. Nor is she an outlier in her view
that vaccines "have been intrinsically injurious from the start of government campaigns promoting their use more than a century ago;"[339] providing an extensive reference list of researchers "who have traveled similar
paths" and reached similar conclusions, she observes:

Each researcher has compiled evidence that US government statements about military, public health, and vaccination program objectives, historical events and scientific, regulatory data, have been
demonstrably false for a very long time, and each researcher's work

has been suppressed and maligned, to prevent widespread public interest in it and access to it.[340]

In late 2023, Watt reminded her readers that the federal government has classified vaccines as "biological products" since 1970, and chillingly described a further reshaping of biological product definitions and oversight that took place in May 2019,[341] when then-FDA Commissioner Scott Gottlieb eliminated the legal requirement for FDA inspectors "to inspect all establishments or facilities producing biological products at least once every two years." Gottlieb also got rid of inspectors' "eight enumerated inspection duties." In one fell swoop, the legal mechanisms through which FDA previously had—at least superficially—regulated biological product manufacturing. The implications are disturbing:

> There is currently no legal requirement for an initial FDA inspection; no minimum interval for subsequent FDA inspections, and there are no legal consequences for compliance failures, such as establishment or product license denial or revocation.[342]

CHAPTER THREE

TWENTY-FIRST CENTURY VACCINE TECHNOLOGIES

As the first two chapters already have shown, examples of medical malevolence—past and present—are not hard to come by. Unfortunately, new technologies and policies added to the arsenal in recent years have opened up further avenues for the medical pharmaceutical killing machine to do its work. Chapter Three focuses on 21st-century mRNA vaccine technologies.

Vaccines have always, and irrefutably, caused illness and death—from the havoc-wreaking smallpox vaccines of the 18th and 19th centuries,[1] to the highly reactive diphtheria/whole-cell pertussis/tetanus (DPT) vaccines used from the mid-20th century through the late 1990s,[2] to the recombinant (genetically modified)[3] and aluminum-adjuvanted HPV jabs,[4] which, until the COVID injections, held the record for being some of the most harmful shots ever licensed.[5] The $5 billion-plus paid out to the injured by the VICP provides one indicator of the extent of harm, albeit a muffled one, given the vast underreporting of vaccine injuries and the VICP's miserly compensation to only a third of petitioners.[6]

In December 2019, the Grand Poobahs of global vaccinology admitted to being aware of the reprehensible safety record. In behind-closed-door confessions caught on camera at a WHO summit, they characterized public and professional confidence in vaccines as "wobbly," and deservedly so.[7] While fretting over the rise in public distrust, the gathered experts essentially acknowledged the lack of relevant vaccine safety science and the fact that monitoring systems permit routine obfuscation of serious adverse events.[8]

At the time, it almost seemed as though the momentous WHO disclosures might presage the death knell for vaccine "business as usual." In January 2020, however, many of the same experts suddenly began drumming up fear about a "coronavirus" threat and started setting the stage for coercive policies designed to stifle the mounting "vaccine hesitancy."[9] With extensive media support,[10] "COVID-19" not only became the means of sending the stunning December 2019 admissions down the memory hole but massaged public opinion into widespread compliance with subsequent mandates to get ethically untenable COVID-19 injections.[11]

Conveniently, COVID also furnished the excuse to debut mRNA technology,[12] which—in conjunction with nanotechnology and various undisclosed and mystery ingredients—has made "vaccines" more unsafe than ever.[13,14,15] As Meryl Nass and others have reasoned, "maybe the vaccines were not made for the pandemic, and instead the pandemic was made to roll out the vaccines."

Modified RNA

DOD's DARPA officially began incubating mRNA technology in the early 2010s,[16] consolidating scientific developments in "RNA biology, immunology, structural biology, genetic engineering, chemical modification, and nanoparticle technologies."[17] In a 2023 paper,[18] U.S. Air Force lieutenant colonel Daniel Schoeni observed that Moderna received its first $25 million grant from DARPA in 2013, cheerfully attributing the "unbelievably rapid pace" of OWS to "the fact that DARPA had already made [these and other] key investments in vaccines pre-COVID." (Schoeni also cutely pointed out that the last three letters of "Moderna" spell "RNA," but he did not dwell on the fact that "mod" is short for "modified," meaning "synthetically designed.")[19]

In using COVID to get mRNA technology "across the finish line,"[20] the U.S. military and its health agency and industry partners[21] positioned the world for a full-steam-ahead, mRNA-dominated pharmaceutical land-scape. On the cusp of the COVID vaccine rollout in December 2020, Jon Rappoport warned that this feat would set the stage for endless genetic-based mRNA injectables (cleverly shoehorned into the CDC's revised def-inition of "vaccines")[22] and launch a highly profitable gold rush of biotech creations that, from a killing machine perspective, are "much faster, easier, and cheaper to produce."[23]

Since 2020, evidence of biotechnology's massive mRNA pivot has been everywhere. For example:

- **Mid-2021:** The news outlet *Fierce Biotech* exulted that pharma giants were "crowd[ing] into mRNA" as the "new frontier for vac-cine development."[24]
- **Late 2021:** The head of Bayer's Pharmaceuticals Division, Stefan Oelrich, celebrated the fact that mRNA vaccine technology and related forms of cell and gene therapy—despite well-documented and even species-threatening dangers—had overcome both regula-tory and public palatability hurdles.[25,26,27,28]
- **Early 2022:** The Cleveland Clinic enthusiastically predicted that a "next generation of mRNA vaccinology" would be one of the "Top 10 Medical Innovations for 2022,"[29] while a Dana-Farber Cancer Institute executive enthused that funding and other resources were "flowing into mRNA vaccine research."[30]
- **Fall 2023:** CNN reported that Pfizer and Moderna were "racing to bring mRNA technology to seasonal flu vaccines," with both also announcing plans to develop triple-whammy mRNA shots for COVID, flu, and respiratory syncytial virus (RSV).[31]

As manufacturers salivate over the prospect of bringing ever more mRNA products to market, they are particularly excited about extending their reach to conditions that previously stumped vaccine developers, such as parasitic infections,[32] Lyme,[33] and Zika.[34] Not coincidentally, mRNA jabs in the pipeline also target some of the most common adverse events aris-ing from the COVID shots (notably cancer and heart problems)—a neat "create-a-problem, develop-a-drug-to-manage-the-problem" trick that perpetuates the poisoning of the population while fortifying the pharmaceutical industry's bottom line.[35,36,37]

In mid-2023, *Fierce Biotech* reported that Merck and Moderna were partnering in a phase 3 clinical trial to test a "personalized" mRNA melanoma vaccine, and that the trial had "stirred up much excitement within the oncology community."[38] Dozens of other clinical trials are experimenting with mRNA vaccines to treat people with "various types of cancer."[39] *STAT* breathlessly asked in late 2022, "mRNA revolutionized the race for a Covid-19 vaccine. Could cancer be next?"[40]

In the cardiac space, UK and U.S. researchers are investigating the use of "exactly the same technology as the Pfizer and Moderna vaccines to inject micro RNAs to the heart," claiming they can use mRNA technology to get heart cells that survive a heart attack to proliferate.[41] At the University of Alabama at Birmingham, cardiac researchers describe their modified mRNA translation system as "SMRTs," although they acknowledge that their not-so-smart "lipid nanoparticle-encapsulated modified RNA" tends to want to head to the liver, not the heart.[42]

The formal inauguration of mRNA technology may be a major feather in the cap of the military-medical-biopharmaceutical cartel, but for the world's citizens, it has represented a horrific and unsubtle ramping up of iatrogenesis.[43] A variety of studies confirm that the COVID vaccines have been responsible for adverse events and fatalities shocking in both their breadth and global scale.[44] In Europe, documents in the European Medicines Agency's possession showed nearly five million adverse events associated with the Pfizer mRNA shots over just a six-month period (December 2021–June 2022).[45] In Korea, roughly 90% of respondents in a nationally representative survey reported adverse events after mRNA vaccination, including menstrual disorders and unexpected vaginal bleeding in over 15% of female respondents.[46] And in the U.S., an online survey of thousands of recipients of the Pfizer and Moderna mRNA jabs found that two-thirds (65%) had experienced adverse effects after one dose, and the likelihood of "severe or very severe adverse effects" significantly increased with further doses.[47]

The *War Room/DailyClout* analysis of 55,000 Pfizer documents released by court order and analyzed by six teams of 3,500 expert volunteers (with

94 reports published as of mid-January 2024) has provided extensive and chilling details about the wide range of debilitating or fatal effects associated with the mRNA shots.[48] In the words of *DailyClout* CEO Dr. Naomi Wolf, the Pfizer documents constitute "a record of a great crime against humanity."[49] The dozens of reports document "neurological events, cardiac events, strokes, brain hemorrhages, and blood clots, lung clots and leg clots at massive scale" as well as "a 360-degree attack on human reproductive capability" and many other types of immediate and long-term damage. One need only look at the titles of some of the *War Room/DailyClout* reports to appreciate the range of harms associated with the mRNA injections (see Table 4).

Table 4. Selected Reports from *War Room/DailyClout* Pfizer Documents Analysis	
Report #	**Report Title**
#13	Adverse events rise in babies breastfed by vaccinated mothers
#33	Pfizer, FDA, CDC hid proven harms to male sperm quality, testes function, from mRNA vaccine ingredients
#43	Blood system-related adverse events following Pfizer COVID-19 mRNA vaccination
#45	Clotting system-related adverse events following Pfizer COVID-19 mRNA vaccination
#46	Serious stroke adverse events following Pfizer COVID-19 mRNA vaccination
#60	449 patients suffer Bell's palsy following Pfizer mRNA COVID vaccination in initial three months of rollout. A one-year-old endured Bell's palsy after unauthorized injection
#61	Histopathology series part 3 – Ute Krüger, MD, breast cancer specialist, reveals increase in cancers and occurrences of "turbo cancers" following genetic therapy "vaccines"
#62	Acute kidney injury and acute renal failure following Pfizer mRNA COVID vaccination. 33% of patients died. Pfizer concludes, "No new safety issue"

(Continued on next page)

Report #	Report Title
#68	Blood vessel inflammation, vasculitis, adverse events occurred in first 90 days after Pfizer mRNA "vaccine" rollout, including one fatality. Half had onset within three days of injection. 81% of sufferers were women
#71	Musculoskeletal adverse events of special interest afflicted 8.5% of patients in Pfizer's post-marketing data set, including four children and one infant. Women affected at a ratio of almost 4:1 over men
#75	mRNA COVID "vaccines" have created a new class of multi-organ/system disease: "CoVax disease." Children from conception on suffer its devastating effects. Histopathology series – Part 4d
#78	Thirty-two percent of Pfizer's post-marketing respiratory adverse event patients died, yet Pfizer found no new safety signals
#79	mRNA COVID vaccine-induced myocarditis at one year post-injection: spike protein, inflammation still present in heart tissue

Moreover, the worst may be yet to come, as Brian Hooker and Margot DesBois suggest in their chapter of the 2023 book *mRNA Vaccine Toxicity* by the group Doctors for COVID Ethics (D4CE):

[T]he available evidence begins to piece together a concerning picture of illness, disability, and death following mRNA COVID-19 vaccination. And these are primarily only the short-term effects, observed within days to six weeks post-injection. It may take **months, years, and decades** for the damage of these toxic biological agents to manifest in chronic cardiac, thrombotic, neurological, immune, reproductive, and other organ dysfunction" [bold added].[50]

In partial confirmation of this somber prediction, an international research group published a study in *Radiology* in September 2023 reporting that people who had received a second COVID vaccine dose within the past 180 days, though asymptomatic, showed significantly increased signs of heart inflammation compared with nonvaccinated patients.[51] A few months earlier, a study in *Frontiers in Cardiovascular Medicine* had noted that one-fourth—a "relevant proportion"—of individuals with mRNA-vaccine-related myocarditis experienced "persisting symptoms" after a median of 228 days, even though their baseline symptoms had not been particularly severe.[52] And in early 2024, a study from Saudi Arabia described heart complications emerging anywhere from one month to over a year after the affected individuals had received mRNA COVID shots.[53]

In a rigorous survey that looked at deaths related to the COVID mRNA jabs, Michigan State University researcher Mark Skidmore estimated the number of U.S. fatalities as of December 2021—that is, just one year's worth of deadly incidents—at 278,000 Americans.[54] (After tens of thousands of downloads, the journal *BMC Infectious Diseases* retracted his article for entirely unconvincing reasons.) Complementing this picture, an analysis by a retired social worker and disabled vet "horrified by all the death and dying around him" found that as of August 2023, U.S. obituaries showed a 62% increase in deaths described as "sudden" or "unexpected" following the rollout of the COVID injections.[55] Canada not only charted a 17% increase in deaths from 2019 to 2022 (despite the population increasing by just 2%), but showed a 475% increase in deaths from "ill-defined and unspecified causes of mortality."[56]

In September 2023, Canadian researcher Denis Rancourt and colleagues published a 17-country study of "COVID-19 vaccine-associated mortality in the Southern Hemisphere"; their devastating analysis revealed a "definite causal link" between "peaks in all-cause mortality and rapid vaccine rollouts."[57] The authors soberly concluded:

> The overall risk of death induced by injection with the COVID-19 vaccines in actual populations . . . is globally pervasive and much larger than reported in clinical trials, adverse effect monitoring, and cause-of-death statistics from death certificates, by **3 orders of magnitude (1,000-fold greater)** [bold added].

In the past, U.S. authorities occasionally have found it politically expedient to respond to public concerns about vaccine-related fatalities by ordering recalls; they recalled polio and swine flu vaccines in less than a year's time after just 10 and 53 reported deaths, respectively.[58] Although FDA did revoke the EUA for the Janssen/Johnson & Johnson (J&J) COVID-19 shot after the CDC admitted to a "plausible" link with blood clots,[59,60] tellingly, the equally risky or even more dangerous Pfizer and Moderna mRNA injections and boosters remain on the market over three years after their deadly premiere, with regulators feigning an unsullied safety record. Addressing this inconsistency, the Substack writer who publishes under the moniker *A Midwestern Doctor* explains:

> At the start of Operation Warp Speed, I hypothesized that a major goal was to get mRNA technology onto the market . . . (but since there were safety challenges with it, nothing short of an "emergency" would be able to break the barrier to human testing). Because of this, I suspected that once vaccine safety concerns emerged, a non-mRNA COVID-19 vaccine would be thrown under the bus to make the mRNA technology look "safe." This is what then happened with the J&J vaccine.[61]

In January 2024, a prominent group of medical researchers, including Dr. Peter McCullough, published a paper in the journal *Cureus* on the COVID-19 mRNA vaccines, "exposing how the industry-sponsored [registrational] trials misled the public," describing an "unacceptably high harm-to-reward ratio," and calling for a global moratorium on the shots.[62] No recalls followed; instead, *Cureus* retracted the article.[63]

In a short documentary released in October 2023, "Cutting Off the Head of the Snake," Dr. Astrid Stuckelberger (a global health expert and long-time consultant for the United Nations [UN] and other international organizations) and retired Swiss banker Pascal Najadi argued that the mRNA shots are bioweapons—"poison injected into humanity"—and tools of democide.[64] In *mRNA Vaccine Toxicity*, D4CE, too, confronts what it views as a "sinister agenda":

> [W]e will make the case that most of the observed severe harm is best understood in terms of these vaccines doing what they are designed to do: the harm is not accidental but rather built into the mRNA technology.[65]

Moreover, D4CE argues, because of the "built-in" and "by design" risks, the public must expect "similar levels of danger and damage with future mRNA vaccines," regardless of the disease or condition such products ostensibly target.

Lipid Nanoparticles (LNPs)

D4CE devotes a chapter of *mRNA Vaccine Toxicity* to the topic of LNPs. In the scientific community's language, LNPs are "sub-micron particles containing ionizable cationic lipids in addition to other types of lipids and encapsulated nucleic acid cargo."[66] Setting this jargon to one side, the immediate thing to understand about LNPs is that they are the biotech industry's trendy Trojan-horse strategy to get mRNA vaccines' bulky "payload" (said to consist of modified-RNA but perhaps containing other or additional components) into cells intact.[67] In other words, manufacturers claim, their LNPs function as a built-in delivery mechanism or "bubble,"[68] "encapsulat[ing] the mRNA constructs to protect them from degradation and promote cellular uptake."[69] Some scientists rave that LNPs used in conjunction with synthetic mRNA "enhance cell uptake and expression by up to 1000-fold compared to naked mRNA."[70]

An additional LNP characteristic—desirable from the standpoint of players who are either heedless of or intent on causing harm—is that LNPs rev up the immune system. Vaccine scientists tamely refer to this as LNPs' "inherent" adjuvant properties.[71] Researchers worried about the "revving-up" warned in 2022 about "overenthusias[m]" for LNP-based vaccines and nanomedicines, due to "pro-inflammatory concerns" with intramuscular, intradermal, and intranasal administration routes; in mice, the researchers noted, "Intranasal inoculation with LNPs led to massive inflammation in the lungs and a high mortality rate . . . in a dose-dependent manner."[72] Thus, it can hardly be reassuring that manufacturers are now eyeing LNP-based nasal and inhalable vaccines—seemingly untroubled by the evidence that this could lead to "sustained pathologic inflammation."[73,74,75]

According to the authors of *mRNA Vaccine Toxicity*, LNPs have a disturbingly "facile" ability to "cross the walls of blood vessels in the brain," ending up in brain tissue; D4CE describes this property as "remarkable, considering that the blood vessels of the brain are generally less permissive to solutes and particles than are those of other organs." Discussing LNPs as a potential feature of future nanomedicines explicitly targeting the brain, the researchers who cautioned the scientific community about LNP "overenthusiasm" concluded:

> Considering the pro-inflammatory nature of the currently available ionizable cationic lipids, notably their **undesirable immune cascade** . . . and of other cationic lipids, the potential application of LNPs for systemic administration must be viewed cautiously. This is important, particularly when targeting biological barriers such as the blood-brain barrier with intravenously administered LNPs . . . to combat neurological diseases and disorders, which could initiate **severe inflammatory reactions in the brain** [bold added].[76]

Another problem with lipid nanoparticles, described in a multipart exposé by Heather Hudson at *A Mother's Anthem* on Substack, is that LNPs actually are not guaranteed to make it all the way into cells; instead, they can remain in circulation or fall apart, leaving their lipid components free to roam the body.[77] One of those components is polyethylene glycol (PEG),[78] a synthetic, nondegradable polymer that is increasingly controversial but remains a pharmaceutical industry darling precisely because of "PEGylation's" professed ability (not always borne out in studies) to keep payloads circulating in the blood longer. Both the Pfizer and Moderna mRNA jabs feature LNP formulations that are PEGylated,[79] meaning, as D4CE describes it, that the PEG "decorates the particle surface."

The inclusion of PEG in the mRNA COVID shots rang alarm bells right from the start due to its known, life-threatening anaphylactic potential.[80] CHD put the FDA on notice about PEG and the risk of hypersensitivity reactions several months before the FDA granted EUA status to Pfizer's and Moderna's mRNA injections; sure enough, anaphylaxis was one of the very first types of adverse events reported.[81,82] The FDA paid no attention to CHD's warnings; this came as little surprise because, for many years, the agency had ignored thousands of reports from parents about their children's severe neuropsychiatric reactions to PEG (such as mood swings, rage, phobias, and paranoia) after taking the PEG-containing laxative MiraLAX (see **The MiraLAX Brush-off**).[83]

The MiraLAX Brush-Off

For decades, tens of thousands of parents have reported that laxatives containing a version of PEG called PEG 3350 caused dramatic neuropsychiatric symptoms in their children, ranging from seizures to psychosis.[84] PEG 3350 is the active ingredient in MiraLAX,[85] originally developed by Braintree Labs and now a Bayer product.

In 2006, the FDA approved switching MiraLAX's status from a prescription drug to an over-the-counter (OTC) drug.[86] Although Braintree developed MiraLAX for short-term use in *adults*, the drug's OTC availability encouraged pediatricians to start recommending it for constipated *children*, sometimes for months or years at a time, heedless of the fact that children "are likely to receive a higher dose per unit body weight than adults."[87] FDA reviewer Dr. Karen Feibus acknowledged, at the time of the OTC status change, "a theoretical concern with long term or frequent repeated use of PEG by consumers;"[88] however, no regulator has ever taken the step of actually assessing the risks to children of prolonged PEG 3350 exposure.

A 2012 citizen petition urged FDA to study both the short-term and long-term pediatric safety of PEG 3350.[89] With its hand forced by the petition, the National Institutes of Health (NIH) in 2014 awarded a $325,000 grant to the Children's Hospital of Philadelphia to assess PEG 3350 safety in children in collaboration with the FDA,[90] but the study apparently went nowhere. The NIH and FDA initially pledged that results would be available within a year, but by late 2022, the study status still showed up online as "recruiting," with a listed completion date of June 2024.[91] In a September 2022 email to a parent whose son had experienced a serious neuropsychiatric reaction to MiraLAX, FDA Public Affairs Specialist Paul Richards offered more prevarication, writing, "While the FDA and CDC cannot predict how long the review of the data and information will take, we will complete our review as expeditiously as possible using a thorough and science-based approach."[92]

There are many other documented and suspected problems with PEG. To name just a few:

- More than seven in ten people may already have some level of sensitization to PEG, predisposing them to adverse reactions and risks that are likely to worsen with each subsequent exposure.[93] In addition to injectable vaccines, PEG is present in other ingested and topical products, including drugs, personal care products, lubricants, gels (such as ultrasound gel), and food additives.
- Some research indicates that PEGylated particles promote tumor growth.[94]
- Studies suggest cross-reactivity between PEG and polysorbates (surfactants/emulsifiers present in consumer products and drugs, including some vaccines, and dangerous in their own right), heightening the risk of hypersensitivity reactions.[95]

The mRNA concoctions that Pfizer and Moderna have marketed as vaccines also include three other lipids inside their LNP carrier systems:[96] cholesterol, the phospholipid DSPC (distearoyl phosphatidylcholine), used to confer "rigidity and stability,"[97] and the cationic (positively charged) synthetic lipids ALC-0315 (Pfizer) and SM-102 (Moderna). In 2021, a letter to the editor of the journal *Allergy* warned that as an LNP component, DSPC could potentially act as an allergen.[98] As for the two synthetic lipids (ALC-0315 and SM-102), which "account for almost half of the total lipid in the vaccine LNPs," D4CE asserts that both contain "unknown amounts of unknown impurities" and that both "can exert toxicity outright."[99] D4CE continues:

> The conclusion is unavoidable that both manufacturers and regulators have acted with gross negligence. This inference is reinforced by the reckless manner in which [the European Medicines Agency] and other regulators brushed aside concerns over lacking quality information pertaining to the novel lipids used by both manufacturers and proceeded with approval.

In studies, SM-102 has shown evidence of being a potent activator of inflammatory responses.[100] Meanwhile, the manufacturer of ALC-0315, Echelon Biosciences, specifies that its lipid is "for research use only,"[101] a fact that in December 2021 prompted European parliamentarian Guido Reil to voice the objection that "Administering a vaccine—particularly to children—which contains unauthorised excipients is illegal, dangerous and unethical"; Reil took the European Commission to task for "distributing a product that is harmful to public health."[102]

According to Heather Hudson, Pfizer internally acknowledged the potential for the LNPs in its COVID mRNA injections to trigger autoimmune and proinflammatory processes "that could, in turn, result in the development of myocarditis and potentially cardiac death."[103] In 2018, Moderna, too, owned up to the serious risks of its LNPs in a corporate prospectus that supported its initial public offering (IPO), describing various problems and acknowledging PEG's known potential to induce hypersensitivity reactions—"immune-mediated toxicities" associated with "severe allergic symptoms with occasionally fatal anaphylaxis."[104,105,106] On p. 33 of the prospectus, Moderna disclosed:

> [T]here can be no assurance that our LNPs will not have undesired effects. Our LNPs could contribute, in whole or in part, to one or more of the following: immune reactions, infusion reactions, complement reactions, opsonization reactions, antibody reactions . . . or reactions to the PEG from some lipids or PEG otherwise associated with the LNP. Certain aspects of our investigational medicines may induce immune reactions from either the mRNA or the lipid as well as adverse reactions within liver pathways or degradation of the mRNA or the LNP, **any of which could lead to significant adverse events in one or more of our clinical trials**" [bold added].[107]

Tracing the history of LNPs, Hudson points out that one of the scientists involved in their development, Janos Szebeni, carried out critical work in Hungary with someone who is now a senior vice president at BioNTech; Szebeni also did LNP-related work for the U.S. military.[108] In a series of papers published from the late 1990s on, Szebeni documented "complement activation-related pseudoallergy" (CARPA) as a frequent problem associated with certain classes of drugs and biologicals administered intravenously, including those containing LNPs.[109,110] Ordinarily, explains Hudson, the complement system "is responsible for immune regulation"; when that system becomes dysregulated, reactions can arise involving "immune-mediated mechanisms that are severe and serious in nature."[111]

As Hudson also notes, conventional allergy testing cannot anticipate CARPA reactions; "prediction of adverse immune consequences of nanomedicines and other 'avant-guard' [sic] drugs," according to Szebeni, requires "new, non-standard toxicity tests" that do not yet exist.[112] Szebeni gauges the frequency of CARPA reactions (5%–45%) to be "much higher than classical anaphylactic reactions to drugs" such as penicillin (<2%).[113] In a 2022 paper, Szebeni and co-authors acknowledged complement activation as a possible contributor to COVID mRNA vaccine reactions.[114]

Russian Roulette Batches

Many observers agree that the COVID jabs have differed across batches and vials—leading to intense Russian-roulette variability in adverse reactions. Diffusing pertinent questions, some biotech industry insiders admit to the inconsistencies but attribute them merely to inadequate quality control; in their stated view, variabilities in manufacturing and handling virtually guarantee that *all vials are not the same* [italics in original].[115] While asserting that vaccines are prone to such "variabilities," these industry partisans profess faith that defective vials are the exception in a sea of "many good-quality vials."

Sasha Latypova agrees about the importance of good manufacturing practices—and regulators' duty to verify manufacturer compliance with those practices—but she parts company with those, including some in the health freedom community, who perceive harmful batches as inadvertent.[116] She states:

> There is a faction . . . that still believe that the FDA/HHS simply haven't seen the data, are too busy and preoccupied, and they are

not intentionally lying and pushing poison onto the public, but just confused and need to really look at the data![117]

When Latypova and others "looked at the data,"[118] examining patterns of adverse event reports across manufacturing lots, they found that "not only were the adverse events high, but the variability of them by batch was absolutely extreme."[119] As Latypova told *Epoch Times* in 2023, "Some batch numbers had two or three [adverse event] reports, and some had 5,000 to 6,000. That should never happen. . . . When you see a variability like this among batches of what is supposed to be good manufacturing practice compliant, it means the product is not, in fact, compliant."[120]

In spring 2023, Danish scientists published evidence reiterating a "batch-dependent safety signal" for the Pfizer shots.[121] According to Latypova, by the end of September, the paper had achieved a ranking of approximately 400 "out of 24 million peer reviewed papers."[122] As *The Defender* summarized in an article titled "'Bombshell' study of Pfizer COVID Vaccine Suggests Some People Got Highly Dangerous Shots, Others Got a Placebo,"

> The Danish researchers . . . found that batches of the Pfizer-BioNTech COVID-19 vaccine could be neatly divided into three groups. Two of the three groups demonstrated higher-than-normal percentages of severe adverse events in recipients. However, for the third group of batches, a total of zero adverse events were reported.[123]

The Defender quoted journalist Kim Iversen's take on the findings: "[E]ither they were actively experimenting on the public or they were covering up for the fact that the vaccines came with numerous side effects."

According to medical-scientific experts like Dr. Mike Yeadon (a former Pfizer senior vice president) and retired neurosurgeon Dr. Russell Blaylock, it is inconceivable that batch-to-batch variability on the scale observed with the COVID shots could be merely random or the result of carelessness. In early 2022, Yeadon emphasized that pharma ordinarily is quite good at "consistent, high-quality, purity manufacturing [from] batch to batch to batch."[124] He then added:

I'm afraid I've come to the conclusion they're doing it [making highly variable batches] on purpose, because they're so professional, and after a year they know this data . . . so the fact that they haven't stopped this tells me that they're at least okay with it. . . . I'm absolutely sure profit alone is not the motivation. . . . This is calibration of a killing weapon.

Pointing to a one to three thousand percent higher incidence of deaths and serious complications in individuals who received COVID vaccine "hot lots" compared to safer lots, Blaylock, too, has speculated that there was "intentional alteration of the production . . . to include deadly batches."[125]

At a House of Representatives "Panel on Injuries Caused by COVID-19 Vaccines" in November 2023, Senator Ron Johnson (R-Wis.) expressed frustration with regulators' insouciance regarding the most dangerous lots:

There was 1 lot of Moderna had 5,297 adverse events reported to the VAERS system. . . . The worst flu vaccine lot ever reported had 137. . . . [A]nd yet the response I get from the CDC or the FDA is even though you show them the data they write back, say "FDA's analysis of counts of serious adverse events reported by COVID 19 vaccine lot numbers showed no unusual concentration reports of a single lot of small group of lots . . . ." With impunity they just say "we see nothing here, so quit asking questions."[126]

Unfortunately, history suggests that CDC and FDA failure to acknowledge "highly reactive lots of vaccine and pull them off the market" may be a matter of long-standing policy (see **Hot Lots: Some History**).[127]

Hot Lots: Some History

Historically, clusters of vaccine injuries or deaths have only very occasionally forced public health officials to acknowledge the possibility of vaccine "hot lots." Researcher Neil Z. Miller describes one notorious example in a paper about vaccines and sudden infant death syndrome (SIDS),[128] looking back at a "hot lot" of DPT vaccine given to 96,000 Tennessee children in the late 1970s. The manufacturer (Wyeth, later acquired by Pfizer), eventually recalled the lot,[129] but not before a number of children died. As Miller outlines the events, an initial investigation found that 11 children had died shortly after DPT vaccination—five within the first 24 hours—and nine of the 11 vaccines had come from the same lot. (Further investigation, according to Miller, identified 23 infant deaths within 28 days of DPT vaccination—over half within 24 hours.) The Institute of Medicine later told the public that samples of the lot in question "were found acceptable with regard to potency and freedom from toxicity" and thereupon dismissed any possibility of a causal relationship.[130]

Starting in 1990, the National Vaccine Information Center (NVIC) played an early and important role in identifying DPT vaccine "hot lots" associated with infant injuries and deaths, enlisting a computer programmer to develop special software that could decode raw VAERS data and then publishing lists of the "hot lots."[131] NBC's Tom Brokaw and Katie Couric broadcast a special show in 1994 about a boy severely brain-injured by a DPT "hot lot."[132]

In 1999, pediatric neurologist Dr. Marcel Kinsbourne shared some thoughts on vaccine "hot lots" with the federal government's Committee on Government Reform, noting that the "well known" phenomenon endangered children:

> However, the manufacturer is protected by law from disclosing the number of doses that derive from a given lot. Therefore, one lacks the denominator of the function which would reveal whether a given lot appears 'hot' because it is more toxic, or because it is the source of more doses.[133]

After the Tennessee DPT disaster, notes Miller, "internal memos by the vaccine manufacturer revealed a new policy of limiting shipments of DPT vaccine so that no geographical location would receive all of the product from a single lot, confounding the ability to trace hot lots that might cause clusters of SIDS cases post-vaccination."[134]

> Specifically, the memo stated, "After the reporting of the SID cases in Tennessee, we discussed the merits of limiting distribution of a large number of vials from a single lot to a single state, county or city health department and obtained agreement from the senior management staff to proceed with such a plan."[135]

"Adulteration" and "Contamination"

Among the evidence assembled by Latypova and Watt since 2021 is compelling information about the dangerous "adulteration" of the COVID mRNA vaccines.[136] To "adulterate," according to the Merriam-Webster dictionary, means "to corrupt, debase, or make impure by the addition of a foreign or inferior substance or element."[137] Other synonyms include "contaminate," "defile," "falsify," "spike," "corrupt," and "pollute."[138]

Under normal pharmaceutical manufacturing circumstances, Latypova emphasizes, if a product is not compliant with current Good Manufacturing Practices (cGMP), regulators consider it "de-facto adulterated, and a potential poison."[139] Legally speaking, however, EUA products are an entirely different matter, as Watt's research shows. Under EUA law 21 USC 360bbb-3(k), Watt reports,

> There are no required standards for quality-control in manufacturing; no inspections of manufacturing procedures; no prohibition on wide variability among lots; **no prohibition on adulteration**; and no required compliance with Current Good Manufacturing Practices [bold added]. EUA products, even though unregulated and non-standardized, "shall not be deemed adulterated or misbranded."[140]

Watt alleges that "Whatever is in the biochemical weapons bearing Pfizer and other pharma labels, is there because US SecDefs [Secretaries of Defense] and their WHO-BIS [Bank for International Settlements] handlers ordered it to be there."[141]

Latypova argues that the COVID vaccination program could not have started or continued without an underpinning of fraud:

After thousands of deaths are recorded with the use of a product and no action is taken by anyone—manufacturer, regulators or law enforcement authorities—the situation should be considered fully intentional. That behavior of flat denial in the face of unequivocal harm signaled the *intent* [italics in original]. . . . **Fraud was the essential part of the Operation Warp Speed from the start**, it was known to the regulators and DOD and pharma, it was expected, it was in fact required, it was and remains essential to this program [bold added].[142]

"If we had honest regulators," Latypova stated in 2022, "the factories would have been raided, shut down, product seized, huge investigations would start. Nothing, obviously, is happening."[143]

These observations are particularly pertinent in light of the detected presence of foreign bodies, including metallic particles (see **Mystery Ingredients**) and apparent plasmid DNA, in selected vials of the COVID mRNA shots examined by independent researchers.[144] Although genetic engineering often uses plasmids ("small rings of DNA" found in bacteria such as *E. coli*) as "starting material,"[145] and Pfizer and Moderna acknowledge that this is the case for the DNA "template" that makes their synthetic mRNA, DNA is not supposed to be present in the final product.[146]

DNA alarm bells first began ringing in 2022. Latypova reported on the findings of an anonymous German scientist, who found "fairly large amounts" of DNA in selected Pfizer and Moderna vials.[147] Latypova and D4CE both emphasize that this manufacturing "deficiency" was known to regulators.[148] Moreover, according to D4CE:

[I]t appears that once vaccine production had commenced, no process quality control data pertaining to the residual DNA content in the mRNA vaccines were ever demanded from or submitted by the manufacturers to the [European Medicines Agency] and other regulators.[149]

In 2023, genomic and DNA sequencing expert Kevin McKernan further confirmed the DNA findings (a development now referred to as

#PlasmidGate). Analyzing anonymously sent vials of Pfizer and Moderna bivalent COVID vaccines,[150] his stated baseline assumption was that he would find "pure mRNA"; instead, he observed levels of DNA "up to 1,000 times higher than deemed to be 'acceptable' by the regulating authorities."[151] When McKernan then assessed Pfizer's monovalent COVID shots, he found DNA present at levels 18 to 70 times higher than regulatory limits.

McKernan has expressed concern about the injection of high concentrations of these "contaminants," which bypass the mucosal defenses.[152] According to McKernan and others, the findings also raise additional concerns:

- The injected plasmids are likely to be "replication competent," meaning "integration into the genome is feasible."[153]
- The lab plasmids contain an antibiotic resistance gene; if not removed, plasmids that later replicate in the human body could make the body resistant to its own innate gut bacteria.[154] (Is it a coincidence that ARPA-H, the Advanced Research Projects Agency for Health, is now planning to spend $100 million to study antibiotic-resistant bacteria?)[155]
- Within the DNA matrix, McKernan also found an undisclosed or "stealth" gene called an SV40 promoter; SV40 promoters are associated with cancer due to their ability to deliver "foreign cargo into the nuclei of cells that are not supposed to divide, but [which] may begin this malignant process due to being hacked by the foreign cargo."[156] McKernan since has noted "really, really potent data" from death records showing "a steady increase in cancer": "That is a clear-cut sign that we have an increase in cancer post-vaccination."[157]

In October 2023, the World Council for Health convened a panel of experts to discuss "Plasmidgate," describing the DNA discoveries as "probably the most important topic of our time," posing a real and shocking threat of "genetically modified humans."[158] As one of the panelists commented, "It's not in any way hyperbolic to talk about the genetic invasion of innocent people without their knowledge."

Watt cautions, however, against letting the hue and cry about "adulteration" and "contamination" become a distraction that pulls attention "away from US kill box laws, DoD, WHO, intentionality and the intrinsic lethality of all mRNA platform technologies."[159] She comments:

> They are trying to shield the mRNA technology and "vaccination" program platforms, and the public health emergency geopolitical and legal platforms from growing public understanding of what's really going on, so that the Monster can keep using "public health emergency" laws, orchestrated "pandemics," "vaccines," and mRNA-platform poisons to sicken and kill many more people for many years to come.

Mystery Ingredients

In September 2021, Japan halted use of 1.6 million doses of the Moderna COVID vaccine and then recalled three batches due to contamination with stainless steel particles.[160] Two Japanese men died after receiving Moderna shots during this period. In April 2022, Moderna had to recall hundreds of thousands of doses distributed in Europe after evidence emerged of contamination with an undisclosed "foreign body."[161] In Germany, scientists also found metallic particles in the Pfizer and AstraZeneca injections, findings cross-validated by investigators in other countries but met with an "eerie silence from global safety and regulatory bodies."[162]

Metallic particles in vaccines are worrisome—and not necessarily a new phenomenon. An earth-shattering 2017 study by Italian researchers Antonietta Gatti and Stefano Montanari, who examined 44 different types of vaccines, found extensive contamination with "micro-, sub-micro- and nanosized, inorganic foreign bodies" with "unusual chemical compositions," including compounds containing lead, stainless steel, chromium, tungsten, and numerous other metals—none declared in the package inserts.[163] The two researchers warned, "The inorganic particles identified are neither biocompatible nor biodegradable . . . and can induce effects that can become evident either immediately close to injection time or after a certain time from administration." They added, "It is a well-known fact in toxicology that contaminants exert a mutual, synergic effect, and as the number of contaminants increases, the effects grow less and less predictable. The more so when some substances are unknown." Gatti is now calling attention to "unknown ingredients" in the COVID injections.[164]

CHAPTER FOUR

ASSISTED SUICIDE AND EUTHANASIA

"Physician-assisted suicide," "assisted suicide," "assisted dying," and "euthanasia," each with slightly different legal connotations, all refer to medical interventions to expedite death, with the main differences being the type of drug protocol used and "who performs the final, fatal act"—the person concerned or a doctor.[1,2,3] In cases of assisted suicide, the individual in question orally takes drugs prescribed by prior arrangement with a doctor; in euthanasia scenarios, doctors directly administer a lethal injection. Some observers also distinguish between the type of person who requests "assisted dying" (individuals who are already dying) versus "assisted suicide" (individuals who are not terminally ill but seek death for other reasons).[4]

Throughout much of human history, assisted death was a "prohibited and morally condemned practice."[5] From that moral standpoint, entities like the North America–based Euthanasia Prevention Coalition argue that irrespective of "whether the person killed has consented to be killed," euthanasia and assisted suicide "should be treated as murder/homicide."[6] Others, too, consider the entirety of such practices to be "medicalized killing"—homicide with a medical "patina"[7]—as well as a back-door strategy to implement eugenics.[8]

Nevertheless, legalization of physician-assisted suicide and/or euthanasia is now a snowballing trend worldwide, exhibiting a "momentum" that "appears unstoppable."[9] Not only are more nations endorsing the practices, but every country or jurisdiction that allows euthanasia or assisted suicide is showing steady year-over-year increases.[10,11] In fact, whatever else one can say about them, "Human Euthanasia Services"—shaped by factors such as economic growth, technology, and political or regulatory risks—have become a "dynamic and promising" global market.[12] The trend provides a disturbing barometer of where the perpetrators of global iatrogenocide could take things if the public does not look beyond the gloss of compassion to consider whether a less friendly "end-of-life" agenda is in play.

The Trendsetters

In many respects, Europe has led the way, with a handful of nations serving as the Western world's assisted-death trendsetters. Switzerland, for example, has permitted assisted suicide (though not euthanasia) for seven decades, including by non-physicians.[13] Playing a pioneering role in normalizing the practice,[14] the Swiss nation even attracts lucrative "suicide tourism" to slick designer-furnished clinics that provide death to those willing to pay.[15,16]

Where euthanasia is concerned, the Netherlands and Belgium have been in the vanguard for two decades. In 2001,[17] the Netherlands became the first country in the world to legalize euthanasia—allowing it for individuals as young as 12—and in 2007, the Dutch nation expanded (for the first but not last time) the range of eligible conditions.[18] Belgium soon followed the Dutch example, legalizing euthanasia in 2003 for adults and emancipated minors, and explicitly defining the practice as "medical treatment."[19] Six years later, Luxembourg joined the ranks of "world leaders" in assisted death, legalizing both euthanasia and assisted suicide.[20] Since legalization, euthanasia has more than tripled in the Netherlands,[21] with almost 60,000 people killed between 2012 and 2021—as of 2017 representing 4.4% of all deaths (or more than one in 25).[22,23] Euthanasia also has become a burgeoning practice in Belgium, going from 1.9% to 4.6% of deaths in less than a decade (2007–2013).[24]

According to reportage in *The Atlantic*, the Dutch directives "are few and broadly drawn," without even a "requirement that a patient be close to death."[25] Moreover, in the Netherlands' permissive euthanasia context, authorities consider doctors to be "virtually always right." However, critical analysis by a French nongovernmental organization, Alliance VITA, suggests that "looser and looser interpretation" of the Dutch law has "trivialized" euthanasia and "created increasingly disputable situations."[26] Alliance VITA describes a wide range of "infringements":

- **Substitution** of euthanasia for palliative care, with euthanasia becoming "the standard way of dying for cancer patients"
- **"Disguised" euthanasia**, with deaths by "deep and continuous sedation" more than doubling between 2005 (8.2%) and 2015 (18%)
- Use of "traveling" doctor-nurse teams and **"death clinics"** that "cater to people whose family practitioners refuse euthanasia"; the teams "willingly accept the most borderline and 'complicated' cases which other doctors do not necessarily believe to be justifiable"
- **Lax government oversight** and virtually no prosecution of overly enthusiastic agents of death
- A continual push to **"broaden the conditions of access,"** including allowing assisted suicide "for people over age 70 requesting it, with no other motive but their age and 'being tired of life'"
- Euthanasia of people with **psychiatric disorders**, autism,[27] or dementia, whose ability to provide "voluntary and well-considered" consent—a legal requirement—is difficult to assess (see **Depressed? Choose Euthanasia!**)

Depressed? Choose Euthanasia!

In 2022, Dutch researchers published a study in *Frontiers in Psychiatry* about the Netherlands' experiences with physician-assisted death for "psychiatric suffering"—an application endorsed by the Minister of Health immediately after passage in 2002 of the country's euthanasia law.[28] Noting that only a handful of other countries (Belgium,

Luxembourg, Switzerland and possibly Canada in 2024) allow physicians to kill individuals with psychiatric disorders, the authors suggested that the growing numbers of Dutch cases—88 in 2020 and 115 in 2021 (versus an average of 27 annually between 2011 and 2014),[29] most often for "depression" or "trauma- or stressor-related disorders"—make the Netherlands "one of the few countries in the world that can offer insights on the practice from real life experience."

The researchers acknowledged a number of facts that make medical killing for reasons of "psychiatric suffering" problematic:

1. Psychiatric disorders "are not in themselves fatal," and persons "potentially have **decades to live** in which they can recover spontaneously" or discover other avenues for help.

2. By definition, psychiatric diagnoses are **subjective** (as we saw in Chapter Two, there are no biological markers), making it "very difficult to give a reliable prognosis."

3. Individuals diagnosed with psychiatric "disorders" often refuse standard treatments (a fact that is not surprising, given how unpalatable most psychiatric "treatment" options are); this poses a Catch-22 to the would-be euthanizer, because someone who has not exhausted **all other options** cannot be said to have "irremediable" psychiatric suffering.

4. Ascertaining **capacity** to request euthanasia and give **informed consent** is challenging, and the stakes—"the choice between life and death"—could not be higher. The Dutch authors also noted concerns "about the rigor of the capacity assessments in the Netherlands, since many reports only [contain] a short statement about the patient possessing decision-making capacity, but [lack] a description on how this was assessed."

Compared with terminal illness, one writer observed in 2019, "the application of the eligibility criteria in psychiatric euthanasia depends much more on doctors' opinions."[30] And, unlike for euthanasia in general, "psychiatric euthanasia is predominantly given to women."

Discussing whether "intractable" mental illness should confer eligibility for euthanasia at all, a Canadian author noted that roughly seven out of ten individuals who receive euthanasia for "mental disorders" are women, and many have "a history of severe sexual or other kinds of abuse."[31] Rather than medically kill such women, the author suggests paying more attention to prevention of gender-based violence.

In 2018, troubled Dutch and British researchers reported that the Dutch "due care criteria" for euthanasia or assisted suicide "are not easily applied to people with intellectual disabilities and/or autism . . . and do not appear to act as adequate safeguards."[32] One of the due care criteria involves determining whether the individual is experiencing "unbearable suffering without prospect of improvement."[33] Other European researchers cautioned in 2020 that granting euthanasia and assisted suicide (EAS) "based on a perception of the patient's illness as being untreatable with no prospect of improvement, could . . . in many cases fail to meet the due care criteria listed in EAS laws."[34] They added, "this practice neglects the individual's potential for having a life worth living."

In a related study published in 2019, the same Dutch and British research group called attention to doctors' pivotal and potentially biased role, observing, "The responsibility for assessing whether the patient's suffering is bad enough and hopeless enough to warrant euthanasia, rests solely with their physician."[35] In their 2018 study, they expressed concerns about physician bias:

Dependency, functional limitations and difficulties with integration in society are often part of conditions which these groups of people live with all their lives. The Dutch cases **raise the possibility that the bar for assessment of intractable suffering is set lower for people with an intellectual disability or autism spectrum disorder than for the general population,** by considering their long term disability as a medical rather than a social condition. **We found no evidence of safeguards against the influence of the**

physicians' own subjective value judgements . . . nor of processes designed to guard against transference of the physicians' own values and prejudices [bold added].[36]

In 2023, this Dutch-British research group described their analysis of 900 publicly available euthanasia case files in the Netherlands; they found that 39 of the euthanized individuals had been diagnosed with autism or an intellectual disability, including 18 individuals (46%) under age 50.[37] Eight (21%) had put forth suffering related to their autism or intellectual disability as the sole reason for their euthanasia request, citing social isolation or coping difficulties; others described their autism or intellectual disability as a major contributing factor. Worriedly, lead author Irene Tuffrey-Wijne commented, "There's no doubt in my mind these people were suffering. But is society really OK with sending this message, that there's no other way to help them and it's just better to be dead?"[38] A Dutch psychiatrist agreed that it was disturbing "that young people with autism viewed euthanasia as a viable solution," and a Canadian expert opined, "Helping people with autism and intellectual disabilities to die is essentially eugenics."[39]

Over a decade ago, just as euthanasia was beginning to take off globally, Canadian palliative care expert Dr. José Pereira wrote that laws and safeguards to prevent "abuse and misuse" of euthanasia and assisted suicide "are regularly ignored and transgressed in all . . . jurisdictions and . . . transgressions are not prosecuted."[40] "Transgressions," as described by Pereira, include:

- Administration of legal substances to individuals who have **not requested** euthanasia
- Euthanasia of people who possess the capacity but have **not given explicit consent** (one out of five euthanized people in the Netherlands)
- Failure to get an **independent second opinion** as required
- Euthanasia **not reported** as required (a conservatively estimated 20% of cases in the Netherlands are not properly reported)
- Performance of euthanasia **by nurses** (not allowed anywhere but Switzerland)

Pereira concluded that legalizing euthanasia and assisted suicide "places many people at risk, affects the values of society over time, and does not provide controls and safeguards." Using the Netherlands as an example of the "social slippery slope," Pereira wrote:

> In 30 years, the Netherlands has moved from euthanasia of people who are terminally ill, to euthanasia of those who are chronically ill; from euthanasia for physical illness, to euthanasia for mental illness; from euthanasia for mental illness, to euthanasia for psychological distress or mental suffering—and now to euthanasia simply if a person is over the age of 70 and 'tired of living . . . .' Denying euthanasia or [physician-assisted suicide] in the Netherlands is now considered a form of discrimination against people with chronic illness, whether the illness be physical or psychological, because those people will be forced to 'suffer' longer than those who are terminally ill. . . . **[E]uthanasia has moved from being a measure of last resort to being one of early intervention** [bold added].

By 2017, the year in which Dutch politicians discussed the option of legalizing euthanasia for "perfectly healthy" seniors,[41] the leader of the Christian SGP party, Kees van der Staaij, was publicly airing concerns—in the *Wall Street Journal*, no less—about his country's "euthanasia culture."[42,43] Other prominent figures agreed that the situation might be "getting out of hand."[44]

"Not So Simple"

In a 2016 study, researchers examined international changes in medical end-of-life practices over time.[45] They broadly defined medical end-of-life practices as (1) withdrawing or withholding treatments that might otherwise prolong life, (2) managing symptoms with drugs, "even if an unintended side-effect may be to shorten life," or (3) intentionally using a lethal drug. Across countries, they found that use of opiates and sedatives had increased significantly between 1990 and 2010, particularly "in situations where the use of lethal drug [sic] is more difficult to justify in legal terms," prompting their mild conclusion that better differentiation "between practices with different legal status is required to properly interpret the policy significance of these changes." Belgian researchers were a bit more direct in a 2018 study, where they argued that evidence "advises against" use of opiates and sedatives, which tend to be associated with "euthanasia cases that remain unreported."[46]

In 2020, a new organization joined the ranks of the U.S. end-of-life community. The American Clinicians Academy on Medical Aid in Dying (ACAMAID), chaired by a doctor who has made a career out of "physician-hastened death,"[47] has the seemingly straightforward mission to make "clinical information about medical aid in dying available to all clinicians."[48] However, as the authors of a powerful article published in the *British Medical Bulletin* wrote in 2022, the assumption "that there exists an easily prescribed drug which consistently brings about death quickly and painlessly" is contradicted by real-life practice; physicians prescribe a "wide variety of lethal drug combinations," and, as a result, "hastening patient death is not so simple."[49]

In assisted suicides, the 2022 article explained, even the seemingly straightforward act of "ingesting sufficiently toxic dosages of the prescribed drugs can prove a significant . . . challenge." Complications are also a concern; the state of Oregon has documented complication rates as high as 15% annually. Still worse, time to death post-ingestion can be "highly unpredictable." Because complications and failures put patients "at risk of distressing deaths," informed choice requires that risks be "understood and clearly explained."

A 2017 report by *KFF Health News* illustrated some of the perils outlined in the 2022 article. At one time, the powerful sedative secobarbital (brand name Seconal) was the "drug of choice" for assisted-suicide practitioners in North America, but after its price doubled (to over $3,000 per dose), clinicians began casting about for alternatives. According to KFF:

> The first Seconal alternative turned out to be too harsh, burning patients' mouths and throats, causing some to scream in pain. The second drug mix . . . led to deaths that stretched out hours in some patients—and up to 31 hours in one case.[50]

Currently, U.S. doctors tend to favor one of two multi-medication protocols: "DDMA" (digoxin, diazepam, morphine, and amitriptyline) or "DDMP" (digoxin, diazepam, morphine, and propranolol). ACAMAID advocates for the addition of phenobarbital to decrease "the upper range of times to death."[51,52] Digoxin and propranolol are both cardiotoxic agents; diazepam is a benzodiazepine, morphine is an opioid, and amitriptyline is a sedative. In countries where doctors perform euthanasia, the cocktail is similar but additionally includes a neuromuscular blocking agent "to prevent respiratory effort and to eliminate muscular spasms which could be interpreted as signs of distress by observing relatives."[53]

Addressing the topic of distress for both the person concerned and their surrounding family members, a study published by Belgian researchers in 2018 disturbingly found that a third of physicians practicing euthanasia used "nonrecommended" drugs to dispatch their patients, rather than adhering to established drug guidelines.[54] As the authors informed readers, incorrect drugs "may lead to traumatic situations such as an extended time to death or awakening of the patient, causing distress for the patient and the attending family and health care providers."

The authors of the 2022 *British Medical Bulletin* study noted that clinicians' opinions are mixed as to "whether death by lethal injection is as peaceful and painless as it may seem," in part due to the challenge of "achieving and ensuring unconsciousness during euthanasia."[55] Because euthanasia protocols include the use of paralytic agents, if an individual happened to remain "cognizant," they would not be able to move a muscle and would, therefore, "appear to be unconscious." Spelling out the implications, the authors observed:

> [M]onitoring consciousness in patients who are undergoing euthanasia and have received paralytic agents is difficult and, unless properly monitored, "there is a risk that vulnerable citizens may be killed by suboptimal, or even cruel, means . . . ." If the person were aware, it has been suggested that the experience of death by lethal injection could be akin to suffocation or drowning.

Even more gruesomely, articles archived by the Euthanasia Prevention Coalition indicate that methods for doling out death by euthanasia include non-pharmaceutical mechanisms, such as dehydration, organ donation, advance directives, and pillows.[56,57]

Killing Kids and Young Adults, Too

In Switzerland, as long as "patients commit the act themselves and helpers have no vested interest in their death,"[58] there is no specified age limit for assisted suicide.[59] *The Guardian* observed in 2014, "technically even a healthy young person could use such services."[60] In fact, according to studies of that time period, one in five assisted suicides in Switzerland involved adults under age 65,[61] including individuals as young as 18.[62] In 2021, Switzerland took steps to legalize a fancy new way for people to kill themselves that is likely to appeal to young people—a space-age "coffin-like capsule with windows" designed so that the decedent-to-be can push a button, flood the interior with nitrogen and supposedly die within 10 minutes.[63]

In the Netherlands and Belgium, persons under age 60 typically have accounted for 12% to 15% of those euthanized.[64,65] In 2015, Belgian doctors made headlines when they agreed to euthanize a healthy 24-year-old woman who had convinced herself of a lifelong "death wish" despite growing up "with a quiet, stable family."[66] The previous year, Belgium became the first country to legalize lethal injection for children as young as one. Implementation of this "radical" provision began in earnest in 2016 and 2017,[67] when Belgian doctors put three children to death, including a nine-year-old,[68] and also euthanized 19 young adults under age 30, followed by another child in 2019.[69] In March 2018, the Colombian government set the precedent in Latin America, legalizing euthanasia in children as young as seven, enabling access for children ages 12 to 14 "even if their parents disagree," and stating that after age 14, "no parental involvement is needed."[70]

Interestingly, at the same time that European and other governments were peddling unwarranted fear-mongering about pediatric COVID-19— used to justify misleading vaccine mandates for kids, deemed genocidal by Thai-German microbiologist Sucharit Bhakdi[71,72]—the Netherlands escalated its kid-focused euthanasia legislation. As a reminder, the Netherlands has allowed euthanasia of children age 12 and up since 2002 and also permits 16- and 17-year-olds to decide to die without parental consent.[73] From 2002 to 2012, assisted dying dispatched five Dutch adolescents, including a 12-year-old,[74] but in late 2020, the Dutch Ministry of Health began taking steps to imitate Belgium and extend "active termination of life" down to age one.[75] After stacking the deck with a favorable commissioned report from the Dutch Society of Pediatrics,[76] Dutch Health Minister Ernst Kuipers renewed the case for under-12 euthanasia in a June 2022 briefing to Members of Parliament.[77] At the same time, university researchers published studies (in medical ethics journals, no less) in favor of "[l]etting go of the age limit or lowering the age limit."[78]

These efforts bore fruit in April 2023, when the Dutch government expanded its euthanasia regulations to cover terminally ill children between one and 12 years old; the government stated, "The end of life for this group is the only reasonable alternative to the child's unbearable and hopeless suffering."[79] Although the expansion set a floor at age one, under a "devil's bargain between medical professionals and prosecutors," Dutch doctors also can kill babies with terminal diagnoses in their first year of life without fear of prosecution, as long as the doctors follow a designated protocol.[80]

Copying Belgium and the Netherlands, the Canadian government has been engaged since 2016 in escalating and normalizing its "medical assistance in dying" program (MAID).[81] In 2021, MAID accounted for 3.3% of Canadian deaths, representing "a growth rate of 32.4% over 2020," and "steady year over year growth" in all provinces, including in the 18- to 45-year age group.[82] Among the first 100 persons to avail themselves of MAID in the province of Ontario that year, over 5% were younger adults aged 35–54 years.[83]

Canada also has signaled its plans to make MAID (currently available only to adults) an option for young people, "starting with those who have terminal illnesses, but eventually extending it to children who want to die for other reasons."[84] Making youth (including "mature minors") eligible for MAID would mean that "before children in Canada can drive vehicles, they may be allowed to consent to physicians taking their lives."[85] In late

2022, a member of the Quebec College of Physicians proposed to the Canadian parliament that babies with "severe deformations or very grave and severe syndromes" also be candidates for assisted suicide.[86] Swimming against the pro-euthanasia tide, a horrified bioethicist-physician objected that this would leave too "much room for parental, physician, personal, social and economic bias."[87]

Freedom to Die—But Not to Live?

In the U.S., state-level legalization of physician-assisted suicide for adults with a terminal illness has been picking up steam since 2013 and is now available to more than one in five Americans[88] in 10 states plus the nation's capital: Oregon (law passed in 1994 and implemented in 1997),[89] Washington (2009),[90] Montana (2009),[91] Vermont (2013),[92] California (2015),[93] Colorado (2016),[94] the District of Columbia (2017),[95] Hawaii (2018),[96] Maine (2019),[97] New Jersey (2019),[98] and New Mexico (2021).[99]

These eleven jurisdictions may well believe their lofty rhetoric about "dignity" and "humane" policies, but their statements about the freedom to die contrast sharply and ironically with their performance during the worst of the COVID period.[100] Specifically, states where physician-assisted suicide is legal were some of the most malevolent in destroying people's livelihoods and using authoritarian measures to suppress constitutionally guaranteed freedoms.[101] A political contradiction was also evident in countries like New Zealand and Spain, where the two prime ministers helped get euthanasia across the finish line (in 2020 and 2021, respectively) but otherwise treated the living in a markedly tyrannical manner.[102,103] Spain's prime minister vigorously enforced lockdowns and restrictions but, without irony, declared that the option of euthanasia made his country "more humane, fairer and freer."[104]

COVID spurred a major trend toward more telemedicine and remote health care in the U.S.,[105] including for assisted death. By March 2020, without missing a beat, ACAMAID was recommending telemedicine "for select aspects of aid-in-dying evaluations,"[106] and by June 2020, doctors were putting the group's recommendations to use nearly from start to finish:

Medically assisted deaths in America are increasingly taking place online, from the initial doctor's visit to the ingestion of life-ending medications. . . . [B]ecause of the coronavirus, volunteers are accompanying patients and families over Zoom, and physicians complete their evaluations through telemedicine.[107]

While doctors may have been enthusiastic about providing remote assisted-suicide guidance, family members apparently found that "dying via telemedicine" could be hard on them. With no physician physically present, the situation forced relatives to assume a more "active" role, including "mixing the life-ending medications themselves." Pre-COVID, most family members had been "glad to outsource this delicate task."[108]

In the state of Vermont, the governor signed a law in 2022 explicitly endorsing telemedicine as a route for "aid in dying"; the bill made it possible for patients to waive the "two in-person consults and . . . 48-hour waiting period" ordinarily required to get a fatal prescription, and also granted health care providers and pharmacists full legal immunity.[109]

Reportedly, the U.S. Drug Enforcement Administration (DEA) is now trying to roll back some of the "online prescribing flexibilities" that took root during the pandemic; protesting, a doctor who has facilitated 80 deaths since 2020 told the Associated Press that telemedicine is a critical assisted-suicide enabler that should not now be "yanked away."[110] A group called "Compassion & Choices" is lobbying to make permanent the end-of-life legislative and regulatory changes implemented during COVID.[111]

Slippery Slope

In Canada, at the same time that the government was rolling out and mandating life-threatening COVID injections in 2021, it passed a consequential modification to the MAID law that eliminated the criterion previously requiring a MAID candidate's death to be "reasonably foreseeable"[112]—rendering decisions about eligibility far more subjective. Even for people with "reasonably foreseeable" deaths, the bill deleted a formerly mandated 10-day "reflection period" and downgraded the requirement for independent witnesses from two to one.[113]

Critics of the broadened policy warned that it would "nurture the country's growing culture of death,"[114] arguing that the availability of assisted suicide would serve as an inducement to consider it while "removing pressure for an improvement in psychiatric and social services."[115] In fact, the relaxation of criteria for hastening death had the predicted result; in 2021, MAID was responsible for the deaths of 219 individuals "whose natural deaths were not reasonably foreseeable," with "neurological" problems cited as the rationale in nearly half (46%) of the cases.[116] Worried observers in Canada and the Netherlands caution that the ranks of assisted suicide candidates are likely to expand beyond the already problematic categories of youth and the mentally ill to other marginalized groups, such as the homeless, the poor, the disabled, those with chronic pain, prisoners "who desperately long for death"—and even "dissenters who the government feels are not fit for society."[117,118,119]

In 2017, the regional director of a U.S. disability-rights group presciently lamented that "assisted-suicide laws inevitably take the lives of innocent people."[120] The director added, "Given that insurers routinely value their bottom lines over patient treatment, and the health care system devalues the lives of disabled people, these laws reduce rather than expand choice." In Canada, cost-benefit analyses have been creeping into the assisted-death calculus, with research and reports issued in advance of the 2021 amendments bragging about how doctor-assisted death could "save millions" (see **Doctor-Assisted Deaths: A Cost-Cutting Measure?**).[121,122] In the Netherlands, even euthanasia's most ardent supporters are concerned that the "financial gutting of the health care sector" will nudge desperate people toward assisted death.[123] Pointing to "the pressure on aging, low-birthrate societies to cut their health care costs," critics also allege that doctors may be promoting assisted suicide as a solution for "sick people seeking a quietus for reasons linked to financial stress."[124]

Doctor-Assisted Deaths: A Cost-Cutting Measure?

In early 2017, barely six months after Canada's legalization of "medical assistance in dying" (MAID), the Canadian Broadcasting Corporation (CBC) started making the case that the MAID program "could result in substantial savings across Canada's health-care system,"[125] citing a just-published report in *CMAJ* (the *Canadian Medical Association Journal*) on the estimated cost savings.[126] The report's first author told

CBC that while there might be "some upfront costs" associated with MAID, he expected "a reduction in spending elsewhere in the system and therefore offering medical assistance in dying to Canadians will not cost the health care system anything extra" but instead would lead to significant savings. Using the example of one Canadian province, his report directed critical attention to the "substantial" and "dramatic" health care costs incurred for those "nearing the end of life," noting that "more than 20% of health care costs are attributable to patients within the 6 months before dying, despite their representing only 1% of the population."[127]

Canadians are not the only ones seeking to frame assisted death as economically beneficial. In a 2020 article in the journal *Clinical Ethics*, two British biomedical ethicists "propose and defend" not one but three economic arguments: the avoidance of "negative" quality-adjusted life years (QALYs); the gain in the "positive" QALYs hypothesized to result if the health care system redirects resources away from the dying toward the living; and (somewhat creepily in light of the growing phenomenon of involuntary organ donation)[128] the enabling of "an additional, highly beneficial source of organs for transplantation."[129] Regarding their third argument, the authors claim that organs obtained after assisted dying are "better from a clinical and economic perspective" compared to organs from individuals who "are denied assisted dying." Though they briefly consider two possible objections—first, that "it is callous to consider assisted dying from the perspective of resource management," and second, that the concept of "negative QALYs" is flawed—the two ethicists maintain that while "the economic costs of denying assisted dying . . . should not be the key driver of any legal change," their theorized costs "should not be ignored."

In 2014, European researchers published a study assessing public acceptance of euthanasia (defined as "terminating the life of the incurably sick") in 47 European countries as of 2008.[130] Unsurprisingly, acceptance was "markedly high" in the three countries where euthanasia is legal (Belgium, the Netherlands, and Luxembourg), but also in Denmark, France, and Sweden, which have not yet enshrined the practice in law. On the other hand, the researchers found evidence of regionally polarized attitudes, "with most of Western Europe becoming more permissive and most of Eastern Europe becoming less permissive." Describing the "poor health care organization and financing" in much of Eastern Europe, the authors hypothesized, "Against this background, it is conceivable that people see euthanasia more as a threat, fearing that it may be used against them to economize on health care costs and free hospital beds."

In the U.S., a 2018 Gallup poll reported "strong" public support for euthanasia—which is illegal nationwide—and slightly lower support for the legal (in eleven jurisdictions) practice of "doctor-assisted suicide.[131] (Gallup hypothesizes that this discrepancy may be "because the [latter] question contains the phrase 'commit suicide,'" whereas the language of the polling firm's euthanasia question "may sound less harsh than committing suicide.") When morality is brought into the picture, a "slim majority" of Americans (54%) consider doctor-assisted suicide "morally acceptable."[132] Back in 2005, a Pew Research Center survey found that Americans' attitudes about physician-assisted suicide were more "deeply divided" and "def[ied] easy categorization."[133]

Looking at European and North American trends, it is clear that cultural conditioning in favor of euthanasia is ramping up worldwide. Describing how medically assisted suicide has become a cherished progressive and liberal value in North America, one writer has dared to ask, "What if a society remains liberal but ceases to be civilized?"[134] In the increasingly unsettled global context, this is a valid question.

In 2020, when the Johns Hopkins Center for Health Security called for administering the experimental COVID vaccines to ethnic minorities and the mentally challenged first as "a matter of justice," Jeremy Loffredo and Whitney Webb characterized (in *The Defender*) the "justice" rhetoric as "odd," pointing to numerous historical examples of medical experimentation and a covert eugenics agenda.[135] Appearing on CHD.TV in late 2022, Canadian physicians Charles Hoffe, Stephen Malthouse, and Chris Shaw likewise speculated that their government's modifications to MAID could well be driven by ulterior and unfriendly motives; as Hoffe commented to his two colleagues, "It is amazing the lengths that the government seems to be going to—to reduce the population."[136] As citizens begin querying their governments' motives for both silently and openly celebrating death, rather than life, many recognize the potential for medically assisted death to become a pretext "to kill the vulnerable."[137]

CHAPTER FIVE

DEADLY MEDICINE: HOW DO THEY GET AWAY WITH IT?

Over the past century, a wide variety of enabling policies, tools, and tactics have made it possible for public and private players to propagate medical mayhem, injuring and killing without incurring widespread public awareness or wrath. The scale of the iatrogenocide since 2020 may be altering that equation.

This chapter dives deeper into factors, both historical and current, that allow the medical-pharmaceutical killing machine to operate with impunity. The non-exhaustive list of factors includes the consolidation of medical authority, the rise of the modern pharmaceutical industry, the sidelining and weaponization of nutrition, the ascendance of an "evidence-based medicine" juggernaut, medical gaslighting, and a growing infrastructure permitting global enforcement.

Consolidation of Medical Authority

Although iatrogenesis has occurred throughout medical history, the consolidation of medical authority and the growing dominance of a chemical-pharmaceutical model of care in the early 20th century made iatrogenic harms feasible on a much wider scale. In his seminal 1982 book,

The Social Transformation of American Medicine,[1] Harvard and Princeton sociologist Paul Starr provides a historical overview of the "rise of a sovereign profession and the making of a vast industry." As Starr delineates, the early 1900s saw a system of standardized medical education and licensing take hold, ensuring that medical authority would be reproduced "from one generation to the next" and transmitted "from the profession as a whole to all its individual members." The public also became vastly more dependent on doctors, who over time began to wield significant gatekeeper power over prescriptions, insurance reimbursement, and medical interventions.

In his chapter titled "The Consolidation of Professional Authority 1850–1930," Starr considers the impact of the milestone "Flexner report," titled *Medical Education in the United States and Canada* and published in 1910 by the non-medically trained Abraham Flexner.[2] The 364-page report, commissioned by the Carnegie Foundation—and, behind the scenes, by an increasingly powerful American Medical Association (AMA)—exhaustively set forth "the ideal of what a medical school should look like."[3] Over the next two decades, the Flexner report became one of several factors that fundamentally reshaped North American medical education, in the bargain helping the AMA's brand of "scientific medicine" solidify its influence.[4]

Key features of post-Flexner medical education were more rigid and uniform curriculum standards (later, Flexner fretted that these "stifled creative work"), a more homogeneous and privileged student body, and a growing concentration of graduates in wealthier parts of the country (prompting the popular press to lament the "vanishing country doctor"). Over the 20-year period following the Flexner report's publication, the number of medical schools fell by 42% (Flexner had called for an even more drastic reduction);[5] this proved to be a welcome development for doctors, "protecting the finances of [the AMA's] member-doctors by limiting the number of new physicians entering the profession."[6]

Perhaps because of his Carnegie and AMA sponsorship, Flexner argued that schools "could be first-rate only if they were well funded."[7] His recommendations became the lodestar guiding health care investments by the Rockefeller Foundation and other major foundations. According to Starr:

> [T]he report was the manifesto of a program that by 1936 guided $91 million from Rockefeller's General Education Board (plus millions more from other foundations) to a select group of medical schools. Seven institutions received over two thirds of the funds from the General Education Board. Though the board represented itself as a purely neutral force . . . its staff actively sought to impose a model of medical education more closely wedded to research than to medical practice. These policies determined not so much which institutions would survive as **which would dominate, how they would be run, and what ideals would prevail** [bold added].

While agreeing that physicians of the era were happy to repress "competing systems of medicine" (such as homeopathy and chiropractic), Starr dismisses the repression as "only a minor and relatively unsuccessful means of advancing the interests of the [medical] profession." Other authors disagree, however, arguing that the AMA, in particular, aggressively sought to "reduce or eliminate the professions that were not aligned with the 'regular' medical paradigm."[8] One avenue for doing so was the AMA's Council on Medical Education, established in 1904, which assigned biased AMA ratings to medical colleges. At the time, graduates of conventional medical schools were failing their medical board examinations at nearly twice the rate of individuals graduating from homeopathic colleges, but the AMA's guidelines nevertheless ensured that homeopathic colleges would get lower ratings.[9] According to homeopathy expert Dana Ullman,

> With the AMA grading the various medical colleges, it became predictable that the homeopathic colleges, even the large and respected ones, would eventually be forced to stop teaching homeopathy or die.[10]

In his report, Flexner, like his AMA sponsors, savagely disavowed healing approaches that he caricatured as "Medical Sects," casting "sectarians" as dogmatic opponents of "scientific medicine." In his view, only a few "sects" (homeopaths, "eclectics," physiomedicals, and osteopaths) were worthy of "serious notice"; even there, he made little effort to hide his disdain for the 32 "sectarian institutions" in existence in the U.S. at the time (including 15 homeopathic colleges), using adjectives like "peculiar," "weak," "uneven," "grimy," "hysterical," "fatally defective," and "utterly hopeless" to describe them (see **Homeopathy Under Attack**). As for chiropractors, Flexner dismissed them outright, labeling them as "unconscionable quacks" guilty of "exaggeration, pretense, and misrepresentation of the most unqualifiedly mercenary character." Chiropractic leaders writing in 2010 observed, "It was apparent that Flexner was not in support of chiropractic or other health care professions and was positioning medicine to be at the center of the health care model."[11]

Homeopathy Under Attack

According to "A Condensed History of Homeopathy" by Dana Ullman, homeopathy was "spectacularly popular" in 19th-century America, including among the rich and famous, a fact that prompted "deep-seated animosity and vigilant opposition" from mainstream medicine and the burgeoning pharmaceutical industry.[12] Ullman notes that orthodox medicine viewed homeopathy as posing a triple threat—"philosophical, clinical, and economic." Homeopathy also challenged the medical profession's reliance on drugs that largely had suppressive effects, not only masking a person's symptoms but often contributing to "deeper, more serious diseases." In response to the perceived competition, the AMA engaged in "incessant and harsh attempts" to destroy homeopathy. As summarized by Ullman, these included:

- Purging local medical societies of physicians who were homeopaths
- Threatening orthodox physicians with loss of their AMA membership "if they even consulted with a homeopath or any other 'non-regular' practitioner"—and regularly enforcing this threat

- Taking actions to "thwart the education of homeopaths"
- "Besmirching" homeopaths' reputation, calling them "immoral," "illegitimate," and "unmanly"

As of 1900, the robust American homeopathic profession boasted of 22 homeopathic medical schools, over 100 homeopathic hospitals, and more than 1,000 homeopathic pharmacies, but by 1923, only two homeopathic colleges remained, and by 1950, there were none. Ironically, though John D. Rockefeller was a strong proponent of homeopathy and used it himself, none of the $300 to $400 million that he gave away in the early 1900s went to homeopathic institutions.

Homeopathy managed to survive, but in more recent times, the FDA has frequently attempted to make access to homeopathic remedies more difficult. The Alliance for Natural Health explains:

The FDA gets a large part of its funding from drug companies, and for this reason sees Big Pharma as its client. Homeopathic medicines, like supplements and other natural products, compete with pharmaceutical drugs, so the FDA tilts the scales in favor of drug companies.[13]

Efforts to tarnish homeopathy's reputation also continue; a 2020 article snarkily described the well-established theoretical basis of homeopathy as "outlandish"[14] and accused the one in ten individuals who use homeopathy worldwide[15] (including the British royal family)[16] of being scientifically illiterate. Another skeptic marveled in 2006 in the *Journal of the Royal Society of Medicine*,

"One might expect that unorthodox medicine . . . would have diminished as a result of the spectacular advances in regular medicine during the second half of the twentieth century, but that does not seem to be the case."[17]

Puzzling over the hostile tone, a UK homeopath commented, "The wiser heads among us realise that all forms of therapeutics have their place and that we should be grateful for the diversity of approach that so adds to the interest of the medical world."[18]

In 1924, Dr. Morris Fishbein (a non-practicing physician) became the AMA's general manager and editor of the influential *Journal of the American Medical Association* (JAMA), serving in those roles for a quarter-century (until 1950).[19] Ullman notes that Fishbein was "an effective advocate for conventional medicine and a vocal critic of unconventional treatments," writing books criticizing "medical quackery," describing chiropractic as a "malignant tumor," and brushing off osteopathy and homeopathy as "cults."[20] Fishbein was particularly adept at using the media to "attack anyone who provided a real or perceived threat to conventional medicine," including via a newspaper column (syndicated in over 200 newspapers) and a weekly radio program (reportedly "heard by millions").[21]

Not coincidentally, according to Ullman, Fishbein's "obsessive" opposition to anything that was not his brand of medicine "provided direct benefits to the physicians he was representing":

> There are . . . numerous stories about Fishbein's efforts to purchase the rights to various healing treatments, and whenever the owner refused to sell such rights, Fishbein would label the treatment as quackery. If the owner of the treatment or device was a doctor, this doctor would be attacked by Fishbein in his writings and placed on the AMA's quackery list. And if the owner of the treatment or device was not a doctor, it was common for him to be arrested for practicing medicine without a license or have the product confiscated. . . . Further, Fishbein wrote numerous consumer health guides, and his choice of inclusion for what works or what doesn't work was not based on scientific evidence.[22]

Fishbein's playbook was not far afield from the tactics used in 2020 to malign therapies such as hydroxychloroquine and ivermectin and harass doctors like Pierre Kory and Meryl Nass who dared to disagree with the CDC's COVID "guidelines."[23,24] One of the medical establishment's key weapons against dissenting doctors in the COVID era has been to have state medical licensing boards suspend or revoke doctors' licenses; like Fishbein, the boards tend to eschew scientific evidence in favor of "consensus opinions."

A related trend includes retraction of published, peer-reviewed studies. Two status-quo-threatening articles retracted since 2020 are "Relative incidence of office visits and cumulative rates of billed diagnoses along the axis of vaccination" by James Lyons-Weiler, PhD, and Dr. Paul Thomas, who concluded that "vaccinated children appear to be significantly less healthy than the unvaccinated;"[25] and "The role of social circle COVID-19 illness and vaccination experiences in COVID-19 vaccination decisions: an online survey of the United States population" by Michigan State professor Mark Skidmore, who reported that one-fifth of his nationally representative sample "knew at least one person who experienced a health problem after COVID-19 vaccination."[26]

To illustrate the perniciousness of "consensus-driven science," Kory cites famous remarks made by writer Michael Crichton in a 2003 speech about the "increasingly uneasy relationship between hard science and public policy":

> Historically, the claim of consensus has been the first refuge of scoundrels; it is a way to avoid debate by claiming that the matter is already settled. . . . Let's be clear: the work of science has nothing whatever to do with consensus. Consensus is the business of politics. . . . In science consensus is irrelevant. What is relevant is reproducible results. . . . There is no such thing as consensus science. If it's consensus, it isn't science. If it's science, it isn't consensus. Period.[27]

Alongside the censorship and attacks on truth-telling doctors, the world has witnessed increasingly horrifying abuses of medical authority against patients since 2020, with many doctors advancing democidal policies through "assisted" deaths engineered without even the pretense of patient or family consent.[28] In the UK—illustrating the dangerous weaponization of euthanasia and medically assisted death discussed in Chapter Four—credible on-the-ground reports by independent British journalist Jacqui Deevoy described the scarcely disguised use of euthanasia as a full-blown medical protocol,[29] documented in her 2021 film *A Good Death?*[30] On paper, the UK does not allow assisted death and promises a jail sentence of up to 14 years for those who help others to die,[31] but observers report that the UK government "and its institutions have been acting as if euthanasia is perfectly legal since 2008."[32]

One of the most damning pieces of evidence assembled by Deevoy and others was UK Health Secretary Matt Hancock's unprecedented acquisition of a two-year supply of the benzodiazepine execution drug midazolam[33] in March 2020, even though the country already had a one-year supply on hand. Reportedly, by October 2020, "there was no midazolam left."[34] UK residents now refer to the death of more than 136,000 elderly residents of care homes since April 2020 as the "Midazolam Murders."[35] One writer soberly calls the intentional use of midazolam a "model for implementing the death of a patient, as determined remotely, without recourse, without the presence of family and cremating of remains to remove any possibility of discovery of false diagnosis or error or malice."[36]

Likewise in the U.S., the past four years have shone a light on hospitals as a locus of lethal medical protocols. In addition to withholding safe and inexpensive treatments and aggressively mandating known-to-be-dangerous COVID shots,[37, 38, 39] the nation's hospitals—immune from liability and salivating over hefty financial incentives[40]—have misused their medical authority by inflexibly adhering to the ventilator-plus-remdesivir COVID "protocols"[41] that are now as notorious as midazolam as a form of medically induced death.[42] According to a white paper produced in mid-2022 by the TN Liberty Network and AJ DePriest, hospitals could earn almost $293,000, on average, by diagnosing and "treating" someone as a "complex" COVID inpatient; a "noncomplex" COVID inpatient earned hospitals roughly $58,000, while a COVID outpatient would only bring in around $2,500.[43]

As mentioned in the Introduction, in a high-profile incident Scott Schara's 19-year-old daughter Grace was allegedly killed by a hospital protocol in 2021 that led to his milestone civil lawsuit (*Schara v. Ascension Health et al.*). Schara's research confirms that "medical murder" is incentivized and profitable, and has become "a mainstream policy position of high-level government and corporate figures in America."[44] Schara also connects the dots to other routes of "medical killing":

By studying the convergence of medical killing across multiple areas—including hospitals, elderly care facilities, the 'jab' agenda, and more—I've noticed a clear pattern. **The medical establishment, backed by coercive government policy and financed and supported by big business, is undertaking a "soft genocide"** [bold added]. They look at what they are doing as "hastening death," but that's just another way of saying "murder." ...[A]ll Americans must protect themselves against this evil—and deadly—agenda.[45]

In the summer of 2023, CHD.TV launched a bus tour that is not only collecting first-hand accounts of vaccine injuries but also documenting how extensively hospitals used—and are still using—the deadly COVID protocols.[46] The COVID-19 Humanity Betrayal Memory Project,[47] an initiative of the FormerFedsGroup Freedom Foundation, is also collecting such stories and, through interviews, has distilled a list of 25 commonalities shared by the victims of deadly hospital protocols (see **Appalling Practices and Deadly Protocols**).[48] The chilling list not only reveals "unthinkable crimes against humanity" and "an ongoing atrocity" but illustrates the variety of tactics deployed under the banner of medical authority. In September 2023, a relative of Sasha Latypova experienced what Latypova described as an attempted (but fortunately unsuccessful) "hospital murder"; her relative and family members experienced many of the practices on the list, including a sneak effort to administer remdesivir in the middle of the night against patient and family member wishes.[49] The experience prompted Latypova to emphasize, "Covid murder wards remain operational in key federally funded locations."[50]

Appalling Practices and Deadly Protocols

In its examination of hundreds of cases of what they describe as "medical murder," the COVID-19 Humanity Betrayal Memory Project has produced an appalling list of 25 of the most prevalent practices and characteristics "associated with the deadly COVID hospital protocols."[51]

1. Isolation
2. Strict adherence to EUA "protocols" ("often forced on victim when refused")

3. Denial of alternative treatments
4. Denial of informed consent (for medications, treatments, intubation, or procedures)
5. Gaslighting
6. Removal of communication devices
7. Dehumanization ("being treated like an animal")
8. Pervasive sense of wrongdoing
9. Vaccination discrimination (including "mocking" and physical abuse)
10. Rapid and intentional oxygen increase (leading to mechanical ventilation)
11. Refusal by providers to communicate with family or advocates
12. Dehydration and starvation (facilitated by administration of diuretics or laxatives)
13. Abuse of physical or chemical restraints
14. Denial of bathroom use in favor of forced catheterization and/or a rectal tube
15. Non-emergency ventilation (typically under false pretenses)
16. Pressure for DNR orders and "shenanigans" such as falsification of paperwork
17. Palliative care pressure
18. Denial of family access during dying and even after death
19. Police or security involvement and threats of arrest
20. Refusal to transfer to another doctor or hospital
21. Hospital-acquired infections or injuries
22. Neglect and lack of basic care
23. Nighttime emergencies to "scare" or "confuse" family members
24. A climate experienced as malevolent
25. Unqualified staff

Ethical individuals might find it difficult to comprehend health care workers' willingness to go along with hospitals' cruel and lethal protocols. In addition to the blunt tool of financial incentives, another possible explanation may have to do with the significant reshaping of the health care workforce that has taken place since 2020. Notably, hospitals' imposition of COVID vaccine mandates beginning in 2021 drove out significant numbers of health care workers who either were fired or chose to quit rather than submit to the experimental jabs—and these likely were some of the industry's most informed professionals. A subgroup of those who remained, willing to comply with the shots, may have been protocol-compliant as well.

Researchers have estimated that over 18% of nurses worldwide refused the jabs in 2021.[52] In various U.S. polls in fall 2021 (in advance of the January 2022 federal mandate for employees at health care facilities participating in Medicare and Medicaid), anywhere from 15% to 30% of nurses reported that they would leave rather than comply with the mandate,[53] and a third of surveyed physicians expressed opposition to the mandate (including some who were "pro-vaccine").[54] As of early 2023, Becker's Hospital Review estimated actual hospital employee losses of 1% to 5% (via resignation or termination) as a result of the mandates,[55] but statistics from Maine—where roughly 10% of health care workers departed around the time of the state's August 2021 mandate[56]—suggest that the workforce impacts may be underreported. Observing the Maine governor's attempts to "hastily rationalize the current acute care shortages as being a decades-long problem" (or a problem caused by the demands of providing COVID care), the Maine Policy Institute noted that 99% of the health care facility staff who left their jobs in August and September 2021 were unvaccinated: "If this were mostly from pandemic burnout and early retirements, why are we just seeing it now?"[57]

In May 2022, without saying a word about the mandated shots, McKinsey issued a report predicting a "dire" nursing shortage by 2025: "Over the past two years . . . nurses consistently, and increasingly, report planning to leave the workforce at higher rates compared with the past decade."[58] Reinforcing McKinsey's "dire" outlook, a survey of health care workers who quit due to COVID vaccine mandates found that one in four had chosen "to leave the field entirely."[59]

Nurse leaders may also be preparing to depart in droves. According to longitudinal survey results released in early 2024 by the American Organization for Nursing Leadership (AONL), 35% of nurse leader respondents reported intending to leave their position within six months—either "maybe" (23%) or definitely (12%)—with a third of those respondents indicating plans to "leave nursing altogether."[60] Of those intending to leave their position, 44% cited as their primary reason the job's negative effects on their health and well-being, and 27% cited inadequate resources and staffing.

When AONL asked survey respondents an open-ended question about "concerns for the future of nursing," several interesting themes emerged in addition to predictable complaints about workload and burnout—with numerous respondents expressing concerns "about the safety and quality of care provided to patients." One of these themes pertained to the perception that **nursing education** has become "diluted" or "subpar." According to respondents, newer graduates arrive on the job "inexperienced" and "unprepared"; have weaker problem-solving, "critical thinking," and communication skills; and display less "professionalism, accountability and responsibility." Specific concerns along these lines included comments about:

- New nurses' lack of "basic nursing skills and critical thinking"
- The amount of time spent "helping [new graduates] with skills they should have learned/acquired in Nursing school"
- New nurses' "increased anxiety" and fear ("They say they did not sign up for this")
- "Watering down of professional practice"
- "[T]he lack of care and concern among younger nurses," "Lack of love for the field of nursing," failure to "grasp . . . the art and heart of nursing," and lack of "passion for the job"

A related concern had to do with **generational differences**. Statements to this effect included:

- "Our millennial and Gen Z nurses don't have the same work ethic as our Baby Boomer and Gen X nurses do."
- "[G]eneration Z are needy and have a hard time thinking outside of the box to solve problems."

- "The younger generation has differing work ethics/values."
- "I am concerned about the lack of professional accountability in the younger generation."
- "The generation[al] lack of caring about patient and families."
- "The younger generation grew up texting each other and don't know how to have a real conversation."

As one respondent concluded,

> I feel we are losing the concept that this is a profession where the whole aspect of patient care needs to be considered, not just a job where we perform a task and consider that to be "what nurses are."

Some of the AONL respondents also made reference to the "use of health care as a commodity to be monetized and profited from"; as one nurse leader put it,

> I am concerned that the overall view of nurses as a commodity . . . versus recognition of attributes that lead to successful outcomes is a significant disconnect.

Others commented that "Nurses seem more financially driven since the pandemic." Interestingly, hospitals' use of travel nurses has exploded since 2020. Whereas "total hours worked by contract or travel nurses in hospitals" was just 3.9% of total hours as of January 2019 (or 4.7% of hospitals' labor expenses for nurses), by January 2022, contract/travel nurse hours represented 23.4% of total hours (or 38.6% of nurse labor expenses).[61]

Although there is no way to pinpoint the extent to which travel nurses constitute some of the manpower carrying out hospitals' deadly protocols, and many travel nurses are undoubtedly diligent and upright professionals, several data points from Trusted Nurse Staffing[62] and other organizations are noteworthy:

- Proportionately more travel nurses—18% vs. 9%—are male compared to regular RNs.[63]

- The top travel nurse specialty in demand, by far, is "med-surg"—these are the very wards where hospitals typically house the patients on whom hospitals are forcibly imposing a diagnosis of "COVID."
- Before 2020, travel nurses averaged $1,800 to $2,600 per week; post-COVID weekly rates soared to as high as $8,000 to $10,000 per week.
- The average annual salary for new travel nurses increased by 25% between 2022 and 2023, with some salaries rising to $200,000 a year.
- New York state—one of the states that behaved most harshly toward health care workers who refused the COVID jab—offers the highest hourly pay for travel nurses and is one of three states (along with California and Texas) with the most travel nurse jobs available. A publicly run New York City hospital network paid one travel nurse contracting company $1.2 billion in fiscal 2022 alone.[64]
- Painter Law Firm, a Texas firm specializing in medical malpractice, pointed out in January 2022 that travel nurse staffing agencies often outsource important oversight and quality assurance functions; because travel nurses are not hospital employees, they typically are "not put through the same hospital processes of training and skills verification that employee nurses must pass."[65]

In 2023, a fed-up Center for Economic and Policy Research argued for "swift and substantial federal and state action . . . to address the new role travel nurse agencies play in the healthcare workforce, especially when profit is the priority of many of these staffing agencies."[66] In 2021, in fact, private equity takeovers of travel nurse agencies hit record levels, with 27 deals—up from 11 in 2020—"well above any year over the previous decade."[67] This reflects a wider trend of exponential growth in private equity acquisitions of hospitals, nursing homes, physician practices, and other health care facilities and agencies—dubbed by some the "financialization" of health.[68] Research indicates that these takeovers correspond to substantial declines in quality of care; according to a December 2023 study, within a few years of private equity acquiring a hospital, the hospital becomes significantly more dangerous for patients, with more incidents suggestive of the types of neglect and incompetence cited by AONL respondents, such as falls and central-line bloodstream infections.[69]

Ultimately, whether one considers the deadly remdesivir protocols,[70] or the sidelining of drugs like ivermectin,[71] or the persistence of medical kidnapping,[72] Starr's long-ago comments about medical authority—and the potentially "severe costs" experienced by patients who refuse "standard" care—resonate now more than ever:

> [P]rofessional authority has become institutionally routine, and compliance has ceased to be a matter of voluntary choice. . . . This social structure is based . . . on the institutionalized arrangements **that often impose severe costs on people who wish to behave in some other way** [bold added].

The Rise of the Modern Pharmaceutical Industry

That the rise of the modern pharmaceutical industry was another major factor involved in the transformation of American medicine is a fact so well established as to almost not be worth mentioning. Yet this historical development was pivotal, helping steer medical practice toward the chemical- and pharma-dominant model that holds sway to this day (now evolving toward a biopharma model). Ironically, enthusiastic paeans to the drug industry's early-20th-century "founding fathers"[73] pay homage to the very same companies (e.g., Pfizer, Merck, Eli Lilly, Bayer) that nowadays often make headlines for bad behavior of various kinds.[74,75]

Starr recounts how, between 1900 and 1920, the AMA took an important step toward cementing a symbiotic medical-pharmaceutical partnership by seizing control over pharmaceutical information. Branding "patent medicine" makers who bypassed doctors and sold their remedies directly to the public as "quacks," the AMA closed its prestigious journal to patent medicine advertisements and established a system that aimed "to withhold information from consumers and rechannel drug purchasing through physicians." As drug makers grew dependent on doctors to market their products, the public in turn became more reliant on professionals to make medication decisions. As mentioned in connection with the discussion of medicine's attacks on homeopathy, the solidifying medical-pharmaceutical relationship also shifted Western medicine's focus toward "palliation"—the easing or suppression of symptoms—and away from efforts to cure the underlying causes of disease or, even better, to prevent them.[76]

The powerful 2021 book *Empire of Pain*,[77] which recounts "the secret history of the Sackler dynasty," illustrates Western medicine's bias toward symptom suppression and adds further color to the historical portrait of melded medical/pharmaceutical interests. In telling the story of OxyContin and the Sackler-family-owned company Purdue Pharma, author Patrick Radden Keefe notably includes extensive background on family patriarch Dr. Arthur Sackler (1913–1987).[78] Though Sackler passed away eight years before the FDA's 1995 approval of OxyContin, he was instrumental in creating the cavalier corporate culture and aggressive marketing tactics that, in Keefe's words, helped OxyContin become "one of the biggest blockbusters in pharmaceutical history."[79]

Though trained as a physician, Arthur Sackler's particular genius was in marketing and medical advertising. Keefe recounts how Sackler "thrived" after taking a second job at the pharmaceutical advertising firm William Douglas McAdams in 1942, to such an extent that he became the firm's president a mere two years later and its owner by 1947. He also acquired a "clandestine stake" in a competing advertising firm, with both agencies together cornering the drug advertising market. The Medical Advertising Hall of Fame posthumously credited Sackler with "shap[ing] the character of medical advertising," claiming that "His seminal contribution was bringing the full power of advertising and promotion to pharmaceutical marketing."[80]

During the postwar period, the efforts of advertising mavericks like Sackler helped trigger a "remorseless migration of medicines from the periphery of the American shopping cart to its center"; between the 1940s and 1961, drug costs became "an important part of the nation's expenditure on health care," rising by 60% and generating enormous profits for the drug industry.[81] The Arthur-Sackler-conceived strategy that subsequently informed the marketing of OxyContin was "to adopt the seductive pizzazz of more traditional advertising—catchy copy, splashy graphics—**and to**

market directly to an influential constituency: the prescribers" [bold added], in essence pitching medicine to doctors "on more or less the same terms as swimwear or auto insurance was marketed to average consumers."

Keefe describes the variety of tactics that Sackler developed and refined to attract doctors' attention, tactics that remain in use to the present day:

- Placing "eye-catching" ads in medical journals—even if "fundamentally deceptive"
- Using "native advertising" (i.e., "paid promotion . . . camouflaged to resemble editorial content")
- Sending hordes of "detail men" ("young, polished sales representatives") to visit doctors' offices and distribute promotional literature and gifts
- Persuading well-known doctors to endorse drugs ("the equivalent, for physicians, of putting Mickey Mantle on a box of Wheaties")
- Having drug companies cite studies that they themselves have funded
- Developing catchy lingo (such as the term "broad spectrum" antibiotic, invented out of whole cloth by Sackler to sound clinical)
- Blaming the victim when drugs turn out to be addictive ("it's not the drugs that are bad; it's the people who abuse them")

These techniques, according to Keefe, made huge successes out of lackluster drugs such as Pfizer's Terramycin antibiotic, and persuaded doctors that two drugs that "did pretty much the same thing," Roche's Librium and Valium, actually were different. Before Valium's rollout, Librium became America's most prescribed drug; leaving Librium in the top five, Valium then achieved the same milestone and became "the first $100 million drug in history" (while also spawning a massive addiction problem).[82] Meanwhile, Roche—which rewarded Sackler with "an escalating series of bonuses"—became "not just the leading drug company in the world but one of the most profitable companies of any kind."

As reported in 2007 in the *Journal of Surgical Research*,[83] the pharmaceutical industry directs over 90% of its very significant marketing outlays (themselves representing at least 30% of industry revenues) at physicians.

"Favored" direct-to-doctor marketing techniques include "continuing medical education," drug detailer visits (estimated at 60 million visits annually), and doctor gifts and payments amounting (as of 15 years ago) to an "astonishing" $10,000 to $15,000 per doctor per year. With none of the parties worried about the blatant conflicts of interest, these strategies pay off:

> Physicians are more likely to prescribe a drug if they had recently attended a sponsored event by the manufacturer; they are more likely to prescribe a drug that is not clinically indicated and have a drug placed on a hospital formulary. . . . Physicians who are given free samples . . . are far more likely to write subsequent prescriptions for that drug.[84]

The obvious conclusion reached by many is that a doctor's "judgment on what represents the best interests of patients may be compromised by a less than cautious relationship with the industry."[85]

The writer *A Midwestern Doctor* describes how the win-win symbiotic relationship between medicine and pharma perpetuates itself:

> [O]ne of the key reasons why doctors are given such a high status in our society . . . is because they are a critical component of the pharmaceutical industry (as their trusted and exclusive prescriptions are the sales mechanism for Big Pharma). Since the pharmaceutical industry has a fairly reliable playbook for grooming doctors to push their drugs, they can expend a relatively small amount of money to ensure a large number of their products are sold—and hence the industry is incentivized to use its massive clout to ensure doctors retain the social status which makes the entire pharmaceutical business model viable.[86]

In more recent times, drug companies have developed additional strategies to capture physicians' loyalties, such as "e-detailing" (incentivized online drug promotion) and "Customer Relationship Management," which allows "[h]ighly refined and comprehensive information about a doctor's prescribing patterns" to be "longitudinally tracked."[87] The latter was a major feature of Purdue Pharma's aggressive and successful marketing of OxyContin to physicians (see **Marketing Takes Over Medicine**). In a 2009 article titled "The Promotion and Marketing of OxyContin: Commercial Triumph, Public Health Tragedy," Virginia physician Art Van Zee explained:

> Drug companies compile prescriber profiles on individual physicians—detailing the prescribing patterns of physicians nationwide—in an effort to influence doctors' prescribing habits. **Through these profiles, a drug company can identify the highest and lowest prescribers of particular drugs in a single zip code, county, state, or the entire country** [bold added]. One of the critical foundations of Purdue's marketing plan for OxyContin was to target the physicians who were the highest prescribers for opioids across the country. The resulting database would help identify physicians with large numbers of chronic-pain patients. Unfortunately, this same database would also identify which physicians were simply the most frequent prescribers of opioids and, in some cases, the least discriminate prescribers.[88]

The horrific result, as described by Keefe and others, was this: "Prior to the introduction of OxyContin, America did not have an opioid crisis. After the introduction of OxyContin, it did."

Not surprisingly, the opioid crisis spawned extensive litigation against Purdue for downplaying and hiding OxyContin's known risks of addiction and overdose. In response, Purdue executives filed for bankruptcy, after first pulling billions out of the company. In September 2021, a judge in the U.S. Bankruptcy Court for the Southern District of New York signed off on a sweetheart bankruptcy deal (and then promptly retired eight years before the expiration of his term), okaying broad and "controversial protections for the Sackler family members that owned the company."[89] In August 2023, however, the U.S. Supreme Court paused the bankruptcy plan. Arguing that the plan constituted an "abuse of the bankruptcy system,"[90] the Solicitor General protested that it "'absolutely, unconditionally, irrevocably, fully, finally, forever and permanently releases' the Sacklers from every conceivable type of opioid-related civil claim—even claims based on fraud and other forms of willful misconduct that could not be discharged if the Sacklers filed for bankruptcy in their individual capacities."

Marketing Takes Over Medicine

As president and co-chairman of Purdue Pharma at the time of OxyContin's launch in 1996, family scion Richard Sackler (nephew of Arthur Sackler) considered the company's "most valuable resource" to be "not the medical staff or the chemists or even the Sackler brain trust but the sales force," according to Keefe's account in *Empire of Pain*. To "sell, sell, sell" OxyContin, the company hired a "phalanx of new recruits" and incentivized them with a bonus program unprecedented in the industry, giving Purdue the reputation of being "a great place to work." Explains Keefe: "Most pharma companies capped what kind of additional bonus you could make as a rep. Purdue didn't. . . . If sales grew, you'd get a bigger bonus. There was no cap." In 2001 alone, Purdue paid out $40 million in bonuses to its 600-plus sales reps—an average of $71,500 annually per rep (on top of an average annual salary of $55K), with the bonuses ranging in amount from $15,000 to nearly $240,000.[91] The company also gave top sales reps extra rewards like all-expenses-paid vacations to Caribbean resorts.

As summarized by Dr. Van Zee (credited with being "among the very first professionals to sound the alarm about the opioid addiction problem"),[92] Purdue's internal sales force more than doubled between 1996 and 2000, as did its physician call list.[93] Incentivized and supported by extensive training and coaching, the sales reps augmented their personal powers of persuasion with:

- A patient starter coupon program (34,000 coupons redeemed by 2001)

- Branded OxyContin swag, including "fishing hats, stuffed plush toys, and music compact discs"
- Instructions to target primary care physicians, despite those generalists not being "sufficiently trained in pain management or addiction issues"
- Aggressive promotion of OxyContin for non-cancer-related pain, which gave a tenfold bump to OxyContin prescriptions for that type of pain between 1997 and 2002 (versus a four-fold increase in prescriptions for cancer-related pain)
- A talking point misrepresenting the risk of addiction as "less than one percent"
- Prescriber profiling data that—coupled with the lucrative bonuses—motivated sales reps to identify the types of doctors willing to operate "pill mills"

According to Keefe, the "carefully scripted phrase that [sales reps] intoned like a mantra" was the unfounded assertion that OxyContin was "the painkiller 'to start with and to stay with.'" For the many individuals who would go on to become addicted, the corollary talking points of this sales pitch were momentous:

For "moderate to severe pain," OxyContin should be the first line of defense. And it was good for acute, short-term pain, as well as for chronic, long-term pain; this was a drug you could use for months, years, a lifetime, a drug "to stay with." From a sales perspective, it was an enticing formula: start early, and never stop.

The Sidelining and Weaponization of Nutrition

Decades ago, an open-minded and curious dentist named Dr. Weston A. Price (1870–1948) wondered why dental decay seemed to go hand in hand with physical degeneration in his patient population, including in younger patients. To try to answer that question, Price and his wife spent 10 years traveling the world to compare two groups: people still eating their traditional diets, and similar populations who had shifted to eating the "displacing foods of modern commerce" (which meant, at the time, items like white sugar and white flour, canned condensed milk, and industrial seed oils).[94] After studying populations inhabiting areas as diverse as the Swiss Alps, the Outer Hebrides, and the South Seas, along with groups like Aboriginal Australians, tribal Africans, and North and South American Indians, Price concluded that nutritional deficiencies were to blame not only for dental problems (ranging from decay and crowded, crooked teeth to facial deformities) but also for many other types of physical degeneration. In contrast, people still eating their traditional diets, "rich in essential food factors," displayed "beautiful straight teeth, freedom from decay, stalwart bodies, resistance to disease and fine characters."[95] For this seminal work, published in his 1939 volume *Nutrition and Physical Degeneration*,[96] Dr. Price earned the nickname the "Isaac Newton of Nutrition."[97]

When Price put what he had learned into practice, he was able to turn malnourished orphans' health around simply by feeding them one nutrient-dense meal a day.[98] Unfortunately, since Price's day, entire populations are now undernourished or malnourished in one way or another, and it takes knowledge and commitment to obtain the nutrient-dense foods that confer radiant health—foods high in vitamins and minerals and especially the fat-soluble vitamins A, D, and K. The list of unhealthy and devitalized foods also has become much longer and virtually ubiquitous.[99] That list includes items like refined and artificial sweeteners;[100] industrially processed oils;[101] white-flour-based breads, snacks, and sweets;[102] toxic extruded and puffed grains;[103] low-quality milk and dairy products;[104] artificial flavors, heavy-metal-containing food dyes,[105] monosodium glutamate (MSG),[106] and other additives (in nearly all processed foods);[107] fake "plant-based meats" and beverages,[108,109] many of them soy-based; foods contaminated with pesticides and herbicides; and foods containing genetically modified ingredients.

Few doctors are likely to tell their patients about the critical importance of the fat-soluble vitamins or warn them about the destructive effects of the denatured and refined foods that make up the bulk of most Americans' diet, for two reasons. First—despite the "intrinsic relationship between food and health" and the epidemics of diet-related diseases that physicians encounter in their practice—doctors know next to nothing about nutrition.[110] Medical students and physicians admit that medical school provides virtually no nutrition education and that they feel "ill-prepared" to counsel patients on nutrition topics.[111]

Second, and far worse than providing no dietary information at all, is the erroneous, dangerous, and industry-friendly dietary propaganda that has replaced the life-giving knowledge that Dr. Price and other early 20th-century nutrition pioneers so carefully gathered and disseminated decades ago.[112] The long-running and willfully unscientific demonization of saturated fat and cholesterol is a prime example. As mentioned in Chapter Two, medical professionals and dieticians who push that party line are causing immense harm, persuading people to remove highly nutritious foods like eggs and butter from their diets while driving them toward the extremely harmful vegetable oils (manufactured from soybean, corn, canola, cottonseed, rapeseed, sunflower, and safflower) that are cheap industry favorites,[113] or toward processed, sugar-laden products billed as "low-fat,"[114] as well as pushing hazardous but highly profitable medications like statins. Knowledgeable experts consider industrially manufactured seed oils to be "chronic metabolic biological poisons" and primary drivers of health problems such as type 2 diabetes, cancer, obesity, heart disease, dementia, and even macular degeneration—but these are all conditions that generate enormous revenues for doctors and drug companies.[115]

Although numerous researchers and studies have refuted the notion that high-fat foods and high cholesterol cause heart disease or that lowering cholesterol improves health, the powerful industries that benefit from this false dogma have made it difficult to get any countervailing messages out. In his book *The Cholesterol Myths: Exposing the Fallacy that Saturated Fat and Cholesterol Cause Heart Disease*, Uffe Ravnskov, MD, PhD quotes Dr. Paul Rosch (1927–2020), an expert on "the role of stress in health and illness" and, notably, on stress's role in cardiovascular disease and cancer. Ravnskov cites Professor Rosch's cut-to-the-chase explanation as to why the cholesterol hypothesis persists despite the lack of "any solid scientific proof" in its favor:

> The cholesterol cartel of drug companies, manufacturers of low-fat foods, blood-testing devices and others with huge vested financial

interests have waged a highly successful promotional campaign. Their power is so great that they have infiltrated medical and governmental regulatory agencies that would normally protect us from such unsubstantiated dogma.[116]

Another area where health care providers are leading the public badly astray concerns salt, a nutrient containing two components (sodium and chloride) "essential to health and happiness."[117] Pointing to the FDA's (mis)guidance telling the public to drastically limit sodium consumption due to putative effects on blood pressure and supposed heart disease risks, the AMA tells its audience that salt is "literally killing people,"[118] a piece of advice that couldn't be further from the truth. (For the majority of individuals, salt intake has no effect on blood pressure whatsoever.)[119] In fact, salt aids "the myriad chemical reactions that support enzyme function, energy production, hormone production, protein transport and many other biochemical processes."[120] Were one to eat a saltless diet, the eventual result would be "gradual desiccation of the body and finally death."[121]

Writing in 2012, molecular biologist and salt expert Morton Satin described the "arbitrary" and "fundamentally flawed" recommendation to lower salt intake as a "blunt fiat" and a "shift away from an evidence-based approach in establishing recommendations to one of subjective inference: opinion."[122] Sally Fallon Morell, president of the Weston A. Price Foundation, additionally notes that "[d]emonization of a substance so vital to our health [as salt] could only happen in a society ignorant of the history of salt." She explains:

> The quest for salt led to the development of the major trade routes in the ancient world. If you look at a map of the world showing the major accessible salt deposits, there you will also see where civilizations developed—in Jordan, the Tigris-Euphrates, the Yellow River of China, the salt swamps in Persia, the deserts of Egypt and the Sahara; in the New World in Central America, the Andes and the Great Lakes; and finally on the seacoasts in areas of abundant sunshine, where salt could be obtained from evaporated sea water.[123]

As with cholesterol, there are conscientious experts who dissent from the "salt myths," but again, they face an uphill battle in combating the entrenched salt dogma and the industries, including medical, that benefit from it (for example, by manufacturing chemical salt substitutes or drugs to treat the problems caused by inadequate salt intake). Bypassing the information embargo, a cardiovascular medicine expert who made it onto ABC News in 2015 declared that the medical community was "stuck in a time warp" with the advice that salt is bad for one's health, stating, "There is no solid evidence to support the current recommendations."[124] Even *Scientific American*, ordinarily not one to break ranks with "consensus" science, suggested a decade ago that "the zealous drive by politicians to limit our salt intake has little basis in science" and admitted that "the correlation between salt intake and poor health" was "tenuous."[125] The magazine also acknowledged that a low-salt diet can have detrimental effects. As Morell has explained:

> "Researchers, politicians, medical professionals and journalists push their no-salt agenda as a surefire way to limit disease when all the evidence points to the opposite—increased health problems in young and old, diminished brain function, increased confusion, and a boon to the food processing and medical industries."[126]

The Evidence-Based Medicine Juggernaut

In the early 1990s, "evidence-based medicine" (EBM) took the medical world by storm, taking off concurrently with the proliferation of clinical practice "guidelines." As an indicator of the sweeping trend, during the 20-plus-year period from 1992 to 2015, the words "evidence-based" appeared in the titles of over 20,000 papers in the National Library of Medicine's PubMed database.[127] The clinical guidelines that accompany EBM, though nominally voluntary, are "treated as law" by the medical system, including insurance and hospitals.[128]

EBM posits that medicine's knowledge base is "rational/technical/linear/predictable rather than contingent/experiential/non-linear/unpredictable."[129] Early proponents marketed EBM as "a new paradigm" that would turn the practice of medicine into an "objective and scientific enterprise."[130] In the process—argued EBM apologists at the AMA—clinical practitioners would necessarily need to "deemphasize intuition, unsystematic clinical experience, and pathophysiologic rationale as sufficient grounds for clinical decision making."[131]

The AMA credits medicine's love affair with quantification and the advent of digital data storage technologies as two factors that helped cement the success of EBM and clinical guidelines, and brags that EBM has become the "gold standard" for clinical practice.[132,133] In a 2007 *British Medical Journal* poll, journal readers seemed to agree, ranking EBM "seventh among the 15 most important milestones that shaped modern medicine."[134] Almost from the beginning, however, EBM also had critics who objected that EBM represents "cookbook medicine," interferes with medical judgment and the art of medicine, sidelines clinical expertise, and fails to "account sufficiently for the complexity of individual cases."[135] In fact, as experts writing for *Harvard Business Review* noted in 2019, the EBM approach "mandates that new information, gleaned from randomized controlled trials and consolidated into clinical practice guidelines, can **and must** be used to improve the quality of care that patients receive" [bold added].[136] Alarmingly, while EBM "may provide guidance to some physicians, others may be nudged to provide treatments that, in the end, are the wrong treatments for their patients."[137]

Illustrating EBM's push toward one-size-fits-all medicine, the *Harvard Business Review* authors further explained:

> Treatments shown to be inferior, *on average*, in randomized controlled trials are assumed by many to be inferior for *all* patients—so much so that keeping a given patient, or a large population of patients, on the inferior treatment is viewed as a departure from evidence-based medicine [italics in original].[138]

However, they added,

> It is possible that *informed* physicians who combine their own clinical experience with up-to-date scientific evidence may have better

outcomes than physicians who unanimously choose treatments shown to be effective, on average, in clinical trials [italics in original].[139]

In 1999, a British general practitioner presciently warned that "[w]idespread adoption of guidelines could . . . result in clinical care informed by guidelines becoming viewed as the norm," with any departure from the guidelines seen "as prima facie evidence of a case to answer."[140] Even the AMA has made a similar point, suggesting that EBM tends to steer physicians toward litigation-avoidant "defensive medicine" because doctors believe that any deviation from EBM-based guidelines will be "immediately considered suspect" unless they can provide "proper justification."[141] Writing in the *World Journal of Clinical Cases* in 2018, researchers made the case that the replacement of clinical reasoning by "guidelines and algorithms" is, in fact, one of several fundamental changes in the "conception of medicine" that are undermining the nature and quality of care that doctors provide.[142] They asked:

> Are patients looking for doctors who rigidly follow algorithms and guidelines? They aren't. Algorithms that transform patient care into a sequence of yes/no decisions do not consider the complexity of medicine and the reasoning inherent in clinical judgment. . . . As much as a recipe book does not guarantee success in cooking, so clinical guidelines cannot guarantee success in diagnosis or treatment. . . . Medicine cannot be, and is not as black and white as protocols and checklists seem to imply.[143]

In a 2019 Reddit post on the topic "medical school does not select for thinkers, it selects for robots," a medical student wrote:

> [D]o you ever realize how we just blindly accept what we are being taught as factually true. . . . [O]ur professors should be telling us the truth but in fact that's not nearly as true as you would want it to be. The amount of times I come out of a lecture and look up information about what I was taught only to find different or varied information. However, we just chose [sic] to believe what we are

taught because that will be the answer on the test. And . . . then we use that information to treat our patients. . . . To even further this . . . we . . . have learned to just accept that EBM is god when in fact EBM can fall under so many fallacies and corruption. . . . [M]y own institution has been under fire for spewing false results and data in research that has built guidelines.[144]

According to "A Midwestern Doctor," the health care industry "has a vested interest in taking doctors out of the medical decision-making equation":

While many doctors will follow the prevailing narratives of the medical field without question, the independent clinical judgement of physician [sic] is an ever-present challenge to these vested interests. There have hence been many converging trends that seek to prevent physician "non-compliance."[145]

For EBM's purposes, "evidence is narrowly defined as having to do with systematic observations from certain types of research,"[146] with randomized controlled trials (RCTs)—and the systematic reviews and meta-analyses that bundle them[147]—ranked as the highest level of evidence.[148] However, critics argue that relying solely on these types of information can be "limiting" and may "restrain the freedom of professionals to use other sources of knowledge"—such as clinical experience and patient preferences—in their decision-making. They explain:

The reliance of EBM on the RCT was useful for acute (mostly single disease) conditions treated with simple interventions, but this approach is not suitable in the current epidemiological context characterized by chronicity and multimorbidity in complex health systems. In particular, EBM has largely disregarded the importance of social determinants of health and local context.[149]

A clinical epidemiologist at Ontario's McMaster University—the institution where EBM concepts first took root—raised other concerns in a 2002 publication in *BMC Health Services Research*:

The expectation of EBM that doctors should keep abreast of evidence from (certain-types-of-health-care-) research raises many issues. First, what is "valid" health care research? Second, what are the "best" findings from this research? Third, when is health care research "ready" for application? Fourth and fifth, to whom and how does one apply valid and ready evidence from health care research?[150]

Another question has to do with which research results get channeled into the EBM and clinical guidelines calculus. In late 2023, clinical research experts lamented that despite a legal requirement since 2007 for clinical trial investigators to report their results to ClinicalTrials.gov within one year of study completion, "less than half of trial results are reported on time" and thousands of results are "missing entirely."[151,152] In addition to drug companies, the culprits who routinely break the law requiring the timely reporting of human trial results include major medical schools, teaching hospitals, and nonprofit organizations.[153] An investigation conducted by *STAT* in 2015 found that even the NIH, responsible for *overseeing* the clinical trial registry, "violated the law the vast majority of the time."[154] So-called "negative results" (results that do not give investigators the answers they want), including embarrassing adverse event data, may be part of the story here—many observers agree that failure to publish "less than favorable results" is a pervasive problem.[155] (The clinical trial reporting law mandates disclosure of adverse event data.)[156]

Questions about the quality and validity of the underlying data sources that form the backbone of EBM and clinical guidelines take on additional resonance when one considers who funds and conducts most RCTs and related research to begin with. More often than not, "the companies that stand to gain the most from an intervention's success fund the studies that investigate them."[157] And, it is increasingly impossible to deny, many of those entities are willing to skew,[158] conceal,[159] or falsify their data,[160] resulting in a biased evidence base embedded with misleading or fraudulent conclusions.

In 2009, a meta-analysis examined studies looking at scientists' perceptions and self-reports of scientific misconduct.[161] The analysis found that over 14% of respondents, on average, reported having observed colleagues fabricating, falsifying, or modifying data or results; up to 72% also reported that colleagues had engaged in "other questionable research practices" (such as "dropping data points" or "changing the design, methodology or results of a study in response to pressures from a funding source"), and up to a third admitted that they themselves had engaged in other questionable research practices. In a similar meta-analysis for the decade from 2011 to 2020, around 3% of researchers acknowledged personal research misconduct (falsification, fabrication, or plagiarism), 13% copped to engaging in other questionable research practices, and 16% and 40%, respectively, were aware of colleagues having committed research misconduct or questionable practices.[162]

In 2023, *Nature's* London-based features editor Richard Van Noorden acknowledged that "faked" or "unreliable" RCTS are not only widespread—with experts estimating that from one-fourth to one-third of trials feature "statistically impossible data"—but also multidisciplinary in scope, spanning areas ranging from "women's health, pain research, anaesthesiology, [and] bone health [to] COVID-19" (see **Flawed RCTs and HPV Vaccines**).[163] Van Noorden warned that this has implications for the clinical guidelines that doctors are supposed to follow:

> [F]aked or unreliable RCTs are a particularly dangerous threat. They not only are about medical interventions, but also can be laundered into respectability by being included in meta-analyses and systematic reviews. . . . **Medical guidelines often cite such assessments, and physicians look to them when deciding how to treat patients** [bold added].

Flawed RCTs and HPV Vaccines

Public health officials have used EBM to justify their unrelenting pressuring of adolescents and young adults to undergo dangerous HPV vaccination.[164] However, in a 2020 study published in the *Journal of the Royal Society of Medicine*,[165] researchers outlined significant flaws in the 12 published RCTs that Cervarix and Gardasil vaccine makers GlaxoSmithKline and Merck used to buttress their assertions about their jabs' efficacy:

- The trials' **questionable methodology** generated uncertainties so significant that they undermined efficacy claims.
- The ages of the women who participated in the trials were **not representative** of the younger adolescents who initially constituted HPV vaccination's primary target group.
- The trials used highly restrictive criteria to **exclude** many potential participants, limiting the studies' "relevance and validity for real world settings."
- The trials used "composite and distant **surrogate outcomes**" that essentially made it "impossible to determine effects on clinically significant outcomes."
- Trial investigators carried out unusually frequent cervical screening of participants, which likely resulted in **overdiagnosis** of low-grade cervical changes and overestimation of the jabs' efficacy.
- Finally, and crucially, no certainty about whether HPV shots prevent cervical cancer was actually possible, because the trials "**were not designed to detect this outcome**, which takes decades to develop."

In numerous countries—including in the UK,[166] Sweden,[167] Australia,[168] and the U.S.—reliance on the HPV jab's murky evidence base has proved disastrous for vaccinated young people, many of whom are now experiencing spiking cervical cancer rates. A current U.S. lawsuit alleges the vaccine can increase the risk of cervical cancer.[169] Numerous cases against Merck are pending in federal court alleging that Gardasil caused a wide range of injuries.[170]

Medical Gaslighting

Every year, Merriam-Webster selects a Word of the Year as a shorthand tool for capturing the zeitgeist. In 2022, following a 1740% increase in the number of people looking up the term online, that word was "gaslighting."[171] Political satirist CJ Hopkins likewise dubbed 2022 "The Year of the Gaslighter," sarcastically stating, "If you were . . . in the process of imposing your new official ideology on the entire planet . . . and you needed the masses confused and compliant, you couldn't ask for much more from your Gaslighting Division!"[172]

Merriam-Webster equates gaslighting with other modern forms of "deception and manipulation," emphasizing gaslighting's "deliberate" nature, the term's utility "in describing lies that are part of a larger plan," and its dual "personal" and "political" applications—with the subcategory of *medical gaslighting* being one conspicuous example. (Seemingly without irony, Merriam-Webster then chose "authentic" as its Word of the Year for 2023.)[173]

Surprisingly, conventional news outlets and medical websites admit to some forms of medical gaslighting, acknowledging that there are occasions when "a physician or other medical professional dismisses or downplays a patient's physical symptoms or attributes them to something else, such as a psychological condition."[174] What the mainstream does not fess up to, however, is the medical cartel's pernicious use of gaslighting to deflect attention away from iatrogenesis—and in particular, from the serious harms caused by liability-free vaccines.[175] In fact, gaslighting is an almost inevitable consequence of the blanket legal immunity enjoyed by vaccine makers;[176] when one does not have to shoulder responsibility for the harms that one's products cause, blaming the victim is a highly convenient alternative.

A favorite technique for gaslighting the vaccine-injured is to concoct amorphous diagnoses—diagnoses like "sudden infant death syndrome" (SIDS), "sudden adult death syndrome" (SADS), "autism spectrum disorder" (ASD), and even "poliomyelitis" (see **The Polio Charade**)—that admit no possible role for toxic compounds in vaccines such as heavy metals or other disclosed (or undisclosed) vaccine components. The neuroimmune disorders labeled as "ASD" offer one telling illustration of how long-standing and multilayered this gaslighting strategy can be. First, and cruelly, many doctors (as well as the industry-friendly arbiters of vaccine injury compensation decisions)[177] gaslight affected families by subjecting parents to the party line that a child's neurodevelopmental crisis can never have anything to do with vaccines, even if vaccination immediately preceded the firestorm. Equally insidiously, officialdom gaslights society as a whole by perpetuating the canard that the dramatic, decades-long rise in conditions labeled as ASD "reflects more awareness . . . rather than a true increase."[178] As informed parents such as JB Handley have explained, such assertions—which fly in the face of biological plausibility, scientific evidence, and parents' direct observations—can only be understood as part of a deliberate playbook "to distract, redirect, and delay,"[179] thereby preventing "inevitable medical injuries from sabotaging business."[180]

The Polio Charade

For decades, public panic about "polio"—ably fomented by alarmist media depictions of children in iron lungs[181]—has served the mass polio vaccination agenda. However, the official story of "poliomyelitis" (where "myelitis" simply refers to inflammation of the spinal cord) is full of more holes than Swiss cheese. There is, and always was, ample evidence showing that various forms of poisoning—whether by lead arsenate and other arsenic exposures, DDT, or vaccines (polio and non-polio)—are credible explanations for the paralytic symptoms and deaths labeled as "polio."[182,183,184]

Historically, myelitis tracks closely with the rollout of pertussis-containing and aluminum-containing pediatric vaccines, and also with other pediatric medical interventions, including other types of injections.[185] Early generations of doctors who noticed the correlation described such cases as "provocation paralysis."[186] Likewise, more recent generations of clinicians have commented on the similarity between "polio" and injection injuries disguised with the vague but important-sounding terminology, "traumatic neuritis."[187] Clinical trial and post-marketing data link "myelitis," "encephalomyelitis," "acute disseminated encephalomyelitis," and "transverse myelitis" to 17 different vaccines on the current pediatric schedule for American children.[188]

In the mid-1950s, significant changes to the diagnostic criteria for "paralytic poliomyelitis" played a major role in misleading and gaslighting the public about the merits and impact of polio vaccination. Whereas previously a "polio" diagnosis had required just 24 hours of paralytic symptoms, after the switch, a doctor could only diagnose it if a patient had experienced at least 60 days of such symptoms![189] As doctors turned to other available diagnoses, "polio" diagnoses plummeted, fostering the erroneous public perception that polio vaccination was eradicating "polio." According to writer Rodney Dodson, the AMA helped orchestrate this deception by instructing physicians to pick one of a number of alternative cover terms for paralysis—for example, acute flaccid paralysis, multiple sclerosis, amyotrophic lateral sclerosis, muscular dystrophy, Bell's palsy, cerebral palsy, or Guillain-Barré syndrome.[190]

In 1962, early public health luminary Bernard Greenberg testified before Congress. Greenberg, the founding chair of the biostatistics department at the University of North Carolina School of Public Health in Chapel Hill, frankly admitted that the victory claimed for the early polio vaccines was entirely undeserved; instead, he said, polio vaccination had "actually increased incidents of polio," even though "misuse of statistical methods had made the opposite seem true."[191]

Dodson sums up the lessons learned from "polio" as follows:

Polio is not a pathogenic infection. Like most established 'plagues,' it was caused by mass poisoning that worked in favor of many industries which made enormous profit via pharmaceuticals, vaccines, and other chemical-based products. Although the Polio story is a tragic one, it's another crucial piece to the puzzle. The list of disease paradigm fraud is long. Turning away from the science fiction novel provided by our corrupt CDC, and understanding well-documented history and geography proves a far better asset in sorting this mess out.[192]

Some insiders who acknowledge the phenomenon of medical gaslighting are hesitant to fault doctors; for this category of critic, "the whole American medical system is to blame."[193] In a Substack "primer" on medical gaslighting, *A Midwestern Doctor* mostly agrees, suggesting that medical gaslighting is, in part, a consequence of medical training: "[N]o well-intentioned doctor wants to harm a patient, and since they often do, the reflexive psychological coping mechanism is to deny the possibility of each injury that occurs."[194] When it comes to vaccine harms, says Midwestern Doctor, physicians also typically lack the training to recognize the complex and subtle manner in which vaccine injury symptoms present, "and as a result, most physicians simply cannot see the large number of vaccine injuries occurring around them."

In a September 2023 think piece about medical gaslighting in *The Conversation*, the article's pharma-friendly author (funded by the Wellcome Trust) reached a similar conclusion, arguing that the health care system and (though she did not name it as such) evidence-based medicine foster gaslighting without any need for malicious intent; suggesting an overlap between gaslighting and misdiagnosis, she wrote:

> [M]isdiagnosis occurs . . . because the symptoms [doctors] observe in the patient . . . are algorithmically out of whack with the standard set of symptoms and characteristics they have been taught to look for and associate with different diseases.[195]

When medical education, medical journals, and a complicit media repeatedly program doctors to believe that vaccine injuries are "one in a million," some physicians' inclination to view parents' reports of vaccine injury as "algorithmically out of whack," while infuriating, may not be terribly surprising.

Events surrounding COVID—and especially the deaths and injuries caused by the experimental jabs—created numerous opportunities for gaslighting. Outside the U.S., the term "gaslighting" has even made it into mainstream media headlines; a flashy headline in the *Daily Mail* in late 2022 declared that the rapper called M.I.A. "slams 'society' for 'gaslighting' her."[196] In Scotland, a news story about the UK Covid-19 Inquiry ("set up to examine the UK's response to and impact of the Covid-19 pandemic")[197] focused on COVID-vaccine-injured individuals' call for "an end to 'gaslighting.'"[198] The story quoted a member of the Scottish Vaccine Injury Group, who stated,

> Our members have not only suffered the massive loss either of a loved one, their health, livelihood, freedom and for some their future, but in addition they have faced disbelief, incredulity and even been ostracised and silenced.[199]

Australian media have pointed to medical gaslighting by the nation's own Department of Health. One story explained that the Department refused compensation to a woman injured by the AstraZeneca jab even though, unusually, 11 doctors had gone on record directly linking her serious health problems to the shots.[200] After quoting some of the 11 experts, the news report described the Department's rejection letter, penned by an anonymous medical officer with undisclosed medical qualifications, who stated, "I am not reasonably satisfied that you have suffered a 'harm' as defined in the policy." On social media, meanwhile, Australian senator Gerard Rennick scolded health authorities in his country and around the world for "gaslighting in regards to injuries and compensation," citing a 7200% increased risk of injury from COVID shots compared to prior vaccines, and telling authorities they "should be ashamed of themselves."[201]

Tragically, as Midwestern Doctor points out, medical gaslighting "is often so powerful that friends and family members of the patient will adopt the reality asserted by those doctors and likewise gaslight the injured patient."[202] This turns out to have been a major theme in a qualitative study that interviewed 14 British individuals diagnosed with "sudden-onset, life-threatening and life-changing" vaccine-induced immune thrombocytopenia and thrombosis (VITT) after receiving COVID injections:

> The government was seen as 'gaslighting' their experiences. As a consequence, all participants reported their condition having been denied by friends and even family, with claims that . . . they

were exaggerating their symptoms: 'they just want to get off the topic—you get treated like an insane antivaxxer!' **Given the degree of trauma and health challenges participants experienced, the denial of their condition was particularly distressing** and had on at least one occasion resulted in complete disconnection from those family members involved [bold added].[203]

The same investigators conducted a related study with family members of COVID-vaccinated individuals who had died from VITT. Several participants reported challenging the information on death certificates, "which did not confirm VITT as cause of death, despite this being stated in medical notes," leaving the relatives ineligible for vaccine injury compensation.[204] One study participant described having a coroner laugh in their face when they disputed the conclusion of death from "natural causes."

As the coroner anecdote illustrates, denial of the true cause(s) of injury and death is but one half of the medical gaslighting equation, with the second and often meaner half pertaining to how the medical system chooses to frame the problem instead. Midwestern Doctor explains:

> Typically, when people have a disabling reaction to a pharmaceutical or a medical procedure (formally known as iatrogenesis), they are either told the reaction is not occurring (i.e., they don't have fatigue and they are just being lazy or trying to get disability), or that the injury was not due to the medication and rather due to psychiatric problems the person has.[205]

Midwestern Doctor highlights two of the most common psychiatric scapegoats—so-called "functional neurological disorder" (FND) and "anxiety"[206]—noting that clinical trial investigators used both to great effect to draw attention away from teen Maddie de Garay's serious vaccine injuries[207] during the dubious clinical trials for Pfizer's COVID shots.[208] (Pfizer continued to conceal adverse events during post-marketing surveillance.)[209] Midwestern Doctor elaborates how the psychiatric gaslighting leads to a vicious cycle for the vaccine-injured:

[O]ne of the greatest issues is neurologic damage subsequently creating psychiatric symptoms as a common side effect from the more
toxic pharmaceuticals. This creates a cycle of circular logic where
the neurologic damage is not recognized, and your psychiatric
symptoms are cited as the cause of your entire illness (which is further worsened by the fact being gaslighted is a traumatic experience
which will often make one appear 'overly emotional').[210]

Long before COVID, there were many historical examples of medical
gaslighting, including Freud's relabeling of the neuropsychiatric disturbances caused by mercury poisoning as "hysteria"; the syphilis diagnosis
given to female factory workers who were actually experiencing radium
poisoning; and the relabeling of military members' anthrax vaccine injuries as "post-traumatic stress disorder."[211] Over a century ago, a doctor
named Alexander Wilder explained that statistical shenanigans could
also serve gaslighting purposes. In his 1899 book titled *The Fallacy of
Vaccination,*[212] Wilder blew the whistle on some of the tactics employed
by physicians to conceal adverse events from smallpox vaccination. He
wrote:

Occasionally . . . a death by vaccination is published, and **immediately the effort is put forth assiduously to make it to be believed
that it was from some other cause** [bold added]. The statistics of
smallpox, purporting to distinguish between vaccinated and unvaccinated persons, are too often not quite trustworthy. Many persons
who have been vaccinated are falsely reported as unvaccinated. Even
when death occurs as the result of vaccination, the truth is concealed and the case represented as scarlet fever, measles, erysipelas
[a bacterial skin infection], or some 'masked' disease, in order to
prevent too close questioning.

Far from excusing his fellow doctors, Wilder saw these tactics as intentional: "Further argument is met by stolid silence, and by an apparent
concert of purpose to exclude carefully all discussion of the matter from
medical and public journals, and to denounce all who object."

The Creation of a Global Enforcement Infrastructure

Chapter Two discussed Katherine Watt's analysis of the extensive legal infrastructure—built decade after decade—that since 2020 has enabled the ramping up of a "biomedical police state kill box system" in the U.S.[213] However, as recent events have made plain, the system is actually global, with harms on a worldwide scale facilitated by, among others, the WHO, one of the leading global proponents of this system.

The WHO, as *Off-Guardian's* Kit Knightly has pointed out, "is the *only institution in the world* empowered to declare a 'pandemic' or Public Health Emergency of International Concern (PHEIC)" and its unelected Director-General "is the *only individual who controls that power*" [italics in original].[214] Knightly reminds us that even before COVID, the WHO "loosened the definition of 'influenza pandemic'" in 2008, making it possible to declare the uneventful "swine flu" of 2009 a "pandemic" and generating "millions upon millions of dollars" for swine flu vaccine manufacturers.

The WHO and its backers are aggressively pursuing a strategy to embed pandemic preparedness, prevention and response in international law and to increase the powers of the organization and its Director-General. To this end amendments were made to the International Health Regulations (IHR), which are "an instrument of international law that is legally-binding on 196 countries," and efforts to negotiate and enact an entirely new pandemic treaty are ongoing.[215] Treaty adoption requires a two-thirds vote by the WHO's 194 member-states (and member-states additionally have to ratify or accept it for it to become binding), but the IHR amendments could be adopted—and thereby become binding to all WHO member-states—by simple majority.[216] In 2022, the U.S. put forth and the WHO members adopted IHR amendments that now leave member-states with far less time (10 months as opposed to 18 months) to opt out of future amendments—and future amendments, once adopted, will take effect in 12 rather than 24 months.[217]

Dr. Meryl Nass (through her organization Door to Freedom and via "Meryl's CHAOS Newsletter")[218,219] is one of the leading voices warning the world about the WHO shenanigans and arguing for the importance of public and legislative pushback against the health organization's attempted "soft coup." Nass and independent journalist James Corbett agree that the Agreement and IHR amendments represent "a 'complete reimagination' of the world's power structure,"[220] and even the European Union openly describes them as "global governance . . . with WHO at its core."[221] Spanish attorney Luis Pardo further explains, "Both texts of the WHO irremediably entail the transfer to the WHO of the power to restrict the fundamental rights of citizens. This is a direct attack on the freedom and health sovereignty of citizens."[222]

Using the COVID "pandemic" as its justification, the WHO professes that the Agreement (see **Proposed Elements of the WHO Pandemic Agreement**) and IHR amendments will allow it to exert better control over "pandemic prevention, preparedness and response." Nass's humorous response: "To paraphrase Ronald Reagan, the words, 'I'm from the WHO, and I'm here to help' should be the most terrifying words in the English language, after what we learned from the COVID fiasco."[223] At Door to Freedom, she characterizes "pandemic preparedness" as a "scam/boondoggle/Trojan horse" designed to get the world's governments to abandon national (and individual) sovereignty by handing over fundamental public health decision-making powers to the WHO Director-General. Nass points out:

> [T]he current draft IHRs include no specific criteria for the Director-General of WHO to declare a . . . PHEIC. A declaration could even be made without the consent of the involved nations. . . . Equally concerning, a PHEIC declaration can be issued for merely the *potential* for a public health emergency, and the emergency powers can be extended beyond the end of the emergency [italics in original].[224]

As Knightly puts it,

> [T]he proposed treaty could allow the [Director-General] of the WHO to declare a state of global emergency to prevent a *potential* pandemic, not in response to one [italics in original]. A kind of pandemic pre-crime.[225]

Nass has outlined other worrisome objectives[226] of the Agreement and/or amendments:

- Transferring billions in taxpayer monies to the WHO, governments, and "favored industries"
- Using public health as the excuse to roll out more and more "rapidly produced" and "liability-free" vaccines and implement vaccine passports, digital IDs, and other forms of central control
- Justifying censorship and propaganda in the name of "public health"
- Committing signatories to implementing and adhering to the WHO's bland-sounding but frighteningly far-reaching "One Health"[227] framework, which offers a "blueprint" for a biosecurity-based model[228] of global governance.[229]

In short, putting the WHO globally in charge would impose dictates that include "vaccine development at breakneck speed, the power to enforce which drugs we may use and which drugs will be prohibited, and the requirement to monitor media for 'misinformation' and impose censorship on media so that only the WHO's public health narrative will be conveyed to the public."[230] As Rob Verkerk of the Alliance for Natural Health International summarizes,

> Fundamental to the proposal is the notion of moving the locus of control from the individual, side-stepping the physician, and putting faceless, unelected and unaccountable bureaucrats in charge of health during times of international public health emergencies and pandemics. . . . If you thought the COVID-19 pandemic response by governments, health authorities, corporations and the media . . . was too heavy-handed, too top-down or too authoritarian, don't imagine the next one will be more even-handed.[231]

Proposed Elements of the WHO Pandemic Agreement

The "negotiating text of the WHO Pandemic Agreement," dated October 30, 2023,[232] includes 17 Articles outlining the WHO's rosy vision for "achieving equity in, for and through pandemic prevention, preparedness and response":

- Pandemic prevention and public health surveillance (Article 4)
- One Health (Article 5)
- Preparedness, readiness and resilience (Article 6)
- Health and care workforce (Article 7)
- Preparedness monitoring and functional reviews (Article 8)
- Research and development (Article 9)
- Sustainable production (Article 10)
- Transfer of technology and know-how (Article 11)
- Access and benefit sharing (Article 12)
- Global supply chain and logistics network (Article 13)
- Regulatory strengthening (Article 14)
- Compensation and liability management (Article 15)
- International collaboration and cooperation (Article 16)
- Whole-of-government and whole-of-society approaches at the national level (Article 17)
- Communication and public awareness (Article 18)
- Implementation capacities and support (Article 19)
- Financing (Article 20)

Behind the jargon and feel-good language about "collaboration and cooperation," the Agreement outlines a variety of open-ended, Big-Brother, sovereignty-destroying activities. These include "science-based actions" like the ones widely abused since 2020 (for example, "infection prevention and control measures" guided by "international standards and guidelines," and strengthening of "integrated" surveillance capacities); control of putative "outbreaks" through implementation of actions—from the national down to the community levels—that "encompass whole-of-government and whole-of-society approaches"; manipulation of perceptions and behaviors through "risk communication and community engagement"; and creation of "human, animal and environmental health workforces" brainwashed into a "pandemic preparedness" mindset.

In July 2023, the White House launched a permanent Office of Pandemic Preparedness and Response Policy (OPPR), housed in the Executive Office of the President, "to coordinate and develop policies and priorities related to pandemic preparedness and response."[233] In keeping with the military hierarchy established during COVID, the OPPR's first director is career military official Major General Paul Friedrichs (ret). Friedrichs previously was the Joint Staff Surgeon at the Pentagon and served as medical advisor to the DOD's COVID-19 Task Force,[234] where he strongly supported COVID vaccine mandates and other destructive countermeasures; he also advised the White House on global health security and biodefense.[235]

Pondering the selection of Friedrichs, Nass and others have questioned whether OPPR is "a health program or a military program."[236] Linking back to the WHO and to the UN's "zero draft" declaration on pandemic prevention, preparedness, and response (PPPR),[237] bioweapons expert and international law professor Francis Boyle suggests that the OPPR is "obviously being coordinated with the U.N. . . . to establish the effective functioning of a WHO globalist worldwide medical and scientific police state here in the United States."[238]

Under the emerging PPPR framework, the UN and WHO envision a global, regional, national *and local* scope of action, where country-level offices like the OPPR and an extensive network of laboratories would have extensive latitude to intervene down to the neighborhood level. To concretely picture what this might mean for the residents of a given community, consider the growing drumbeat for "sewage surveillance," occurring at a time when more parents than ever are questioning childhood vaccines like polio vaccines.[239] In the summer of 2022 in New York's Rockland County, officials claimed to have found genetic fragments in wastewater samples and attributed a case of paralysis in an adult (classifying it as "polio") to the fragments because they purportedly matched up to oral polio vaccine components.[240] Officialdom's theory is that individuals who receive oral polio vaccine in other countries (because the U.S. does not administer that type of polio vaccine) pose a "shedding" risk that can leave detectable vaccine-related genetic material in stool samples and sewage.[241] On this tenuous basis, New York's governor declared a state disaster.[242]

Conveniently, the governor's declaration "open[ed] up more vaccine resources" to go after New York counties where, from the state's perspective, polio vaccination rates were too low.[243] This included deputizing "midwives, pharmacists, emergency medical workers and other health care workers" to administer polio vaccines and boosters to all and sundry— children and babies who had not yet started the polio vaccination series; adults who were either unvaccinated, partially vaccinated or "unsure whether they received the vaccine"; and individuals (such as health care providers and wastewater treatment workers) deemed to be at "increased risk."

That summer, health authorities in London used a similar pretext— sewage samples flagged as suspicious[244]—to aggressively push polio boosters citywide for one- through nine-year-olds, even though there were zero cases of disease.[245] UK officials described the vaccination campaign as "a precautionary measure." Around the same time in Israel,[246] authorities pronounced "polio" the cause of a three-year-old's paralysis. Multiple Israeli cities suddenly began reporting "traces" of polio in sewage,[247] prompting media hype about "spread" and exhortations for children and teens (ages 7–17) to get the oral polio vaccine.[248]

Not coincidentally, New York (City and State), London, and Jerusalem were among the jurisdictions that imposed the harshest COVID-19 restrictions and pushed the COVID vaccines the hardest—showing themselves to be willing agents for tyranny disguised as a health emergency.[249] This influential triumvirate's provocative sewage claims and alarmism about a possible polio resurgence—along with the push for widespread polio vaccination—looked like nothing less than a dress rehearsal for future wastewater-driven coercion.

An October 2020 explication of "sewage surveillance" gives credit to an early-2000s study in Helsinki as one of the first to effectively use sewage to snitch on putative polioviruses,[250] but in reality, the experiment made intentional use of an oral polio vaccine:

> Scientists flushed a polio vaccine down a toilet 20 kilometres [about 12 miles] away from a wastewater treatment plant. The researchers then collected wastewater samples . . . over four days, and showed they could still detect the vaccine after 800 million litres of wastewater had passed through the system.[251]

The researchers did not ask what it was about the hardy vaccine that seemingly and scarily allowed it to resist millions of flushes, but a 2021 study showing bioaccumulation of "rotavirus vaccine" in oysters suggests the question might be pertinent.[252] Instead, public health authorities decided to add wastewater monitoring to their disease (not vaccine) surveillance toolkit.

Wastewater experts admit that their "surveillance" techniques are fallible and far from reliable. Observing in mid-2022 that "wastewater surveillance" was suddenly "all over the news," Tufts University researchers cautioned that analysis of wastewater "is a chemically and biologically complex process" involving "multiple steps that are difficult to standardize and that require systematic controls," with wastewater often containing "compounds that can interfere" with the principal method used to spot supposed pathogens.[253] That "principal method," since the 1990s, is the very same polymerase chain reaction (PCR) technique[254] abused to such an extent with COVID as to be roundly denounced as "useless."[255] The Tufts author also noted "privacy and ethical concerns," describing (and then glossing over) the potential for misuse if officials happen to link wastewater data with identifiable genetic or personal data—linkages that could be facilitated by concurrent social media analysis or intelligence from geographic information systems.[256]

In a paper published in August 2021, 70 international experts bemoaned the lack of "harmonized" quality assurance and quality control procedures, admitting that false or "inconclusive" results could conceivably cause "policymakers, public health officials, and the public to lose confidence" in the utility of wastewater monitoring.[257] However, these concerns have not deterred environmental microbiologists from pronouncing sewage monitoring the "next frontier" in the fight against polio, specifically,[258] nor CDC officials—using supposed coronaviruses as the excuse—from enthusiastically and more broadly positioning wastewater surveillance as a "new frontier for public health."[259,260]

Water scientists who can see which way the funding winds are blowing are rushing to endorse sewage monitoring despite the pitfalls that they readily acknowledge, with Tufts itself calling for the practice's further development and expansion and supporting Harry-Potter-style "constant vigilance." Ominously, others have helpfully pointed out how officials could use wastewater analyses to justify "isolation practices"[261] and implementation of stay-at-home directives, masking, social distancing, or other "mitigation measures"[262]—never mind that COVID conclusively demonstrated the arbitrariness, lack of science, and tyranny behind such measures.[263]

One clue that public health officials are not truly concerned about sewage and what it can tell them is the lack of attention to addressing "the health issues associated with long-term simultaneous exposure to a large number of pharmaceutical products"—including the active ingredients and metabolites of non-steroidal anti-inflammatory drugs, cardiovascular drugs, antidepressants and antipsychotics—"known to partially survive the conventional process of wastewater treatment."[264] Nor does the unsexy upgrading of sewage systems—part of classic water, sanitation, and hygiene interventions—command much interest.[265] Instead, the global cheerleaders of "pandemic preparedness" seem intent on using dubious lab techniques to "find" something in the wastewater that can justify their imposition of draconian public health dictates in any neighborhood they choose to persecute.

CHAPTER SIX

WHY DO THEY DO IT?
MONEY, PRESTIGE,
AND CONTROL

As laid out in the first five chapters, it is utterly disingenuous to dismiss iatrogenesis—or iatrogenocide—as inadvertent or accidental. Despite the good intentions of many, Western health care not only tolerates but frequently incentivizes harmful outcomes, and never more so than with medicine's stepped-up weaponization since March 2020. As political economist Toby Rogers reminds us—commenting on a film "about an entire society that ignored the genocide all around them"—"under the right conditions, lots of people are willing to participate in great evil."[1]

At the same time, because such a wide variety of players is involved in the medical-pharmaceutical killing machine—ranging from individual clinicians to corporations, governments, and the military—it is not possible to do more than speculate about the various motivations fueling the iatrogenic carnage. This chapter offers a few thoughts on what some of the drivers may be, ranging from banal pecuniary considerations, to power and prestige, to depopulation and control agendas. It goes without saying that these are not mutually exclusive.

Customers for Life

Money can be an overly simplistic explanation for complex economic, cultural, and sociopolitical phenomena, but there is no doubt that the medical-pharmaceutical complex is highly motivated to create customers for life and frequently does so with great success. As amply illustrated in the earlier chapters, one medical or pharmaceutical intervention tends to beget another, creating problems (explained away as "side effects" and "complications") that necessitate further lucrative medical tinkering.

Pharmaceutical "Gold Mine"

The skyrocketing use of prescription medications across all age groups furnishes a telling illustration of the medical pharmaceutical cartel's cradle-to-grave hold on many residents of the U.S. and other Western nations (see **Illustrating "Cradle to Grave" Customers Through Art**). Americans, in particular, are the global drug industry's "gold mine," according to *Axios*, with U.S. sales of the 20 top-selling drugs accounting for nearly two-thirds (64%) of the world's total sales of those drugs.[2] In the U.S., lobbying by the powerful Pharmaceutical Research and Manufacturers of America (PhRMA) trade group (the biggest spender on federal lobbying of any industry)[3] also helps ensure that Americans pay more for drugs "than the rest of the world combined."[4] In December 2023, leading vaccine and drug makers, including Pfizer and Sanofi, announced plans to raise prices in the new year on more than 500 drugs.[5]

Illustrating "Cradle to Grave" Customers Through Art

In 1998, a Wellcome Trust-funded "medical-art" group in the UK called Pharmacopoeia began using art to engage audiences "in the debate around our relationship with medical treatments," exploring "the tension between the dependence of our society on pharmaceuticals and the ambivalence we often feel towards them."[6]

One of Pharmacopoeia's subsequent projects, a British-Museum-commissioned installation titled "Cradle to Grave"[7]—measuring 14 meters long (approximately 46 feet)—used "Pill Diaries" made from fabric, packaging, and actual pills to communicate "the story of an average man and woman told through the medication they have taken in their life." At the time, in the early 2000s, the artists were astonished to learn that the British health system prescribed an estimated average of 14,000 drug doses "to every man and woman in the UK in their lifetime," a number that did not even include over-the-counter drugs.

The artists wrote that their installation, which laid out pills "in the exact sequence in which they would be taken," focused on "the Western biomedical approach to ill health with its reliance on medicines, which we take in ever increasing amounts as we move from birth, childhood and adulthood into old age and eventually death." Foreshadowing globalists' push two decades later for vaccine passports and other digital monitoring tools,[8] the artists commented that the UK's electronic medical records system already permitted "accurate" documentation of "nearly everyone's . . . medical history and prescribing from birth onwards."

Common medications featured in "Cradle to Grave" included childhood vaccines, hormones (such as contraceptives and hormone replacement therapy), and drugs for a wide range of chronic conditions—asthma, hay fever, indigestion, high blood pressure, back pain, depression, arthritis, diabetes, stroke, cancer, and more. The artists also used photographs, documents, and other objects to illustrate the medicalization of "ordinary life" and old age.

Table 5, using data from early 2019, summarizes U.S. adults' pattern of escalating prescription drug use over their life course. Citizens' readiness to medicate over a lifetime is somewhat astonishing in light of annual polls that consistently identify the pharmaceutical industry as "the most loathed sector in America"[9] (with health care not far behind).[10]

Table 5. U.S. Adults' Use of Prescription Medications (as of February 2019)

Age Group	Taking > 1 medication (%)	Taking > 4 medications (%)
18–29 years	38%	7%
30–49 years	51%	13%
50–64 years	75%	32%
65 years or older	89%	54%

Source: Kirzinger A, Neuman T, Cubanski J, Brodie M. Data note: prescription drugs and older adults. KFF, Aug. 9, 2019. https://www.kff.org/health-reform/issue-brief/data-note-prescription-drugs-and-older-adults/

How does all the pill (and injection) peddling translate into dollars and cents? An investigation published in *JAMA* in 2020 reported that for the period from 2000 to 2018, large pharmaceutical companies were significantly more profitable than 357 publicly traded nonpharmaceutical companies in the S&P 500 Index,[11] benefiting from a 39% higher gross profit margin (defined as the money made "after accounting for the cost of doing business").[12] The year 2022—when the percentage of Americans willing to give a "very" or "somewhat" positive rating to the pharmaceutical industry dropped from 31% to 25%[13] (falling still further in 2023 to an all-time low of 18%)[14]—turned out to be an "especially profitable year,"[15] with the industry pulling in domestic revenues of roughly $364 billion.[16]

The company leading the profitable pack that year was Pfizer, an entity that ranks among the top three "most-fined drug companies" ever.[17] As one publication sarcastically remarked about Pfizer in mid-2022, "It's been a good pandemic for a company that was, until recently, the least-trusted company in the least-trusted sector in the United States."[18] Pfizer's 2022 revenues of $100.3 billion set records "for both itself and the industry as a whole"[19] and were 23% higher than the record it had already set in 2021. As the trade rag *Fierce Pharma* proudly reported, "For the first time in biopharma history, a company has topped the $100 billion mark in annual revenue."[20] In terms of *net* income, Pfizer's year-over-year growth from 2021 to 2022 was 42.5%.[21]

Pfizer's COVID mRNA jabs and Paxlovid drug accounted for nearly three out of five (57%) of the 2022 dollars the company raked in; without that "windfall," its revenues would have been "right in line with . . . pre-pandemic sales."[22] Ironically, Pfizer's "windfall" largely came from the same U.S. taxpayers who so mistrust pharma. As of March 2023, according to KFF, U.S. government (i.e., taxpayer) spending on COVID-19 jabs exceeded $30 billion, with the federal government not only "incentivizing" the shots' development but "guaranteeing a market."[23] Focusing on the lavish support for Pfizer and Moderna, in particular, KFF noted:

- The federal government's initial purchases of **100 million doses** from each of the two companies occurred in mid-2020, "months before any COVID-19 vaccine was yet authorized or had even completed clinical trials."
- Pfizer and Moderna received **80% of all federal funds spent on COVID shots**, and their shots accounted for 97% of all U.S. doses administered.
- Government purchases of Pfizer and Moderna doses cost taxpayers **$25.3 billion.**
- By March 2023, the federal government had made six different bulk purchases from Pfizer and five from Moderna, paying more per dose over time; Pfizer benefited from a **56% increase in price per dose**, while Moderna enjoyed a **73% increase** compared to initial prices.[24]

In May 2023, Brad Setser (a senior fellow at the Council on Foreign Relations, hardly a radical organization) testified before the Senate Committee on Finance about "tax avoidance by American pharmaceutical companies," discussing drug companies' offshoring of profits and production, which Setser deemed detrimental to both "the U.S. Treasury and the strength and resilience of the U.S. biopharmaceutical industrial base."[25] Proving the first point, eight of pharma's largest companies (again including Pfizer) paid just 2% in U.S. taxes for 2022, whereas American workers paid an average tax rate of 24.8% that year.[26] (For insights into Setser's second point about the U.S. industrial base, see next section, **Power, Prestige, and Perks**.) Suggesting that the COVID shots constitute "a particularly egregious example of public investment turned to private profit," a writer for the international business news publication *Quartz* commented on the boondoggle in multiple countries, stating,

> It's almost as if these states—and their citizens—are paying for these vaccines twice over: once to bankroll much, or nearly all, of the research itself, then again to buy back the products of this public-funded research.[27]

Vaccine-Related Illness: A Key Driver of Pediatric Profits

"Pediatrics" means "healer of children," but conventional, modern-day pediatrics (whether doctors acknowledge it or not) might more aptly be termed as the medical specialty that "makes children sick"—and keeps them coming back for more. This may seem harsh to some, but the evidence is incontrovertible that vaccination—one of the primary activities that pediatricians engage in during their busybody schedule of at least 10 "well-child visits"[28] in the first two years of a child's life (see Table 6)—is responsible for iatrogenic harms on a vast scale.[29] (The CDC's 2024 schedule includes 77 doses of 18 different vaccines for children from birth through age 18.)[30] The book *Vax-Unvax: Let the Science Speak* by Robert F. Kennedy Jr. and Brian Hooker, PhD conclusively shows that whereas public health, without irony, bills vaccines as "prevention," they are anything but.[31]

Table 6. How the CDC Schedule for Vaccines and "Other Immunizing Agents" Matches Up to "Well-Child" Visits	
Age at Well-Child Visit	**Vaccines**
Week 1 (3 to 5 days)	HepB#1 (if not given on day of birth); RSV-mAb (depending on maternal RSV vaccination status)
1 month	HepB#2 (recommended at 1 month or 2 months)
2 months	DTaP#1, Hib#1, IPV#1, PCV#1, rotavirus#1 (plus HepB#2 and RSV-mAb if not already given)
4 months	DTaP#2, Hib#2, IPV#2, PCV#2, rotavirus#2 (plus RSV-mAb if not already given)

Age at Well-Child Visit	Vaccines
6 months	COVID-19 ("1 or more doses"), DTaP#3, HepB#3, Hib#3 (if 4-dose series), influenza#1 (if available), IPV#3, PCV#3, rotavirus#3 (if 3-dose series) (plus RSV-mAb if not already given)
7 months	Influenza#2
9 months	If not already given: COVID-19, HepB#3, influenza#1, IPV#3
10 months	If influenza#1 given at 9 months: influenza#2
12 months	HepA#1, Hib#3 or Hib#4 (depending on series), MMR#1, PCV#4, varicella#1 (plus HebB#3, influenza#1, and IPV#3 if not already given)
15 months	DTaP#4 (plus catch-up on shots not already given)
18 months	HepA#2 (plus catch-up on other shots, including influenza and COVID-19)
2 years	Influenza (plus catch-up on shots not already given)

Key: DTaP=Diphtheria, tetanus, and acellular pertussis; HepA=Hepatitis A; HepB=Hepatitis B; Hib=*Haemophilus influenzae* type b; IPV=Inactivated polio vaccine; MMR=Measles, mumps and rubella; PCV=Pneumococcal; RSV-mAb=Respiratory syncytial virus (monoclonal antibody)

Source: https://www.cdc.gov/vaccines/schedules/hcp/imz/child-adolescent.html#table-1

"Healthy children," Brian Shilhavy at *Health Impact News* has pointed out, "are a terrible business model for pediatricians,"[32] whereas children who suffer one or more of the approximately 400 types of adverse events or illnesses related to vaccination[33] are cash cows. As mentioned in Chapter Five, the *International Journal of Environmental Research and Public Health* retracted a 2020 paper by James Lyons-Weiler, PhD, and Dr. Paul Thomas because its powerful take-home message that unvaccinated children were far healthier than vaccinated kids was too paradigm-threatening.[34] Five days after the paper's publication, the state of Oregon also suspended Dr. Thomas's license to practice medicine. The two authors reported on 10 years of retrospective data from Dr. Thomas's Oregon-based pediatric practice; with children in the practice having received some or many vaccines or no vaccines, the data presented "a unique opportunity to study the effects of variable vaccination on outcomes." The researchers' findings were, as they noted, "compelling," with "strikingly clear" age-specific differences in the "health fates"—and the "accumulation of human pain and suffering"—of vaccinated compared to unvaccinated children:

> We have found higher rates of office visits and diagnoses of common chronic ailments in the most vaccinated children in the practice compared to children who are completely unvaccinated.

The "common chronic ailments" that Drs. Lyons-Weiler and Thomas refer to—again, "significantly increased" in the vaccinated group—included "fever, otitis media, conjunctivitis, sinusitis, breathing issues, anemia, gastroenteritis, and weight/eating disorder," as well as asthma and allergic rhinitis in the youngest one-third of patients and eczema in the oldest one-third of patients. Although the pediatrics profession claims that vaccines themselves are an "expense" rather than a money-maker,[35] the ill health that so often follows the barrage of early vaccinations results in the very conditions that are conventional pediatricians' bread-and-butter.

The corporate- and pharma-funded trade group American Academy of Pediatrics (AAP)[36] also offers advice to pediatricians regarding how to "manage" vaccination costs. AAP recommends "maximizing opportunities"[37] by, for example, ruthlessly vaccinating at "sick visits" (despite package inserts for vaccines such as the MMR listing "febrile illness" as a contraindication)[38] and using combination vaccines—associated with an increased odds of sudden infant death[39]—that allow the doctor to code "for each component of the vaccine," garnering more reimbursement than if they were to vaccinate with individual doses.

When Dr. Lyons-Weiler and his co-author Dr. Russell Blaylock carried out a follow-up to the retracted study, titled "Revisiting Excess Diagnoses of Illnesses and Conditions in Children Whose Parents Provided Informed Permission to Vaccinate Them,"[40] their findings unequivocally supported the earlier conclusions, estimating that "vaccination increases the need for visits to the doctor for vaccine-related health outcomes **at a rate of 2.56 to 4.98 new chronic-illness-related visits per unit increase in vaccination**

per year" [bold added].[41] Commenting on the implications—disadvantageous for children and their families but advantageous for pediatricians and other medical specialties—Dr. Lyons-Weiler remarked:

> That translates into far more chronic illness in vaccinating children than in those not vaccinating, a disease burden that is not considered in risk:benefit considerations when it comes to vaccine policies and laws.[42]

The "Cradle" Often Begins Prenatally

To further understand the corporate rewards of a cradle-to-grave business plan, it is particularly revealing to look at the "cradle" end of the spectrum, which, more than ever before, starts with a prenatal assault. Judging by trends over the past few decades, many different arms of the medical-pharmaceutical cartel appear emboldened to behave as if it is open season on pregnant women and their babies. The reckless abandon with which modern medicine is willing to intervene and experiment prenatally provides a telling—and disturbing—clue that agendas other than health and well-being are in play.

Consider the steady encroachment of pharmaceuticals into the pregnancy market. According to the CDC, about 5.4 million U.S. pregnancies are "exposed to medications" each year.[43] Once upon a time (and notably in the aftermath of the thalidomide debacle of the late 1950s and early 1960s,[44] briefly mentioned in Chapter Two), medical ethicists classified pregnant women as a "vulnerable population" and believed that studies should steer clear of experimenting on them due to the "unacceptable" possibility of fetal risks.[45] In the mid-1980s and early 1990s, the iatrogenic agenda-setters encouraged a reversal of that consensus, and suddenly it became fashionable to advocate for clinical trial "inclusion" of pregnant women.[46]

Still, for more than nine in 10 medications currently taken during pregnancy, information about potential fetal risks remains unavailable, meaning that "testing" instead plays out in unacknowledged, real-life experiments.[47] Researchers have long warned that physicians do not even adequately counsel pregnant women about *known* medication risks.[48] Considering that drugs taken during pregnancy can cause birth defects, developmental disabilities, heart anomalies, and other serious direct and indirect impacts, a baby exposed to pharmaceuticals in utero stands a good chance indeed of becoming a "customer for life."[49,50,51]

Unfortunately, many pregnant women unquestioningly accept the chancy medicines, biologics, and procedures hawked by medicine and pharma[52]—including during the developmentally pivotal first trimester. Among the reasons analysts offer to explain the explosive rise in the drugging of pregnant women is the fact that today's women are sicker themselves, entering pregnancy with more preexisting chronic conditions than women of three or four decades ago.[53] Reflecting what is becoming a concerning multigenerational cycle of iatrogenically caused ill health, an online survey (published in 2022) of U.S. pregnant women recruited through the WebMD.com website, nearly half of whom were in their first trimester, found that 48% reported at least one chronic health condition.[54] Topping the list of afflictions (self-reported) were mental health issues (32%), chronic pain (13%), and digestive or respiratory conditions (8%–9%). Eight in 10 respondents reported taking medication—prescription, over-the-counter, or both—during pregnancy.

Less recent but equally staggering longitudinal data (spanning a 33-year period beginning in 1976) highlight the growth of the prenatal drug market over time. In an urban study that recruited pregnant women in Boston and Philadelphia, 94% of the women, as of 2006–2008, took at least one prenatal med, taking an average of 4.2 drugs (and a range of 1 to 28!) over the course of their pregnancy.[55] From 1976 to 2008, the proportion of women who took four or more medications prenatally more than doubled, going from 23% to 50%. For the 2004–2008 period, flu shots topped the list of medical products most commonly taken during the first trimester, as women swallowed the false claims of mouthpieces like WebMD that the shots are safe "at any time" during pregnancy (never mind well-documented risks of miscarriage, birth defects, and developmental disabilities).[56,57] Other popular first-trimester meds, according to the same longitudinal study, included antibiotics, asthma inhalers, thyroid medication, drugs for gastrointestinal issues, and antidepressants. These types of studies show that when pharmaceutical companies begin medicating women *before* pregnancy, they can reasonably expect to retain that market *during* and *after*, as well as folding at least some of the babies into their medication pipeline.

The issue of transgenerational drug effects has prompted mounting concern.[58] For example, if women take opioids or benzodiazepines during pregnancy, the drugs can cause babies to develop dependence; after birth, the babies are likely to experience withdrawal symptoms—known as "neonatal abstinence syndrome" (NAS)—that sometimes last as long as six months.[59] This can lead to a medical cascade involving further medications and lengthy hospital stays in neonatal intensive care. Estimates are that NAS may affect two to seven out of every 1,000 babies born.[60]

As of 2015, an estimated 9% of pregnant women had received opioid prescriptions, and doctors had handed out benzo prescriptions to 3%.[61] The U.S. witnessed a 67% overall increase in benzodiazepine prescriptions between the mid-1990s and 2013; as CNBC later reported in 2018, benzos are a market where "there's plenty of money to be made."[62] By 2022, market analysts were cheering "[r]ising cases of mental disorders" and "increasing [benzo] prescriptions . . . across the world" as a "huge growth opportunity," rating North America as "the most favorable regional market for benzodiazepine drug manufacturers over the coming years."[63] Benzo market leaders once again include Pfizer, along with multinationals like the Swiss firm Roche, India's Torrent Pharmaceuticals, Ireland's Mallinckrodt Pharmaceuticals, and Israel's Teva Pharmaceuticals.[64]

Researchers have warned for years that "parental ingestion of abused drugs [can] influence the physiology and behavior of future generations **even in the absence of prenatal exposure**" [bold added], and may set the stage for a later propensity toward substance use by offspring.[65] Looking just at the risk window from 90 days *before* conception, a 2023 study emphasized that benzodiazepine exposure could increase the likelihood of adverse pregnancy outcomes.[66] Dutch researchers similarly warned in 2020, "The relatively common use of benzodiazepines with possible risks for both mother and (unborn) child is worrying and calls for prescription guidelines for women, *starting in the preconception period*" [italics added].[67]

Prenatal cannabis use represents another growing and worrisome trend to which both industry and government are turning a fruitful blind eye (see **Of COVID and Cannabis**). Despite convincing evidence of harm, pregnant women's use of cannabis is surging, with women reportedly turning to the drug for relief of stress and anxiety.[68] The prevalence of self-reported "medical" and "nonmedical" use of cannabis among pregnant women more than doubled from 2002 to 2017, with past-month use jumping from an estimated 3% to 7% of pregnant women overall, and, during the first trimester, going from 6% to 12%.[69]

Cannabis has known adverse effects on fetal growth and neurodevelopment, particularly when women regularly partake in the first and second trimesters.[70] In 2023, researchers reported that pregnant women who used cannabis early in pregnancy (during formation of the placenta) had significantly worse pregnancy outcomes compared to non-exposed women: more stillbirths, preterm births, small-for-gestational-age babies, and hypertensive disorders of pregnancy.[71] Of perhaps even greater concern, however, are the long-term reproductive implications of adolescents' exploding cannabis use (in addition to other major risks such as psychosis and suicide).[72] As Tufts and University of Pennsylvania researchers soberly reported in 2014:

> Marijuana use by adolescents may be particularly problematic as developing systems may be more vulnerable to the impact of exogenous cannabinoids. . . . Thus, cannabinoid exposure during this critical period could result in **lasting modifications in female reproduction**, and might impact future offspring [bold added].[73]

Of COVID and Cannabis

As of April 2023, use of marijuana for medical and recreational purposes was legal in 38 and 24 U.S. states, respectively (and counting).[74] State-level legalization and positive media spin are, in the view of critics, encouraging commercialization of "habituation [and] addiction," which, in turn, is giving rise to health and social problems that some deem "catastrophic."[75]

According to a 2021 study published in *JAMA*, the stressful events of 2020—the declaration of a pandemic and the imposition of destructive public policies—spurred pregnant women to greater cannabis use.[76] Kaiser Permanente researchers, based in California, noted that cannabis retailers experienced record sales in 2020 due to the state's categorization of the retailers as an "essential business." Rates of biochemically confirmed prenatal cannabis use among pregnant women in Northern California increased significantly, rising by 25% compared to the 15 months before March 2020.

The press likes to paint medical cannabis and the cannabis industry (estimated, by 2026, to be worth $57 billion globally and $42 billion in the U.S.)[77] as a threat to the pharmaceutical industry. However, major pharmaceutical multinationals like AbbVie, Sanofi, Merck, and Pfizer already constitute seven of the top 10 companies with medical cannabis patents in the U.S., and Visual Capitalist argues that "Big Pharma will inevitably enter the space themselves."[78] *Forbes* reported in late 2021 on Pfizer's acquisition of a company developing cannabinoid-type therapeutics,[79] and in 2023, cannabis industry watchers noted comparable mergers and acquisitions in Europe.[80] Some are even posing the question, "Is cannabis the high-growth sector that Pharma has been waiting for?"[81]

In August 2023, HHS—which as of 2019 was the 10th top holder of medical cannabis patents in the U.S.[82]—signaled the possibility of a game-changing shift in federal policy, recommending a reclassification of marijuana from a drug with a "high potential for abuse and no accepted medical use" to one with "a moderate to low potential for dependence and a lower abuse potential." Reuters discreetly noted that this "could allow major stock exchanges to list businesses that are in the cannabis trade,"[83] and would also encourage big pharma to dive into "more ownership of the medical cannabis industry."[84] It is a pretty safe bet that HHS will pay closer attention to its patents and what its corporate partners want than to researchers sounding the alarm about cannabis's significant public health downsides, even though HHS's own Surgeon General issued an advisory in 2019 recommending that adolescents and pregnant women abstain from cannabis use.[85] As the lead author of the California Kaiser Permanente study noted in another *JAMA* publication, "Despite the evidence of negative effects, particularly on vulnerable populations, the balance of cannabis regulation . . . favors industry-friendly regulations rather than true public health protections."[86] She also warned, "The cannabis industry is not held accountable for misleading or inadequate information regarding the safety of cannabis during pregnancy."

Invisible Hazards

Prenatal ultrasound trends provide a final illustration of the potential hazards for offspring—and the rewards for a relentless medical model— when clinicians and parents-to-be ignore (or become inured to) iatrogenic risks.[87] Ultrasound has become such a taken-for-granted component of modern maternity care that many American women undergo four to five ultrasounds per pregnancy.[88] (In countries such as Israel, the average number of prenatal ultrasound procedures is six to eight.)[89] Ultrasound relies on non-ionizing radiation (the form of radiation characterizing microwave ovens and wireless devices); because it is not the ionizing radiation used in X-rays and CT scans (which the public has learned to treat with caution), the ultrasound industry gets away with perpetuating the discredited whopper[90] that ultrasound "can be safely repeated again and again without posing risks to patients."[91] However, even the complacent Environmental Protection Agency (EPA) admits that non-ionizing radiation "cause[s] atoms to vibrate, which can cause them to heat up."[92]

The fact is, there is a large and still-growing body of evidence showing that prenatal ultrasound may predispose babies for profit-generating postpartum problems. As long ago as 1993, researchers documented an association between prenatal ultrasound exposure and subsequent speech delays.[93] The book by Jim West titled *50 Human Studies: A New Bibliography Reveals Extreme Risk for Prenatal Ultrasound*[94] discusses the concept of "toxic synergy," suggesting that ultrasound can be "a powerful synergist, capable of initiating fetal vulnerabilities to subsequent toxic exposure." The implications are profound: "[T]he risk of subsequent exposure to vaccines, birth drugs, antibiotics and other environmental stressors would be raised by prenatal ultrasound, not in addition, but as a multiplier."

Beyond standard 2D ultrasound, optional 3D and 4D ultrasound (where 4D creates a "live video effect, like a movie")[95] has become a "booming baby business" that is luring a growing number of parents into stockpiling footage "purely for entertainment" and scrapbook purposes.[96] Some medical groups pay lip service to the idea that too much exposure to these types of ultrasound may be harmful to babies, but most doctors reportedly say yes to 3D and 4D ultrasounds "as a courtesy to women who want them."[97] Opportunists are now pitching elective ultrasound as a "profitable and fulfilling business venture" for those seeking to generate a "substantial income."[98] Describing the trend back in 2015, FOX News observed that so-called "keepsake shops" and their ultrasound machines are entirely unregulated, with no training or certification required.[99]

In recent years, a "revolution" in point-of-care ultrasound (POCUS) also has taken place, with the emergence of portable devices that can "fit seamlessly in a clinician's back scrub pocket."[100] As a sign of the times, a company called Exo (founded in 2015) incorporates artificial intelligence (AI) into a hand-held ultrasound device the firm released in September 2023, which "uses AI to capture scans" that connect to a smartphone.[101] Using an accompanying (and proprietary) mobile app, clinicians can then "slot into a hospital's electronic health record software to add the scans to patient records," and they have the option of storing the scans on Exo's cloud server.[102] A number of observers have cautioned that such devices come with patient confidentiality and safety concerns that have yet to be adequately studied or regulated.[103] Reminding its audience that ransomware attacks affected over 18 million patient records in the U.S. in 2020 alone, Exo admitted (one year before the release of its product) that POCUS comes with "overlooked cyber risks" that can "present unforeseen circumstances or create gaps for cyber exploitation."[104]

Meanwhile, an Israeli startup has developed a do-it-yourself ultrasound device—now approved in both Israel and Europe—that pregnant women can use in tandem with their smartphone; good "for up to 25 ultrasound

exams," the developers tout it as something with the potential to "revolu-
tionize . . . the frequency at which pregnant women check the fetus."[105]
However, enthusiasts also emphasize that the invention's purpose is *not* to
check for "defects": "This is a device with the sole purpose of calming a
woman down, giving her a chance to see her fetus, plugging in when she
is stressed, and understanding that everything is fine."[106] Women using
such devices may be unaware of the research linking prenatal cell phone
use with pregnancy risks such as shorter pregnancy duration and preterm
birth as well as behavioral problems in offspring.[107,108]

Power, Prestige, and Perks

The medical pharmaceutical killing machine could not function without a cadre of functionaries, health care providers, researchers, and other
high-level foot soldiers willing to play-act the charade of scientific and
regulatory legitimacy that has proven so effective at disguising iatrogenesis
and keeping iatrogenocide largely hidden from public view. Some of these
individuals clearly are comfortable "violat[ing] the rights and well-being
of others" (a trait of psychopaths),[109] while others may simply put above
all else the power, prestige, and material perks that flow, at least in the
short term, from the corrupting system that Catherine Austin Fitts calls
"tapeworm economics," where the parasite consumes its host.[110]

In and Out They Go

The revolving door between federal agencies and the industries they regulate is one of the most infamous and in-your-face facilitators of the dangerous and unproductive system.[111] Notably, two-thirds of pharma lobby-
ists are former government employees.[112]

In the top tier of the revolving door, involving agency heads, is the
well-known example of Julie Gerberding, who became the first woman to
sit at the helm of the CDC (2002–2009). After a stint at San Francisco
General Hospital in the mid-1980s and 1990s, where she promoted the
mainstream, pharma-enriching narrative about AIDS, Gerberding's CDC
tenure included presiding over the agency's response to over 40 other purported public health crises—"including anthrax bioterrorism, SARS, and
natural disasters"—and advising other governments on "urgent issues

such as pandemic preparedness."[113] Drawing on her AIDS and SARS street cred, Gerberding also helped ramp up COVID concern as early as January 2020 by granting interviews that emphasized rapid and "pretty intense" spread, "hot spots" and "clusters"—and vaccines as the solution.[114]

Robert F. Kennedy Jr. has frequently reminded the public of other significant cartel-friendly actions carried out by Gerberding during her stint at CDC, including some that turned out to be very lucrative for Merck.[115] Claiming that she "silenc[ed] and punish[ed] whistleblower Dr. William Thompson when he told her that CDC bigwigs were destroying data linking Merck's MMR to autism" and that she ignored serious and glaring risks of the Gardasil vaccine, some alleged that Gerberding's actions laid the groundwork for ongoing vaccine injuries and deaths in children, adolescents, and young adults all over the world. (Although Gardasil became an instant "blockbuster," the more than 200 and rising lawsuits currently consolidated in multi-district litigation may now give Merck a well-deserved run for its money.)[116]

Far from suffering any ill consequences for industry-friendly actions or inactions at the CDC on her watch, Gerberding left CDC in January of 2009 and moved on to plum positions at Merck in January 2010, first serving as head of Merck Vaccines and then as "chief patient officer" and executive vice president in charge of corporate social responsibility.[117] In 2020 and 2021 alone, these advantageous roles yielded Gerberding approximately $20 million from the sale of some of her Merck shares.[118] Sending the revolving door spinning back in the other direction, Gerberding's latest soft landing (as of May 2022) is as CEO of the Foundation for the National Institutes of Health (FNIH), the foundation established by Congress in 1990 to "convene public and private partnerships between the NIH, academia, life science companies, and patient advocacy groups."[119] At the time of the FNIH announcement, NIH's Acting Director gushed about Gerberding that it was "hard to imagine a more accomplished and respected public health leader" for the job.

Over the years, Scott Gottlieb has also attracted scandalized attention from watchdog groups like Public Citizen as a flagrant case study of the "dangerous, rapidly revolving door between federal regulatory agencies like the FDA and regulated industries."[120] According to Public Citizen,

Gottlieb "was entangled in an unprecedented web of close ties to Big Pharma" well before becoming FDA Commissioner in May 2017. After remaining at FDA for only two years, Gottlieb then "firmly reestablished his financial links to the pharmaceutical industry" by stepping onto Pfizer's board of directors (a post paying "more than $330,000 in cash and company stock"). Commenting on Gottlieb's inaction while at FDA when a Pfizer drug aroused safety concerns, Public Citizen quoted a bioethicist as stating that Gottlieb's subsequent appointment to Pfizer's board "sounds like a reward for a job well done." *The Hill*, which reports on "the intersection of politics and business," pointed out that Gottlieb's actions during his brief stint at FDA included lowering the number of inspections at domestic and foreign drug manufacturing facilities that produce drugs for the U.S. market, and speeding up experimental and generic drug approvals, "directly benefit[ing] Big Pharma bottom lines."[121] (Upon announcement of his departure from FDA, *The Hill* cutely quoted pharma CEOs as saying "we're going to miss him.")

Since joining the Pfizer board, Gottlieb has been a frequent and shameless cheerleader on CBS and CNBC for Pfizer's COVID injections—promoting booster shots and condoning jabs for young children[122,123]—as well as supporting censorship of contrary viewpoints.[124] In late 2023, Texas Attorney General Ken Paxton filed a lawsuit against Pfizer for the drugmaker's "false, misleading and deceptive claims" about the COVID jab, also explicitly mentioning Gottlieb's censorship efforts.[125] In Pfizer's response to the suit, notes the Brownstone Institute, the company does not bother to refute the allegations that Gottlieb and Pfizer "conspired to censor the vaccine's critics."[126]

Belonging to the Club

Lower down on the status totem pole are the motley crew of "experts" who perform rubber-stamping functions for groups like the FDA's Vaccines and Related Biological Products Advisory Committee (VRBPAC) and the CDC's Advisory Committee on Immunization Practices (ACIP). Although the two agencies bill the members of their committees as independent authorities in their respective fields,[127] the conflicts of interest flagged in databases such as the Centers for Medicare & Medicaid

Services' Open Payments[128] and elsewhere indicate that the individuals who people such committees actually are thoroughly beholden to pharmaceutical, military, philanthropic, and academic paymasters, with stock options, grants, patents, and prestige clearly ranking well ahead of patient safety.

The 17–0 vote by VRBPAC members in October 2021 in favor of giving Pfizer's COVID jab to children ages 5 to 11 furnishes a clear example. (See the article titled "17 Pharma Henchmen Who Voted to Experiment on Your Kids—and How to Shun Them" in *The Defender*, which provides a detailed explication of each of the 17 members' unsubtle conflicts of interest.)[129] Teeing up the committee's sham deliberations toward a predetermined outcome, the Biden administration had already purchased tens of millions of pediatric doses and released a detailed pediatric vaccination plan,[130] while the CDC had already issued vaccination guidelines for the age group under consideration. Buttressed by these top-down directives, the VRBPAC members apparently then felt free to ignore the over 140,000 dissenting comments submitted to FDA by the public.[131] Two of VRBPAC's "yes" votes came from CDC officials—hardly a disinterested or independent contingent. Most of the others who voted "yes," including 11 "temporary" voting members and an "acting" committee chair, were affiliated with leading schools of medicine and public health heavily reliant on government grants and funding from entities like the Bill & Melinda Gates Foundation, a major funder of U.S. universities.[132]

A week later, the 14 supposedly "independent" members of ACIP repeated the farce, voting unanimously to foist the dangerous COVID shots on 5- to 11-year-old children. Like their VRBPAC counterparts, all 14 had deep ties to pharma, with careers hinging on promoting the nation's destructive one-size-fits-all vaccination agenda.[133] (See article in *The Defender* from November 2021 titled "14 ACIP Members Who Voted to Jab Your Young Children—and Their Big Ties to Big Pharma" for details on their conflicts.)[134] Condemning the ACIP meeting, Toby Rogers stated that it "was not a scientific review. It was banal bureaucrats announcing plans for a Blitzkrieg and the bought white coats were cheering them on."[135] The powerful institutions with which the ACIP members were affiliated include the nation's top universities and leading pediatric hospitals. Without exception, all the universities at which ACIP members had appointments—Brown,[136] Drexel,[137] Harvard,[138] Michigan State,[139] Ohio State,[140] Stanford,[141] University of Maryland,[142] University of Washington,[143] Vanderbilt,[144] and Wake Forest University[145]—had mandated the COVID shots at the time, while the universities' pediatric hospitals stepped forward to play a frontline role as vaccination sites.[146]

Complicity and Two-Way Streets

In 2021, authors Rosemary Gibson and Janardan Prasad Singh published the book *China Rx: Exposing the Risks of America's Dependence on China for Medicine*.[147] The book calls attention to China's dominant global role as a supplier of "active pharmaceutical ingredients" (APIs)—the components of medications (either prescription or over-the-counter) that produce a drug's intended effects. (Ingredients such as binders, dyes, and preservatives, supposedly "inactive," are called "excipients.") China supplies APIs to the U.S. both directly and indirectly. As an example of an indirect pathway, India is the fifth largest source of U.S. pharmaceutical imports, but China supplies nearly 70% of India's APIs.[148]

China Rx highlights numerous problems with U.S. overreliance on foreign suppliers, but one of the most unsettling is the largely unverifiable quality and safety—and sometimes fatal fallout—of drugs made with imported APIs. A chapter titled "The Perfect Crime" shows how FDA complaisance not only leaves these problems largely unchallenged but helps mask them. For example:

- Though federal regulations require drug manufacturers, distributors, and packagers to report problems to the FDA (reporting by health care providers and pharmacists, in contrast, is "voluntary"), manufacturers frequently **"underreport and misreport"**—and the FDA knows it.
- In reports submitted to FDA, the agency will only count deaths if the submitter has checked the "death" box, but not if deaths get reported with **other wording** (such as by mentioning an "autopsy" or stating that the patient "passed away").

- A former FDA employee told the *China Rx* authors that "agency employees don't change or correct reports submitted, because of concern about liability if a manufacturer's **stock price** drops after a lot of deaths are associated with a product."
- FDA data are not "fully used for the public's benefit," nor are they **"transparent and searchable."**
- FDA often "doesn't inform doctors or the public" about what its data reveal, collaborating with the industry "to keep problems . . . **out of public view.**"
- In theory, FDA is supposed to assure the safety "of more than $1 trillion worth of food, drugs, and other products coming from a vast and complex web of suppliers," yet it is given a "pittance" budget of approximately $2.6 billion annually from Congress (an additional $2 billion comes from the companies it regulates). This chronic underfunding leaves it **hamstrung.**
- In 2020, Gibson and Singh also note, the FDA recalled its **overseas inspectors** "to protect them from the pandemic," leaving the agency without the ability to perform even its usual perfunctory inspections of overseas drug manufacturing facilities.

Although *China Rx* criticizes the "pivot east" of U.S. behemoths like Pfizer, Merck, and Johnson & Johnson that began around the 2010s—deleting U.S. jobs and factories and offshoring research and development to China just as China began declaring ambitious plans to evolve its industry from "Made in China" drugs to "Designed in China" biotech products—the U.S.-China pharmaceutical trade is nevertheless a "booming" two-way street. On the one hand, as outlined in an April 2023 report by the Atlantic Council (a major U.S. think tank), U.S. imports of Chinese products ("defined by the US tariff code to include packaged medicaments, vaccines, blood, organic cultures, bandages, and organs") have grown by a whopping 485% since 2020, encompassing a sizable share of the U.S. antibiotics market and a rapidly growing market share for heart drugs and cancer treatments.[149] On the other hand, U.S. exports of "immunological products" are a significant trade counterbalance, with U.S. companies controlling "more than 65%" of that market in China.

According to *Forbes*, immunological medicines and plasma and related products are the United States' ninth most valuable export, and China's purchases account for up to 19% of the total.[150] In fact, the U.S. supplies 70% of the world's human blood plasma, obtaining it (under sometimes dubious circumstances) from an estimated 20 million Americans annually.[151] (A blogger describes the U.S. as "the Saudi Arabia of blood plasma and plasma products, with . . . annual exports valued in the billions of dollars.")[152]

Depopulation and Central Control?

Admittedly, democide is a grim matter, not least if it's accomplished through stealthy iatrogenic means. As a cogent critic of what she considers post-2020 democidal policies, Latypova often fields questions about why Western governments would want to maim and kill their own citizens.[153] To answer that question, she cites the analysis of the *Solari Report*'s Catherine Austin Fitts,[154] who appears regularly on CHD.TV's popular show, *Financial Rebellion*.[155]

Balancing the Books

As a former investment banker and former Assistant Secretary of Housing at the U.S. Department of Housing and Urban Development (HUD), Fitts has unique insights about governance, finance, and the secretive but dominant economic and governance role played by central banks and, in the U.S., by the privately owned member banks of the New York Federal Reserve, banks that "operate—even control—a significant part of the U.S. federal balance sheet and related accounts," including acting as depository for the U.S. government.[156] Her clear-sightedness about these matters includes an in-depth understanding of the $21-plus trillion (and counting) that went "missing" from the U.S. government between the years 1998 and 2015, a breathtaking "financial coup d'état" that drew Fitts' attention almost as soon as it began to occur. After Fitts spent years crying in the wilderness about the coup, professor Mark Skidmore at Michigan State University joined her to painstakingly piece together more of the details[157] (see the extensive documentation available at the *Solari Report*'s "Missing Money" website).[158]

Fitts explains her early warnings to the public:

> After trillions of dollars started to go missing from the U.S. government, I began in 2000 to warn Americans that our retirements and social safety nets depended on simple mathematical formulas. If we continued to permit trillions to be stolen, **then the financial books would be balanced by other methods** [bold added]. These would include curtailing or inflating away financial and health benefits, implementing delayed retirement ages, intentionally lowering life expectancy, or some combination thereof.[159]

Dismally matching Fitts' expectations, average U.S. life expectancy "ceased" to increase after 2010,[160] and began falling precipitously in 2020, going from 78.8 years in 2019 to 76.1 years in 2022.[161] In the spring of 2023, the Harvard School of Public Health described the decline in U.S. life expectancy and the escalating death rates in young people as "shocking."[162] The dropoffs put America noticeably behind residents of wealthy peer nations, whose populations generally enjoy a life expectancy of 80 years or more. The life expectancy figures for U.S. males are even worse—73.2 years versus 79.1 years for women—with researchers blaming the widening gap on (iatrogenic) factors like opioid overdoses, among others.[163] For men, the U.S. has "a higher rate of avoidable deaths . . . measured as death before the age of 75 . . . than any comparable country."[164]

Still more alarmingly, in 2021—the year in which the COVID shots rolled out—maternal mortality reached a new high, as did mortality among U.S. children and adolescents.[165] In a 2023 paper titled "The New Crisis of Increasing All-Cause Mortality in US Children and Adolescents," the worried authors concluded, "A nation that begins losing its most cherished population—its children—faces a crisis like no other."[166] The paper's lead author, Dr. Steven Woolf (who received his MD 40 years ago and has a long career in pediatrics), morosely commented,

This is the first time in my career that I've ever seen [an increase in pediatric mortality]—it's always been declining in the United States for as long as I can remember. Now, it's increasing at a magnitude that has not occurred at least for half a century.[167]

In early 2024, KFF reported, "U.S. life expectancy is still well below pre-pandemic levels and continues to lag behind life expectancy in comparable countries."[168] Moreover, studies show that booming immigration can confound interpretation of overall life expectancy. In a 2021 study, life expectancy for the foreign-born in the U.S. was roughly "7 and 6.2 years longer [for men and women, respectively] than the average lifespan of their U.S.-born counterparts"; the researchers concluded that immigrants are making "an outsized contribution to national life expectancy."[169] In addition, *Fortune* reported at the close of 2023, "U.S. population increase in 2023 was driven by the most immigrants since 2001—and immigration will be the 'main source of growth in the future.'"[170]

DOD: Covering Its Tracks?

In their analysis of the "missing money," Fitts and Skidmore established that while one trillion disappeared from Fitts' former agency, HUD, the bulk of the money ($20 trillion) went missing from DOD. As the entity responsible for 95% of the heist, DOD would be motivated to carry out a "military maneuver" against the people to whom the trillions are owed, Fitts suggests.

Commenting on a March 2023 podcast interview between Latypova and Robert F. Kennedy Jr. about the DOD's orchestration of the COVID "biological weapon project,"[171] Fitts has taken pains to reiterate the take-home message:

[W]e are not dealing with just bad manufacturing practices or even "mere" pharmaceutical fraud. We are dealing with a much bigger premeditated crime that goes beyond the imagination of most.[172]

Notably, just as the "missing money" story acquired Dr. Skidmore's involvement and began to gain some traction, it suddenly became impossible to continue tracking the back-door outflows. In October 2018, the federal government (with concurrence from both sides of the aisle and the executive branch) implemented an obscure accounting policy ("Statement of Federal Financial Accounting Standards 56" or "Standard 56") that essentially took the federal finances dark, conveying a message that the government and its contractors would no longer respect constitutional financial management laws and transparency.[173] As summed up in the single article about Standard 56 published in the mainstream press (by journalist Matt Taibbi in *Rolling Stone*), the policy conveniently "allows federal officials to fake public financial reports" and also "expressly allows federal agencies to refrain from telling taxpayers if and when public financial statements have been altered."[174]

With the United States' massively "out-of-whack balance sheet" (Latypova's words) having built up an intolerable head of steam thanks to the missing trillions,[175] many of the promises made to voters (such as entitlements) may "never be fulfilled," according to Fitts. Stated another way, "We are at the end of the financial musical chairs game."[176] Aided by helpful tools of obfuscation such as Standard 56, the "criminal cabal" at the top (including the private owners of central banks) appears to have decided, as Fitts and Latypova argue, to "restructure" the balance sheet. Rather than admit to the theft of trillions or make restitution by returning the assets acquired with those trillions, one of the cabal's major goals seems to be depopulation. In that context, medicine and pharma are an effective and time-tested means of downsizing voters who make inconvenient demands on budgets or on those who govern.

It is a truism that populations that are weak and sick are likely to be more obedient and less able to think critically or challenge tyrannical governments. In his 2023 book *The Indoctrinated Brain: How to Successfully Fend Off the Global Attack on Your Mental Freedom*,[177] German scientist Michael Nehls makes the case that the measures adopted during and since 2020 (including the lockdowns, masking, "social distancing," and jabs) represent a "targeted neuropathological attack"[178] on the part of the brain, the hippocampus, that is central to memory, learning, and emotional processing—the very brain capacities that help people recognize threats and develop the wisdom to avoid them.[179] Having spent years demonstrating that conditions like Alzheimer's are preventable—showing that factors like a nutrient-rich diet, physical activity, a "rich social life," and a sense of life purpose are hugely hippocampus-protective[180]—Nehls was well positioned to recognize how the COVID-era assault on these lifestyle factors dampened the public's courage, weakened cognitive abilities, and augmented compliance.

Nehls' thesis may be useful to those who hope to withstand and defeat iatrogenic depopulation efforts, helping an informed public to recognize the wider dystopian agenda. *If* the public does not resist and push back, that agenda is likely to include, according to Fitts, an aggressive effort to eliminate financial transaction freedom and grab land and other real assets.[181,182] Latypova describes the installation of "a global system of totalitarian control" as a strategy to "utterly impoverish and terrorize the survivors" (further disabling the hippocampus); she semi-humorously paraphrases the cabal's mindset as, "we need all the real assets and resources for ourselves, and fewer of you plebs around."[183]

Family Wealth Under Attack

Fitts has often explained how medical malfeasance provides a clever strategy and alibi for the theft of family wealth. Unfortunately, the mesmerizing spell that modern medicine seems to cast over so many citizens prevents most people from grasping the significant financial ramifications of iatrogenic injuries and deaths for individuals, households, and society as a whole.[184] For example, in addition to generating weighty medical expenses, such events often deprive primary earners of their ability to work or (as men's drastically lower life expectancy indicates) their life. The premature death of male heads of household also leaves in the lurch surviving wives, who generally have a lower earnings history and "patchier careers because of caregiving, both for children and often for elders."[185]

In the U.S., unpaid medical bills are the leading cause of bankruptcies, directly causing 67% of them, with "medical problems that lead to work loss" accounting for 44% of the total.[186] Medical debt adversely affects household finances in other ways, too, often resulting in "lawsuits, wage and bank account garnishment, [and] home liens."[187] Moreover, medical debt outweighs all other forms of debt—"credit card debt, personal loans, utilities and phone bills" combined. As of 2022, nearly one out of six (17%) adults with medical debt "had to declare bankruptcy or lose their home because of it."[188] Nearly a quarter of a million Americans per year turn to crowdfunding for help with medical bills, but research suggests that only 12% of campaigns reach their target, while an estimated 16% raise no funds at all.[189] Admitting even pre-COVID that medical need was a "gigantic" category on his GoFundMe crowdfunding platform, the company's CEO Rob Solomon argued in 2019, "We shouldn't be the solution to a complex set of systemic problems."[190]

It is challenging to tease out the specific financial impact of the injuries, disabilities, and deaths caused by the COVID injections, but *Politico* may have been hinting at the growing dimensions of the problem in its September 2023 reportage about personal medical debt, which the news outlet rated as having attained "startling heights."[191] Whereas 19% of U.S. *households* were medically indebted in 2017,[192] a 2021 survey found that more than a third of U.S. *adults* (37%) were carrying medical debt, and 13% of those owed $20,000 or more.[193] Five percent of Americans with medical debt owe more than $100,000.[194] When the Consumer Financial Protection Bureau (CFPB) tallied up the medical debt on consumer credit scores in mid-2021, it amounted to $88 billion,[195] and more comprehensive estimates place collective medical debt in the U.S. at $200 billion.[196]

In December 2023, Toby Rogers spelled out the mechanisms of the COVID-facilitated family wealth grab in a potent synopsis on Substack that is worth quoting at length:

> The Pharma Industrial Complex works like this: picture a 60-year-old woman. She has one million dollars in a 401(k) and another million dollars in equity in her house that she shares with her husband. If Pharma can convince this woman to get a Covid shot every year it has a chance to steal her entire net worth. The shot itself only generates about $100 in revenue for Pharma. But soon thereafter she develops an autoimmune disorder, myocarditis, or has a stroke. She's in and out of the hospital and sees an endless string of specialists who order hundreds of useless tests and prescribe a growing quantity of high-priced medicines. She slowly gets worse and

now additional chronic conditions develop (diabetes, cancer, and/
or dementia). At first her insurance covers some of the costs. Then
the government picks up some costs. But as her health continues
to deteriorate the family spends down their savings. They cash out
her 401(k) and then her husband's as well. Eventually the husband
sells the house to pay for more treatments that don't work. After a
decade of illness she dies and the family says, "she received the best
care." But now they are in debt with no assets. [. . .] Now multiply
this crime times the 270 million Americans . . . and the 5.5 billion
people around the world who got Covid shots and you start to realize the enormity of what we are up against. This is the largest crime
in human history, it's ongoing, and Pharma's goal is literally to steal
the stored wealth of the entire industrialized world.[197]

Three and a half years earlier, Fitts had sought to give the public an advance
warning about the financial risks of the COVID injections. Offering
insights gleaned from her decade of work as an investment advisor to
individuals and families, Fitts wrote:

Many of my clients and their children had been devastated and
drained by health care failures and corruption—and the most common catalyst for this devastation was vaccine death and injury. . . . I
got quite an education about the disabilities and death inflicted on
our children by what I now call "the great poisoning." I had the
opportunity to repeatedly price out the human damage to all concerned—not just the affected children but their parents, siblings,
and future generations—mapping the financial costs of vaccine
injury again and again and again.[198]

Likening the current environment to wartime, Fitts emphasizes that
proper navigation tools and risk management strategies are essential for
staying out of the medical-pharmaceutical killing machine's crosshairs.[199]
As she stated in her May 2020 warning,

When you help a family with their finances, it is imperative to understand all their risk issues. Their financial success depends on successful mitigation of all the risks—whether financial or non-financial—that they encounter in their daily lives. Non-financial risks can have a major impact on the allocation of family resources, including attention, time, assets, and money.[200]

CHAPTER SEVEN

LIFESAVING FACTS
FOR YOU AND THOSE
YOU LOVE

In 1961, President Dwight D. Eisenhower coined the famous term, the "military-industrial complex," foreseeing the likelihood of a "disastrous rise of misplaced power" wielded by a military conjoined with private industry.[1] Since that time, numerous commentators in the medical-pharmaceutical space have riffed on Eisenhower's astute phrasing, adapting it to warn the public about the "medical-industrial"[2] or "pharma-industrial"[3] complex, or—reflecting private equity's incursions and medicine's transmogrification into a tech-driven military tool—the "finance-energy-medical-military-IT" complex.[4,5,6]

The oppressive policies and over-the-top propaganda disseminated since 2020 by this weaponized public-private juggernaut have done a great deal to wake people up to the risks of outsourcing responsibility for their health to entities that plainly do not have their interests at heart. Joining the ranks of those who were clear-eyed (or at least asking questions) about medical risks at the outset of the alleged pandemic are new battalions of citizens who have learned some ugly truths the hard way—by suffering or witnessing injuries and deaths caused by the COVID jabs and other so-called countermeasures.[7]

In a forthright January 2024 Note to his Substack subscribers, Jon Rappoport summed up the broad dimensions of the problem:

> The medical cartel, at the highest level, aims to debilitate, exhaust, cripple, and kill human will power and stamina and strength. . . . We are in a war. Whether or not we want to be.

Plummeting Trust

When oppression disguised as "public health" inevitably circles back around, there are indications that more people will recognize the deadly policies and posturing for what they are. Facing facts and adding words like "iatrogenocide" and "democide" to one's vocabulary may be uncomfortable or unpleasant, but doing so is an essential step toward rejecting medical tyranny, dodging threats, and crafting positive solutions.

As already mentioned, the public does not hold the pharmaceutical and health care *industries* in terribly high regard. Gallup's 2023 Honesty and Ethics poll provides yet another marker of the growing public mistrust, showing that the percentage of Americans willing to rate the honesty and ethical standards of the *professionals* who practice in these fields as "very high" or "high" is also falling.[8] Since 2019, the perceived trustworthiness of medical doctors, psychiatrists, dentists, pharmacists, and nurses has declined noticeably (see Table 7); however, nurses remain the most trusted of the 23 professions included in the survey. (Members of Congress and senators rank at the bottom alongside car salesmen and admen, with all four of the lowest categories receiving "very high" or "high" ratings from only 6% to 8% of Americans.) Respondents who are college graduates appear more likely to trust health professionals' honesty and ethics, a pattern that is interesting to ponder given that the college-educated were far more likely and willing to get COVID shots than those with less education.[9]

Table 7. Percentage of Americans Rating Honesty/Ethical Standards as "Very High" or "High"		
Profession	**2023**	**2019**
Nurses	78%	85%
Dentists	59%	61%
Medical doctors	56%	65%
Pharmacists	55%	64%
Psychiatrists	36%	43%

There are other encouraging signs that members of the public, both domestically and internationally, are wising up and are subjecting official narratives to closer scrutiny. This was evident in New York State in January 2024, when Governor Kathy Hochul used the pretext of wintry weather to issue stay-at-home orders and a county-wide travel ban to one million New Yorkers. The majority of citizen comments on social media, according to freedom-defending attorney Bobbie Anne Cox, were "negative, logical rebuttals to [Hochul's] power grab."[10]

To cite a few other examples of the growing repudiation of medical narratives:

- During the 2022–23 school year, U.S. officials charted "the highest vaccination exemption rate ever reported" among kindergartners—3% (up from 2.5% in 2019–20)—with an exemption rate four times higher (12%) in states like Idaho.[11] The vast majority of exemptions were religious or philosophical rather than medical. In tandem with this trend, the proportion of kindergartners receiving the most heavily promoted childhood vaccines (DTaP, MMR, polio, chickenpox) fell in over two dozen states compared to the prior school year. Officials admit that sentiments about the COVID shots "may have caused a decrease in confidence in vaccination overall."[12] In the UK, too, MMR coverage, at 85%, is the lowest it has been in a decade.[13] Officials in both nations are trotting out a favorite propaganda weapon—measles hysteria—to hound citizens back into the MMR fold,[14] blatantly misstating history's documented lesson that vaccination had nothing to do with 20th-century declines in measles mortality.[15]

- Facing a "glut," *Politico* reported in December 2023 that EU countries had discarded "at least" 215 million doses of COVID shots (an average of 0.7 doses per capita)—admitting that the number was probably a substantial underestimate because EU governments "are reluctant to reveal the scale of the waste."[16] Although the estimated hit to taxpayers (more than four billion euros) is nothing to celebrate, the falling demand is a noteworthy development. *Politico* also pointed to the efforts of Romanian prosecutors to remove immunity protections for the former prime minister and two previous health ministers; the prosecutors claim "excessive vaccine purchases caused more than €1 billion in damages to the state."

- In December 2023, a consumer group in Malaysia called for the immediate recall of the COVID shots. Apparently willing to withstand potential attacks on his reputation, one of the press conference speakers was a high-level toxicologist with a lengthy history of advising the WHO and involvement in international drug trials, who stated, "Our people are dying. . . . And now the world has come to know that this vaccine is not safe, vaccines are not safe; they should be stopped immediately."[17] As mentioned in an earlier chapter, an unfairly retracted 2024 paper by a prominent group of U.S. medical researchers called for a global moratorium on the shots.[18]

New Threats on the Horizon

The need for citizen vigilance—and noncompliance—has not abated, with the despots who ran amok during the purported pandemic showing that they have no plans to let up. In New York State (one of tyranny's consistent trendsetters), the governor's actions indicate that she and her dictatorial cronies have even worse plans up their sleeve than county-wide weather lockdowns—and intend to continue using manufactured public health threats as a wedge to implement their dystopian policies. Bobbie Anne Cox has had ample opportunity to size up the New York leadership's authoritarian designs, having spearheaded a lawsuit (*Borrello v. Hochul*), initially successful, against a chilling quarantine camp regulation promulgated by the governor and New York's Department of Health (DOH). After Cox achieved a trial court victory against Hochul in July 2022, five governor-appointed judges from the state's Appellate Division (Fourth Judicial Department) overturned the decision in November 2023, dismissing the lawsuit and allowing the governor to proceed with the quarantine regulation. Cox, who is now appealing, explains what is at stake:

> [T]he language in the reg makes it crystal clear that the DOH can pull you from your home (and your life) and, with the force of police, hold you anywhere they deem appropriate, including "other residential or temporary housing" . . . [T]he reg says they don't have to prove you are sick, they can hold you for however long they want, and there is no way for you to get out of lock up or lock down (unless you get a lawyer and sue them)!!!![19]

Another disturbing indicator of medicine's intensified weaponization is the Biden administration's creation, in March 2022, of a new biomedical research agency, the Advanced Research Projects Agency for Health (ARPA-H), "modeled after the US military's 'high-risk, high-reward' . . . DARPA."[20] The ARPA-H director reports to the HHS Secretary, but the agency emphasizes that it is "distinct and independent" from the NIH, operating as a funding agency "with its own culture, policies, and processes."[21] In its second year (fiscal year 2023), ARPA-H's budget had already increased by 50%.[22]

In an investigative report published just months before the agency's formal launch, journalist Whitney Webb highlighted a number of causes for concern, including ARPA-H's "direct connection" to DARPA, its manifest interest in building a "total surveillance dragnet," and its merged national-security-plus-health-security focus, which could allow "any decision or mandate promulgated as a public health measure" to be "justified as necessary for 'national security.'"[23] To accomplish biosurveillance goals, Webb suggested that the agency is likely to rely on DARPA "mainstays"— "biotechnology, supercomputing, big data, and artificial intelligence"— including promoting "wearables" to monitor and collect "pre-diagnostic medical data" and "behavioral indicators." Elaborating on the critical role played by Big Tech, Webb added:

> "[The creation of ARPA-H] combines well with the coordinated push of Silicon Valley companies into the field of health care, specifically . . . companies that double as contractors to US intelligence and/or the military (e.g., Microsoft, Google, and Amazon). During the COVID-19 crisis, this trend toward Silicon Valley dominance of the health-care sector has accelerated considerably due to a top-down push toward digitalization with telemedicine, remote monitoring, and the like.

Tying together New York's push for quarantine regulations and ARPA-H's biosurveillance thrust, Webb's warning that "the era of digital dictatorships is nearly here" does not seem far-fetched. As she commented:

> Making mandatory wearables the new normal . . . for monitoring health in general would institutionalize quarantining people who have no symptoms of an illness but only an opaque algorithm's determination that vital signs indicate 'abnormal' activity.

In Webb's view, ARPA-H at its worst could turn into a "technocratic 'precrime' organization with the potential to criminalize both mental and physical illness as well as 'wrongthink.'"

Refocusing on Health, Not Medicine

Aided and abetted by Big Medicine and Big Pharma (along with Big Food, Big Ag, and others), a multipronged "Great Poisoning" (the term used by Fitts) has been underway for far too long, as attested to by the oft-cited (and now probably understated) statistic derived from the 2007 National Survey of Child Health that at least 54% of American children have one or more chronic health problems.[24] However, with greater awareness and understanding of the risks—and of the pain and suffering that iatrogenesis causes, both directly and indirectly—there is no reason why an informed public cannot say "no more" and reverse direction. The fact demonstrated by Dr. Paul Thomas's practice data that unvaccinated children interface with the health care system far less often than vaccinated children is a potent clue that parents, in particular, have the power to reshape future generations' trajectory and protect them from a cradle-to-grave downward spiral.

Clearly, factors beyond vaccination (such as nutrition and other toxic exposures) also influence health and play into the frequency of doctor visits, but as Drs. Lyons-Weiler and Thomas point out:

> Lifestyle differences between the vaccinated and unvaccinated groups . . . cannot explain the large difference in outcomes, and if they do, then it would be objective to conclude that **everyone should adopt the lifestyle followed by the unvaccinated if they want healthier children**. That lifestyle choice includes, for many families, avoiding some or all vaccines, and thus, the lifestyle choice concern is inextricably linked to vaccine exposure [bold added].[25]

Learning about iatrogenic dangers and how to avoid them is a vital step toward ensuring the radiant health that should be every child's birthright. Redirecting time and dollars away from the "health care" system and toward health-supporting lifestyle choices can make a big difference. This includes eating real foods that are nutrient-dense; eliminating or minimizing exposure to glyphosate and other pesticides; drinking water free of fluoride, heavy metals, and other toxins; spending time in nature; being physically active; reducing phone and screen time; and creating the conditions for regenerative sleep (such as turning off Wi-Fi at night, keeping smartphones out of bedrooms, or returning to hard-wired Internet solutions).

These types of lifestyle choices are equally important for those who have already experienced iatrogenic harm. In addition, there are numerous non-pharmaceutical pathways that hold out the possibility of either facilitating recovery (partial or full) or at least improving quality of life. Many people have experienced remarkable healing—even from the most serious of chronic conditions—by stepping off the medical cartel's treadmill and exploring options such as:

- Healing diets (for example, the GAPS, ketogenic, or carnivore diets)[26,27,28]
- Detoxification and cleansing
- Homeopathy
- Herbal medicine
- Non-Western approaches such as Traditional Chinese Medicine or Ayurveda
- The Wim Hof Method (involving breath work and cold exposure)[29]
- Coherence healing techniques such as those taught by Dr. Joe Dispenza[30]
- And many more

On a broader level, all of us must find ways to reckon with the growing technocratic-totalitarian threat, by protecting and strengthening the freedom to make and implement paradigm-changing life choices. This requires the qualities that make pushback possible—commitment, courage, and a willingness to think and act outside the box. In a 2016 Foundation for Economic Education (FEE) article titled "A totalitarian state can only rule a desperately poor society," the author suggests that free and productive societies need people who not only have ingenuity and creativity but are willing to "go[] against the grain or traditionally accepted view of things" and "stand up to the existing order."[31]

It would be easy to become discouraged by the encroachment of authoritarian measures such as quarantine camps and digital biosurveillance, but, as Sasha Latypova and Catherine Austin Fitts both emphasize, it is important not to "succumb to [a] mentality of doom and victimhood," which only plays into the cabal's hands. Latypova offers this advice:

Do not get into [a] "we are doomed" mindset. We are not doomed. We have solutions, we can work with what we have and build resilient support networks. Most importantly, as [we] shake the sleepers out of their stupor, the cabal loses more and more of its willing foot soldiers. Do not comply.[32]

Fitts describes the process of moving forward as "turtling forth," stating that "Forward action is a 'force multiplier.'"[33] Notwithstanding the potential for iatrogenocide and (to use Katherine Watt's term) the "medical mercenaries" who might commit it, Fitts reminds us that "There is a wonderful world full of good people and good things," and "anything is possible when you leave hopelessness behind."

ENDNOTES

Introduction

1   Elizabeth Lee Vliet, MD, "Lethal Connections: 'Complete Lives' Morphs into 'COVID Protocol' in America's Hospitals," *Association of American Physicians & Surgeons*, Oct. 26, 2021, https://aapsonline.org/lethal-connections -complete-lives-morphs-into-covid-protocol-in-americas-hospitals.

2   Robert F. Kennedy Jr., "'60 Minutes' – Swine Flu 1976 Vaccine Warning," Children's Health Defense, Apr. 1, 2020, https://childrenshealthdefense. org/video/60-minutes-swine-flu-1976-vaccine-warning.

3   Jon Rappoport, "Dispatches from the War: Epidemics Are Staged on Television," *Jon Rappoport's Blog*, Apr. 12, 2021, https://blog.nomorefake news.com/2021/04/12/dispatches-from-the-war-epidemics-are-staged-on -television.

4   John Cumbers, "The COVID-19 Pandemic Has Dramatically Changed How Biopharma Does Business," *Forbes*, May 19, 2021, https://www.forbes .com/sites/johncumbers/2021/05/19/the-covid-19-pandemic-has-dramatically -changed-how-biopharma-does-business.

5   Jon Rappoport, "COVID: The Medical Cartel Destroying Millions of Lives Is Nothing New," *Jon Rappoport's Blog*, Sep. 10, 2021, https: //blog.nomorefakenews.com/2021/09/10/covid-the-medical-cartel -destroying-millions-of-lives-is-nothing-new.

6   Pearce Wright, "Ivan Illich," *The Lancet* 361, no. 9352 (2003): 185, doi: 10.1016/S0140–6736(03)12233–7.

7   Toby Rogers, "Who Is 'They': Mapping the Institutions and Actors Involved in the Global Iatrogenocide," *uTobian*, Aug. 31, 2022, https://tobyrogers .substack.com/p/who-is-they.

8   Toby Rogers, "How Then Shall We Think about the Economy?" *uTobian*, Jun. 30, 2023, https://tobyrogers.substack.com/p/how-then-shall-we-think -about-the.

9   Children's Health Defense Team, "Pfizer Works to Fast-Track More Vaccines for Pregnant Moms, Despite Mounting Evidence Rushed COVID

Shots Harmed Babies," *The Defender*, Nov. 9, 2022, https://childrenshealth defense.org/defender/pfizer-fast-track-vaccines-pregnant-moms.

10   Toby Rogers, "How Then Shall We Think about the Economy?" *uTobian*, Jun. 30, 2023, https://tobyrogers.substack.com/p/how-then-shall-we-think -about-the.

11   Mike Capuzzo, "Landmark Lawsuit Alleging Medical Battery Killed 19-Year-Old with Down Syndrome Will Go to Trial," *The Defender*, Nov. 20, 2023, https://childrenshealthdefense.org/defender/grace-schara-lawsuit -medical-battery-trial.

12   Andrew Lohse, "Press Release: 'Medical Murder' Outpaces Heart Disease and Cancer, Becoming America's #1 Cause of Death," *DailyClout*, Sep. 22, 2023, https://dailyclout.io/press-release-medical-murder-outpaces-heart -disease-and-cancer-becoming-americas-1-cause-of-death.

13   Toby Rogers, "The Iatrogenocide Accelerates," *uTobian*, Oct. 18, 2022, https://tobyrogers.substack.com/p/the-iatrogenocide-accelerates.

14   Jon Rappoport, "COVID: The Medical Cartel Destroying Millions of Lives Is Nothing New," *Jon Rappoport's Blog*, Sep. 10, 2021, https://blog .nomorefakenews.com/2021/09/10/covid-the-medical-cartel-destroying -millions-of-lives-is-nothing-new.

15   Ibid.

Chapter 1

1   Kevin M. Rice et al. "Environmental Mercury and Its Toxic Effects," *J Prev Med Public Health* 47, no. 2 (2014): 74–83, doi: 10.3961/jpmph. 2014.47.2.74.

2   Merinda Teller, "Mercury's Poisonous Persistence in the Medical Armamentarium," *Wise Traditions in Food, Farming and the Healing Arts*, May 4, 2018, https://www.westonaprice.org/health-topics/environmental -toxins/mercurys-poisonous-persistence-in-the-medical-armamentarium.

3   "Mercury in Drug and Biologic Products," U.S. Food & Drug Administration, Mar. 28, 2018, https://www.fda.gov/regulatory-information/food-and-drug -administration-modernization-act-fdama-1997/mercury-drug-and -biologic-products.

4   David Kirby, *Evidence of Harm: Mercury in Vaccines and the Autism Epidemic: A Medical Controversy* (New York: St. Martin's Press, 2005), https://archive .org/details/evidenceofharmme00kirb_0.

5   David Kirby, Foreword to *The Age of Autism: Mercury, Medicine, and A Man-Made Epidemic*, by Dan Olmsted and Mark Blaxill (New York: St. Martin's Press, 2010), ix–xi, https://archive.org/details/ageofautismmercu0000olms.

6   Eleanor McBean, *The Poisoned Needle: Suppressed Facts about Vaccination* (1957), 10, https://archive.org/details/the_poisoned_needle_mcbean.

7   Robert L. North, "Benjamin Rush, MD: Assassin or Beloved Healer?" *Baylor University Medical Center Proceedings* 13, no. 1 (2000), doi: 10. 1080/08998280.2000.11927641.

8   Albert Wilking, "The Heroic Age of Medicine," *Mercury Free Kids*, Jul. 29, 2015. https://www.mercuryfreekids.org/mercury101/p/mercury-and-heroic -age-of-medicine.html.

9   Marisa Sloan, "Following Lewis and Clark's Trail of Mercurial Laxatives," *Discover*, Jan. 29, 2022, https://www.discovermagazine.com/planet-earth /following-lewis-and-clarks-trail-of-mercurial-laxatives.

10   Albert Wilking, "The Heroic Age of Medicine," *Mercury Free Kids*, Jul. 29, 2015. https://www.mercuryfreekids.org/mercury101/p/mercury-and -heroic-age-of-medicine.html.

11   Robert L. North, PhD, "Benjamin Rush, MD: Assassin or Beloved Healer?" *Baylor University Medical Center Proceedings* 13, no. 1 (2000), doi: 10.1080/08998280.2000.11927641.

12   Eglė Sakalauskaitė-Juodeikienė, "'Heroic' Medicine in Neurology: A Historical Perspective," *Wiley Online Library*, Nov. 20, 2023, doi: 10.1111 /ene.16135.

13   Alexander Horacio Toledo, "The Medical Legacy of Benjamin Rush," *Journal of Investigative Surgery* 17 (2004): 61–63, doi: 10.1080/0894193 0490427785.

14   Robert L. North, PhD, "Benjamin Rush: Medical Quackery in the Yellow Fever Epidemic of 1793," *Brewminate*, May 26, 2020, https://brewminate .com/dr-benjamin-rush-medical-quackery-in-the-yellow-fever-epidemic -of-1793.

15   Nicholas E. Davies et al., "William Cobbett, Benjamin Rush, and the Death of General Washington," *JAMA* 249, no. 7 (1983): 912–915, doi:10.1001 /jama.1983.03330310042024.

16   Grace Holley, "How George Washington Died: Bloodletting, Enemas, and the Dangers of 18th Century Medicine," *History Daily*, May 30, 2020, https://historydaily.org/how-george-washington-died-bloodletting -enemas-18th-century-medicine.

17   Robert L. North, PhD, "Benjamin Rush, MD: Assassin or Beloved Healer?" *Baylor University Medical Center Proceedings* 13, no. 1 (2000), doi: 10.1080/08998280.2000.11927641.

18   Tess Lanzarotta and Marco A. Ramos, "Mistrust in Medicine: The Rise and Fall of America's First Vaccine Institute," *American Journal of Public Health* 108, no. 6 (2018): 741–747, doi: 10.2105/AJPH.2018.304348.

19   Claudia Huerkamp, "The History of Smallpox Vaccination in Germany: A First Step in the Medicalization of the General Public," *Journal of Contemporary History* 20, no. 4 (1985): 617–635, https://www.jstor.org /stable/260400.

20   Tess Lanzarotta and Marco A. Ramos, "Mistrust in Medicine: The Rise and Fall of America's First Vaccine Institute," *Am J Public Health* 108, no. 6 (2018): 741–747, doi: 10.2105/AJPH.2018.304348.

21   Alexander Wilder, *The Fallacy of Vaccination* (New York: The Metaphysical Publishing Company, 1899), https://archive.org/details/101229606.nlm .nih.gov.

22   Robert L. North, PhD, "Benjamin Rush: Medical Quackery in the Yellow Fever Epidemic of 1793," *Brewminate*, May 26, 2020, https://brewminate.com/ dr-benjamin-rush-medical-quackery-in-the-yellow-fever-epidemic-of-1793.

23   Andrew G. Shuman et al., "Bleeding by the Numbers: Rush Versus Corbett," *The Pharos*, Sep. 1, 2014, 9–16, https://papers.ssrn.com/sol3 /papers.cfm?abstract_id=2974410.

24   Ibid.

25   Nicholas E. Davies et al., "William Cobbett, Benjamin Rush, and the Death of General Washington," *JAMA* 249, no. 7 (1983), doi: 10.1001 /jama.1983.03330310042024.

26   Andrew G. Shuman et al., "Bleeding by the Numbers: Rush Versus Corbett," *The Pharos*, Sep. 1, 2014, 9–16, https://papers.ssrn.com/sol3/papers.cfm ?abstract_id=2974410.

27   Ibid.

28   Ibid.

29   Aaron Levin, "The Life of Benjamin Rush Reflects Troubled Age in U.S. Medical History," *Psychiatric News*, Jan. 29, 2019, doi: 10.1176/appi. pn.2019.2a23.

30   Suzanne Humphries, MD, and Roman Bystrianyk, *Dissolving Illusions: Disease, Vaccines, and the Forgotten History* (self pub., 2013), 74, https: //dissolvingillusions.com.

31   "Benjamin Rush and the State of Medicine in 1803," *Gateway Arch, National Park Missouri*, Apr. 10, 2015, https://www.nps.gov/jeff/learn /historyculture/medrush.htm.

32   Grace Holley, "How George Washington Died: Bloodletting, Enemas, and the Dangers of 18th Century Medicine," *History Daily*, May 30, 2020, https://web.archive.org/web/20201020002259/https://historydaily.org /how-george-washington-died-bloodletting-enemas-18th-century-medicine.

33   "The FDA's Drug Review Process: Ensuring Drugs Are Safe and Effective," *U.S. Food & Drug Administration*, Nov. 24, 2017, https://www.fda.gov /drugs/information-consumers-and-patients-drugs/fdas-drug-review -process-ensuring-drugs-are-safe-and-effective.

34   Jared Staver, "Can Medical Malpractice Be Criminal?" *Staver Accident Injury Lawyers, P.C.* (blog), n.d., https://www.chicagolawyer.com/blog/can -medical-malpractice-be-criminal.

35 *Medpagetoday.com*, https://www.medpagetoday.com/search?q=investigative +roundup.

36 Daren Nicholson et al., "Medication Errors: Not Just a 'Few Bad Apples,'" *Journal of Clinical Outcomes Management* 13, no. 2 (2006): 114–115, PMID: 16862227.

37 Lucian L. Leape, "Errors in Medicine," *Clinica Chimica Acta* 404, no. 1 (2009): 2–5, doi: 10.1016/j.cca.2009.03.020.

38 Lucian L. Leape et al., "Perspective: A Culture of Respect, Part 1: The Nature and Causes of Disrespectful Behavior by Physicians," *Academic Medicine* 87, no. 7 (2012): 845–852, doi: 10.1097/ACM.0b013e318258338d.

39 Robert Kaplan, "The Clinicide Phenomenon: An Exploration of Medical Murder," *Australas Psychiatry* 15, no. 4 (2007): 299–304, doi: 10.1080 /10398560701383236.

40 Erin Tilley et al., "A Regulatory Response to Healthcare Serial Killing," *Journal of Nursing Regulation* 10, no. 1 (2019): 4–14, doi: 10.1016/S2155– 8256(19)30077–8.

41 Kaiser Health News, "List: 769 Hospitals Fined for Medical Errors, Infections, by CMS," *HealthCare Finance,* Jan. 5, 2017, https://www.healthcarefinance news.com/news/list-769-hospitals-fined-medical-errors-infections-cms.

42 Lena Groeger, "Big Pharma's Big Fines," *ProPublica*, Feb. 24, 2014. https: //projects.propublica.org/graphics/bigpharma.

43 "Violation Tracker Industry Summary Page," *Good Jobs First, Violation Tracker*, https://violationtracker.goodjobsfirst.org/industry/pharmaceuticals.

44 Children's Health Defense Team, "Pharma's Criminal Business Model— and How the U.S. Government Benefits From It," *The Defender*, Sep. 9, 2022, https://childrenshealthdefense.org/defender/big-pharma-criminal-business -model-us-government.

45 Sy Mukherjee, "So Your Biopharma Got Fined Millions by the Feds— Where Does That Money Go?" *BioPharma Dive*, Nov. 25, 2015, https: //www.biopharmadive.com/news/so-your-biopharma-got-fined-millions-by -the-fedswhere-does-that-money-go/407168.

46 Sophia Putka, "Is DOJ Underusing Authority to Hold Pharma, Device Execs Accountable?" *Medpage Today*, Sep. 19, 2022, https://www.medpage today.com/special-reports/features/100799.

47 Children's Health Defense Team, "Pharma's Criminal Business Model— and How the U.S. Government Benefits from It," *The Defender*, Sep. 9, 2022, https://childrenshealthdefense.org/defender/big-pharma-criminal -business-model-us-government.

48 Sy Mukherjee, "So Your Biopharma Got Fined Millions by the Feds— Where Does That Money Go?" *BioPharma Dive*, Nov. 25, 2015, https: //www.biopharmadive.com/news/so-your-biopharma-got-fined-millions-by -the-fedswhere-does-that-money-go/407168.

49   "Twenty-Seven Years of Pharmaceutical Industry Criminal and Civil
     Penalties: 1991 through 2017," *PublicCitizen*, Mar. 14, 2018, https://www
     .citizen.org/article/twenty-seven-years-of-pharmaceutical-industry-criminal
     -and-civil-penalties-1991-through-2017.

50   World Orders Review, "A Good Death? [Midazolam]," Bitchute, Dec. 10,
     2021, https://www.bitchute.com/video/Jgc6smKYJiXq.

51   Elizabeth Lee Vliet, "Lethal Connections: 'Complete Lives' Morphs into
     'COVID Protocol' in America's Hospitals," Oct. 26, 2021, https://aapson
     line.org/lethal-connections-complete-lives-morphs-into-covid-protocol-in
     -americas-hospitals.

52   Jeremy Loffredo, "British Funeral Director Tells RFK, Jr. about Fear,
     Propaganda and COVID Vaccine Deaths during Early Days of Pandemic,"
     The Defender, Nov. 24, 2021, https://childrenshealthdefense.org/defender
     /rfk-jr-podcast-john-olooney-fear-propaganda-covid-vaccine-deaths.

53   Defender Staff, "Where's the Emergency? 18 Congress Members Demand
     Answers as FDA Looks to Approve COVID Shots for Kids Under 5,"
     The Defender, Jun. 8, 2022, https://childrenshealthdefense.org/defender
     /congress-members-fda-approve-covid-shots-kids

54   Gavin de Becker, "Seeing is Believing: What the Data Reveal about
     Deaths Following COVID Vaccine Rollouts around the World," *The
     Defender*, Jan. 9, 2023, https://childrenshealthdefense.org/defender/covid
     -vaccine-deaths-cause-unknown.

55   Defender Staff, "Holocaust Survivor Vera Sharav Premiers 'Never Again Is
     Now Global' on CHD.TV," *The Defender*, Jan. 24, 2023, https://childrens
     healthdefense.org/defender/never-again-vera-sharav-chd-tv.

56   Robert Kaplan, "The Clinicide Phenomenon: An Exploration of Medical
     Murder," *Australas Psychiatry* 15, no. 4 (2007): 299–304, doi: 10.1080/
     10398560701383236.

Chapter 2

1   Editors of Encyclopedia Britannica, "Cassandra," in *Encyclopaedia Britannica*,
    last rev. Jan. 9, 2018, https://www.britannica.com/topic/Cassandra-Greek
    -mythology.

2   Andrea Carlino, "Petrarch and the Early Modern Critics of Medicine,"
    Journal of Medieval and Early Modern Studies 35, no. 3 (2005): 559–582,
    doi: 10.1215/10829636–35–3–559.

3   Ibid.

4   Natasha Hobley, "Big Pharma's Influence in Shaping the U.S. Medical
    Model," *The Vaccine Reaction*, Jan. 22, 2023, https://thevaccinereaction.org
    /2023/01/big-pharmas-influence-in-shaping-the-u-s-medical-model.

5   Ihor B. Gussak et al., *Iatrogenicity: Causes and Consequences of Iatrogenesis
    in Cardiovascular Medicine* (New Brunswick, NJ: Rutgers University Press,
    2018), https://www.jstor.org/stable/j.ctt1q1cr8b?turn_away=true.

6   Andrew D. Auerbach et al., "Diagnostic Errors in Hospitalized Adults Who Died or Were Transferred to Intensive Care," *JAMA Internal Medicine* 184, no. 2 (2024): 164–173, doi: 10.1001/jamainternmed.2023.7347.

7   Sophie Putka, "Diagnostic Errors Common in Hospital Deaths, ICU Transfers, Study Suggests," *Medpage Today*, Jan. 8, 2024, https://www .medpagetoday.com/special-reports/features/108175.

8   David E. Newman-Toker et al., "Burden of Serious Harms from Diagnostic Error in the USA," *BMJ Quality and Safety* 33, no. 2 (2023), doi: 10.1136/ bmjqs-2021–014130.

9   "Iatrogenesis in Pediatrics," *AMA Journal of Ethics* 19, no. 2 (2017): 737– 842, https://journalofethics.ama-assn.org/issue/iatrogenesis-pediatrics.

10 Dr. Avitzur, "E-Iatrogenesis, Cloning, Pseudo-Histories and Other Unintentional Pitfalls of Electronic Health Records: an Interview with James Bernat, MD," *Neurology Today*, Apr. 4, 2013, https://journals.lww .com/neurotodayonline/Fulltext/2013/04040/E_Iatrogenesis,_Cloning, _Pseudo_Histories_and.10.aspx.

11 Mark Meckler and Kim Boal, "Decision Errors, Organizational Iatrogenesis, and Errors of the Seventh Kind," *Academy of Management Perspectives* 34, no. 2 (2020), doi: 10.5465/amp.2017.0144.

12 Deirdre K. Thornlow, Ruth Anderson and Eugene Oddone, "Cascade Iatrogenesis: Factors Leading to the Development of Adverse Events in Hospitalized Older Adults," *International Journal of Nursing Studies* 46 (2009), doi: 10.1016/j.ijnurstu.2009.06.015.

13 Michael J. Saks and Stephan Landsman, "Use Systems Redesign and the Law to Prevent Medical Errors and Accidents," *STAT*, Aug. 4, 2021, https://www.statnews.com/2021/08/04/medical-errors-accidents-ongoing -preventable-health-threat.

14 "Iatrogenesis," *Merriam-Webster Dictionary*, accessed Jun. 24, 2024, https: //www.merriam-webster.com/dictionary/iatrogenesis.

15 Trisha Torrey, "Iatrogenic Events During Medical Treatments," *VeryWellHealth*, Nov. 24, 2021, https://www.verywellhealth.com/what-is -iatrogenic-2615180.

16 Michael J. Saks and Stephan Landsman, "Use Systems Redesign and the Law to Prevent Medical Errors and Accidents," *STAT*, Aug. 4, 2021, https://www.statnews.com/2021/08/04/medical-errors-accidents-ongoing -preventable-health-threat.

17 Ibid.

18 R. Smith, "Limits to Medicine. Medical Nemesis: The Expropriation of Health," *Journal of Epidemiology and Community Health* 57, no. 12 (2003): 928, doi: 10.1136/jech.57.12.928.

19 "Iatrogenesis," Hartford Institute for Geriatric Nursing, 2023, https://hign .org/consultgeri/resources/protocols/iatrogenesis.

20   Deborah C. Francis and Jeanne M. Lahaie, "Iatrogenesis: The Nurse's Role in Preventing Patient Harm," in *Evidence-Based Geriatric Nursing Protocols for Best Practice*, 4th ed., (New York: Springer Publishing, 2012), https://repository.poltekkes-kaltim.ac.id/674/1/Evidence-Based%20Geriatric%20Nursing%20Protocols%20for%20Best%20Practice%20(%20PDFDrive.com%20).pdf.

21   David W. Bates et al., "The Safety of Inpatient Health Care," *New England Journal of Medicine* 388 (2023): 142–153, doi: 10.1056/NEJMsa2206117.

22   Ivan Illich, *Limits to Medicine. Medical Nemesis: The Expropriation of Health* (New York: Penguin Books, 1977), 23, https://archive.org/details/limitstomedicine00illi.

23   Mirenda Teller, PhD, "Fake Meat and Other Fake Foods: Synthetic Biology Wolves in 'Sustainable' Sheep's Clothing," *Wise Traditions in Food, Farming and the Healing Arts*, Feb. 8, 2022, https://www.westonaprice.org/health-topics/fake-meat-and-other-fake-foods.

24   John M. Travaline, "The Moral Dangers of Technocratic Medicine," *Linacre Q* 86, no. 2–3 (2019: 231–238), doi: 10.1177/0024363919858463.

25   Christine Buttorff, Teague Ruder, and Melissa Bauman, "Multiple Chronic Conditions in the United States," *RAND*, May 26, 2017, doi: 10.7249/TL221.

26   Pearce Wright. "Ivan Illich," *The Lancet* 361, no. 9352 (2003): 185, doi: 10.1016/S0140–6736(03)12233–7.

27   Richard Smith, "Time to Assume That Health Research Is Fraudulent until Proven Otherwise?" *The BMJ Opinion* (blog), Jul. 5, 2021, https://blogs.bmj.com/bmj/2021/07/05/time-to-assume-that-health-research-is-fraudulent-until-proved-otherwise.

28   Tae Ik Chang et al., "Polypharmacy, Hospitalization, and Mortality Risk: A Nationwide Cohort Study," *Sci Rep* 10, no. 18964 (2020), doi: 10.1038/s41598–020–75888–8.

29   Aseem Malhotra, "Why Modern Medicine Is a Major Threat to Public Health," *The Guardian*, Aug. 30, 2018, https://www.theguardian.com/society/2018/aug/30/modern-medicine-major-threat-public-health.

30   Doron Garfinkel, IGRIMUP, "Overview of Current and Future Research and Clinical Directions for Drug Discontinuation: Psychological, Traditional and Professional Obstacles to Deprescribing," *European Journal of Hospital Pharmacy* 24, no. 1 (2017): 16–20, doi: 10.1136/ejhpharm-2016–000959.

31   Christina Charlesworth et al., "Polypharmacy Among Adults Aged 65 Years and Older in the United States: 1988–2010," *The Journals of Gerontology Series A*, Aug. 2015, doi: 10.1093/gerona/glv013.

32   Ashley Kirzinger et al., "Data Note: Prescription Drugs and Older Adults," *KFF*, Aug. 9, 2019, https://www.kff.org/health-reform/issue-brief/data-note-prescription-drugs-and-older-adults.

33 "National Ambulatory Medical Care Survey: 2018 National Summary Tables," National Center for Health Statistics, 2018, 32, https://www.cdc.gov/nchs/data/ahcd/namcs_summary/2018-namcs-web-tables-508.pdf.

34 Andrea Sonnenberg, "Polypharmacy Killed My Son. He's Not Alone," *TIME*, May 19, 2023, https://time.com/6280929/polypharmacy-dangers-essay.

35 Children's Health Defense Team, "Gaslighting Autism Families: CDC, Media Continue to Obscure Decades of Vaccine-Related Harm," *The Defender*, Dec. 17, 2021, https://childrenshealthdefense.org/defender/autism-cdc-media-vaccine-related-harm.

36 Children's Health Defense Team, "'Profiles of the Vaccine-Injured': New CHD Book Exposes Life-Changing Impact of Vaccine Injuries," *The Defender*, Oct. 17, 2022, https://childrenshealthdefense.org/defender/profiles-of-the-vaccine-injured-new-chd-book-vaccine-injuries.

37 "Biography: The Life of Robert S. Mendelsohn, Medical Heretic [1926–1988]," *The People's Doctor*, accessed Jun. 24, 2024, https://thepeoplesdoctor.net/biography.

38 Robert S. Mendelsohn, MD, *Confessions of a Medical Heretic* (New York: McGraw-Hill, 1979).

39 Robert S. Mendelsohn, MD, *How to Raise a Healthy Child . . . in Spite of Your Doctor* (New York: Ballantine Books, 1987).

40 "Biography: The Life of Robert S. Mendelsohn, Medical Heretic [1926–1988]," *The People's Doctor*, accessed Jun. 24, 2024, https://thepeoplesdoctor.net/biography.

41 Richard Smith, "The Most Devastating Critique of Medicine since Medical Nemesis by Ivan Illich in 1975," *The BMJ Opinion* (blog), Feb. 13, 2019, https://blogs.bmj.com/bmj/2019/02/13/richard-smith-most-devastating-critique-medicine-since-medical-nemesis-ivan-illich.

42 Deleted128562, "Don't Go into Medicine If Critical Thinking Is Important to You," *The Student Doctor Network (SDN)*, Jul. 17, 2013, https://forums.studentdoctor.net/threads/dont-go-into-medicine-if-critical-thinking-is-important-to-you.1020875.

43 Ray Moynihan, "Too Much Medicine?" *BMJ* 342, no. 7342 (2002): 859–860, doi: 10.1136/bmj.324.7342.859.

44 American Heart Association, "High Blood Pressure Redefined for First Time in 14 years: 130 Is the New High," *AHA Newsroom*, Nov. 13, 2017, https://newsroom.heart.org/news/high-blood-pressure-redefined-for-first-time-in-14-years-130-is-the-new-high.

45 Ahlia Sekkarie et al., "Abstract 9858: Trends in Lipid-Lowering Prescriptions by Statin Intensity—United States, 2017–202[2]," *Circulation* 146, no. suppl_1 2022, doi: 10.1161/circ.146.suppl_1.9858.

46 Timothy Aungst, "What Are High Intensity Statins?" *GoodRx Health*, Jan. 19, 2022, https://www.goodrx.com/classes/statins/high-intensity-statins.

47   Pantelis A. Sarafidis, Angeliki I. Kanaki, and Anastasios N. Lasaridis, "Statins and Blood Pressure: Is There an Effect or Not?" *J Clin Hypertens (Greenwich)* 9, no. 6 (2007): 460–467, doi: 10.1111/j.1524–6175.2007.06625.

48   Sally Fallon Morell, "Our Seniors: Dumping Ground for Drugs," *Wise Traditions in Food, Farming and the Healing Arts*, Oct. 18, 2018, https://www.westonaprice.org/health-topics/health-issues/our-seniors -dumping-ground-for-drugs.

49   Hilda Labrada Gore, "Episode 397: The Problem with Statins with David Diamond," Dec. 12, 2022, in *Wise Traditions Podcast*, produced by Hilda Labrada Gore, podcast, 00:37:56, https://www.westonaprice.org/podcast /the-problem-with-statins.

50   Michael O. Schroeder, "Death by Prescription: By One Estimate, Taking Prescribed Medications Is the Fourth Leading Cause of Death among Americans," *U.S. News*, Sep. 27, 2016, https://health.usnews.com/health -news/patient-advice/articles/2016–09–27/the-danger-in-taking -prescribed-medications.

51   David Kupelian, "70 Million Americans Taking Mind-Altering Drugs," *WND*, Feb. 9, 2014, https://www.wnd.com/2014/02/70-million-americans -taking-mind-altering-drugs.

52   "Robert Koch Biographical," The Nobel Prize, https://www.nobelprize.org /prizes/medicine/1905/koch/biographical.

53   Stefan H. Kaufmann, "Koch's Dilemma Revisited," *Scandinavian Journal of InfectiousDiseases* 33, no. 1 (2000): 5–8, doi: 10.1080/00365540175006 4004–1.

54   Ibid.

55   Shoichi Takekawa et al., "History Note: Tragedy of Thorotrast," *Japanese JournalofRadiology*33(2015):718–722,doi:10.1007/s11604–015–0479–1.

56   Julian G. West, "The Accidental Poison That Founded the Modern FDA," *The Atlantic*, Jan. 16, 2018, https://www.theatlantic.com/technology /archive/2018/01/the-accidental-poison-that-founded-the-modern -fda/550574.

57   Kathryn C. Zoon and Robert A. Yetter, "The Regulation of Drugs and Biological Products by the Food and Drug Administration," in *Principles and Practice of Clinical Research*, 2nd ed. (2007): 97–107, doi: 10.1016/ B978–012369440–9/50011–6.

58   Carol Ballentine, "Sulfanilamide Disaster," U.S. Food & Drug Administration, last updated Jan. 31, 2018, https://www.fda.gov/about-fda /histories-product-regulation/sulfanilamide-disaster.

59   Stephen Phillips, "How a Courageous Physician-Scientist Saved the U.S. from a Birth-Defects Catastrophe," University of Chicago Medicine, Mar. 9, 2020, https://www.uchicagomedicine.org/forefront /biological-sciences-articles/courageous-physician-scientist-saved-the-us -from-a-birth-defects-catastrophe.

60 Kendall Nelson, "Polio Vaccines: Medical Triumph or Medical Mishap?" *Wise Traditions in Food, Farming and the Healing Arts,* Aug. 3, 2019, https://www.westonaprice.org/health-topics/vaccinations/polio-vaccines -medical-triumph-or-medical-mishap.

61 Children's Health Defense Team, "What Polio Vaccine Injury Looks Like, Decades Later," Children's Health Defense, Sep. 5, 2019, https: //childrenshealthdefense.org/news/government-corruption/what-polio -vaccine-injury-looks-like-decades-later.

62 Children's Health Defense Team, "Think the FDA Is Looking out for Your Health? History Tells a Different Story," *The Defender,* Oct. 6, 2021, https://childrenshealthdefense.org/defender/fda-regulatory-capture -revolving-door-jobs.

63 "The True Story of Thalidomide in the US," *USA Thalidomide Survivors,* accessed Jun. 24, 2024, https://usthalidomide.org/our-story-thalidomide -babies-us.

64 Ayesha Rascoe and Matthew Schuerman, "Jennifer Vanderbes on Her Book 'Wonder Drug,'" NPR, Jun. 25, 2023, https://www.npr.org/2023 /06/25/1184198904/jennifer-vanderbes-on-her-book-wonder-drug.

65 "Medicine: Off the Market," *TIME,* Apr. 27, 1962, https://content.time .com/time/subscriber/article/0,33009,896123,00.html.

66 "Drugs: Triparanol Side Effects," *TIME,* Apr. 3, 1964, https://content.time .com/time/subscriber/article/0,33009,939476,00.html.

67 Daniel Steinberg, "Thematic Review Series: The Pathogenesis of Atherosclerosis. An Interpretive History of the Cholesterol Controversy, Part V: The Discovery of the Statins and the End of the Controversy," *J Lipid Res* 47, no. 7 (2006): 1339–1351, doi: 10.1194/jlr.R600009-JLR200.

68 Jon Rappoport, "Mental Disorders Do Not Exist," *Jon Rappoport's Blog,* Oct. 30, 2018, https://blog.nomorefakenews.com/2018/10/30/mental-disorders -do-not-exist-outside-matrix.

69 Ralph Lewis, MD, "Is ADHD a Real Disorder or One End of a Normal Continuum?" *Psychology Today* (blog), Jan. 6, 2021, https://www .psychologytoday.com/us/blog/finding-purpose/202101/is-adhd-real-disorder -or-one-end-normal-continuum.

70 Brian J. Piper et al., "Trends in Use of Prescription Stimulants in the United States and Territories, 2006 to 2016," *PLoS One* (2018), doi: 10.1371/journal. pone.0206100.

71 Juho Honkasilta and Athanasios Koutsoklenis, "The (Un)Real Existence of ADHD—Criteria, Functions, and Forms of the Diagnostic Entity," *Frontiers in Sociology* 7, no. 814763 (2022), doi: 10.3389/fsoc.2022.814763.

72 "Number of U.S. Children Taking Psychiatric Drugs," *Citizens Commission on Human Rights International,* Jan. 2021, https://www.cchrint.org /psychiatric-drugs/children-on-psychiatric-drugs.

73   Thomas P. Shellenberg et al., "An Update on the Clinical Pharmacology of Methylphenidate: Therapeutic Efficacy, Abuse Potential and Future Considerations," *Expert Review of Clinical Pharmacology* 13, no. 8 (2020): 825–833, doi: 10.1080/17512433.2020.1796636.

74   Brigette S. Vaughan and Christopher J. Kratochvil, "Pharmacotherapy of ADHD in Young Children," *Psychiatry (Edgmont)* 3, no. 8 (2006): 36–45, PMID: 20963194.

75   Melissa L. Danielson, "A National Profile of Attention-Deficit Hyperactivity Disorder Diagnosis and Treatment among U.S. Children Aged 2 to 5 Years," *Journal of Developmental & Behavioral Pediatrics* 38, no. 7 (2017): 455–464, doi: 10.1097/DBP.0000000000000477.

76   Judith Stewart, "What Are the Brands of Methylphenidate?" *Drugs. com*, updated Feb. 14, 2024, https://www.drugs.com/medical-answers /brands-methylphenidate-3510739.

77   Matthew Smith, "Ritalin at 75: What Does the Future Hold?" *The Conversation*, Sep. 18, 2019, https://theconversation.com/ritalin-at -75-what-does-the-future-hold-121591.

78   Ibid.

79   Sean Estaban McCabe et al., "Prescription Stimulant Medical and Nonmedical Use among US Secondary School Students, 2005 to 2020," *JAMA Network Open* 6 no. 4, (2023): e238707, doi: 10.1001/ jamanetworkopen.2023.8707.

80   Melanie Wolkoff Wachsman, "Study: One in Four Teens Has Abused Stimulant Medications for ADHD," *ADDitude*, updated Aug. 11, 2023, https://www.additudemag.com/adhd-drug-abuse-stimulant-medications -misuse-teens.

81   David B. Clemow and Daniel J. Walker, "The Potential for Misuse and Abuse of Medications in ADHD: A Review," *Postgraduate Medicine* 126, no. 5 (2014): 64–81, doi: 10.3810/pgm.2014.09.2801.

82   CCHR International, "FDA Finally Adds 'Addiction' to Black Box Warning on ADHD Drugs," *Citizens Commission on Human Rights International*, May 16, 2023, https://www.cchrint.org/2023/05/16/fda -finally-adds-addiction-to-black-box-warning-on-adhd-drugs.

83   Ibid.

84   Nortin M. Hadler, *The Last Well Person: How to Stay Well Despite the Health-Care System* (Montreal, Quebec: McGill-Queen's University Press, 2004).

85   Nortin M. Hadler, *Worried Sick: A Prescription for Health in an Overtreated America* (Chapel Hill, NC: University of North Carolina Press, 2008).

86   Geoff Watts, "Barbara Starfield," *The Lancet* 378, no. 9791 (2011): 564, doi: 10.1016/S0140–6736(11)61281–6.

87   Barbara Starfield, "Is US Health Really the Best in the World?" *JAMA Network* 284, no. 4 (2000): 483–485, doi: 10.1001/jama.284.4.483.

88 Steven H. Woolf, MD, "Falling Behind: The Growing Gap in Life Expectancy between the United States and Other Countries, 1933–2021," *American Journal of Public Health* 113, no. 9 (2023), doi: 10.2105/AJPH.2023.307310.

89 "Lucian L. Leape," *Harvard School of Public Health*, accessed Jun. 24, 2024, https://www.hsph.harvard.edu/profile/lucian-l-leape.

90 Lucian L. Leape, "Error in Medicine," *JAMA Network* 272, no. 23 (1994): 1851–1857, doi:10.1001/jama.1994.03520230061039.

91 Domenico Paparella et al., "Acute Iatrogenic Complications after Mitral Valve Repair," *Journal of CardiacSurgery* 37, no. 12 (2022): 4088–4093, doi: 10.1111/jocs.17055.

92 Charlotte Demoor-Goldschmidt and Florent de Vathaire, "Review of Risk Factors of Secondary Cancers among Cancer Survivors," *British Journal of Radiology* 92, no. 1093 (2018), doi: 10.1259/bjr.20180390.

93 Joette Calabrese, "My Dad and the Dead Cardiologists," *Wise Traditions in Food, Farming and the Healing Arts*, Jan. 21, 2014, https://www.westonaprice.org/health-topics/my-dad-and-the-dead-cardiologists.

94 Sally Fallon and Mary G. Enig, "What Causes Heart Disease?" *Wise Traditions in Food, Farming and the Healing Arts*, Mar. 1, 2001, https://www.westonaprice.org/health-topics/modern-diseases/what-causes-heart-disease.

95 George V. Mann, ScD, MD, "Diet-Heart: End of an Era," *New England Journal of Medicine* 297, no. 12 (1977): 644–650, doi: 10.1056/NEJM197709222971206.

96 James Delingpole, "The Big Fat Lie about Cholesterol," *The Spectator*, Jun. 21, 2014, https://www.spectator.co.uk/article/the-big-fat-lie-about-cholesterol.

97 Uffe Ravnskov, *The Cholesterol Myths: Exposing the Fallacy That Saturated Fat and Cholesterol Cause Heart Disease* (Washington DC: New Trends Publishing, 2003).

98 "Dr. Zoë Harcombe Ph.D.," accessed Jun. 24, 2024, https://www.zoeharcombe.com.

99 Tim Boyd, "Modern Diseases: Cardiovascular Disease," Oct. 6, 2016, https://www.westonaprice.org/modern-diseases.

100 Children's Health Defense Team, "Read the Fine Print, Part Two—Nearly 400 Adverse Reactions Listed in Vaccine Package Inserts," Children's Health Defense, Aug. 14, 2020, https://childrenshealthdefense.org/news/read-the-fine-print-part-two-nearly-400-adverse-reactions-listed-in-vaccine-package-inserts.

101 Defender Staff, "Utah Woman Is First to Sue Merck Alleging Gardasil HPV Vaccine Caused Cervical Cancer," *The Defender*, Apr. 26, 2023, https://childrenshealthdefense.org/defender/caroline-cantera-merck-gardasil-hpv-vaccine-lawsuit-cervical-cancer.

102 Dr. Chris Flowers, "Report 11: Pfizer Vaccine—FDA Fails to Mention Risk of Heart Damage in Teens," *DailyClout*, Apr. 7, 2022, https://dailyclout.io /pfizer-vaccine-fda-fails-to-mention-risk-of-heart-damage-in-teens.

103 Brucha Weisberger, "Cancer Resulting from Covid Vaccines: Causal Mechanisms, Case Studies, Doctors' Reports—and NIH Coverup," *In G-D's Army There's Only Truth*, May 11, 2023, https://truth613.substack. com/p/cancer-resulting-from-covid-vaccines.

104 J. A. Johnson and J. L. Bootman, "Drug-Related Morbidity and Mortality and the Economic Impact of Pharmaceutical Care," *American Journal of Health System Pharmacy* 54, no. 5 (1997): 554–558, doi: 10.1093/ajhp/ 54.5.554.

105 Saul N. Weingart et al., "Epidemiology of Medical Error," *Western Journal of Medicine* 172, no. 6 (2000): 390–393, doi: 10.1136/ewjm.172.6.390.

106 Jon Rappoport, "An Exclusive Interview with Dr. Barbara Starfield: Medically Caused Death in America," *Jon Rappoport's Blog*, Dec. 9, 2009, https://blog.nomorefakenews.com/2009/12/09/an-exclusive-interview -with-dr-barbara-starfield-medically-caused-death-in-america.

107 Ibid.

108 Lucian L. Leape, "Errors in Medicine," *Clinica Chimica Acta* 404, no. 1 (2009): 2–5, doi: 10.1016/j.cca.2009.03.020.

109 John T. James, "A New, Evidence-Based Estimate of Patient Harms Associated with Hospital Care," *Journal of Patient Safety* 9, no. 3 (2013): 122–128, doi: 10.1097/PTS.0b013e3182948a69.

110 Defender Staff, "As COVID Vaccine Injuries Pile Up, It's Worth Remembering: Medicine Is a Leading Cause of Death in the U.S," *The Defender*, Aug. 24, 2022, https://childrenshealthdefense.org/defender /covid-vaccines-medicine-leading-cause-death-united-states.

111 Ariana Eunjung Cha, "Researchers: Medical Errors Now Third Leading Cause of Death in United States," *Washington Post*, May 3, 2016, https: //www.washingtonpost.com/news/to-your-health/wp/2016/05/03/researchers -medical-errors-now-third-leading-cause-of-death-in-united-states.

112 "About Me," *GaryNull*, accessed Jun. 25, 2024, https://garynull.com/about -me.

113 Lucian L. Leape, "Error in Medicine," *JAMA* 272, no. 23 (1994): 1851– 1857, doi: 10.1001/jama.1994.03520230061039.

114 John T. James, "A New, Evidence-Based Estimate of Patient Harms Associated with Hospital Care," *Journal of Patient Safety* 9, no. 3 (2013): 122–128, doi: 10.1097/PTS.0b013e3182948a69.

115 Lucian L. Leape, "Error in Medicine," *JAMA* 272, no. 23 (1994): 1851– 1857, doi: 10.1001/jama.1994.03520230061039.

116 Michael Nevradakis, PhD, "Exclusive: Woman Diagnosed with Vaccine-Induced Transverse Myelitis after Pfizer Shots," *The Defender*, Jan. 31, 2023, https://childrenshealthdefense.org/defender/danielle-baker-pfizer-shots -transverse-myelitis.

117 Eleanor Chelimsky, *FDA Drug Review: Postapproval Risks, 1976–85* (Washington, DC: United States General Accounting Office, April 1990), https://www.gao.gov/assets/pemd-90–15.pdf.

118 Timothy F. Jones et al., "Neurologic Complications Including Paralysis after a Medication Error Involving Implanted Intrathecal Catheters," *Am J Med* 112, no. 1 (2002): 31–36, doi: 10.1016/s0002–9343(01)01032–4.

119 "How Often Do Surgical Instruments Get Left Inside Patients?" *Golden Law Office*, Mar. 12, 2020, https://goldenlawoffice.com/medical-malpractice/how -often-do-surgical-instruments-get-left-inside-patients.

120 "Wrong-Site, Wrong-Procedure, and Wrong-Patient Surgery," *Patient Safety Network*, Sep. 7, 2019, https://psnet.ahrq.gov/primer/wrong-site-wrong -procedure-and-wrong-patient-surgery.

121 Kristina Fiore, "Wrong-Eye Surgery Suit; Baby-Killer Nurse's Life Sentence; $1.4B COVID Relief Fraud," *Medpage Today*, Aug. 24, 2023, https://www .medpagetoday.com/special-reports/features/106028.

122 Biao Cheng et al., "Iatrogenic Wounds: A Common but Often Overlooked Problem," *Burns & Trauma* 7 (2019), doi: 10.1186/s41038–019–0155–2.

123 Annie Waldman, "In the 'Wild West' of Outpatient Vascular Care, Doctors Can Reap Huge Payments as Patients Risk Life and Limb," *ProPublica*, May 24, 2023, https://www.propublica.org/article/maryland-dormu-minimally -invasive-vascular-medicare-medicaid.

124 Arun K. Thukkani and Scott Kinlay, "Endovascular Intervention for Peripheral Artery Disease," *Circulation Research* 116, no. 9 (2015): 1599–1613, doi: 10.1161/CIRCRESAHA.116.303503.

125 Niveditta Ramkumar et al., "Adverse Events after Atherectomy: Analyzing Long-Term Outcomes of Endovascular Lower Extremity Revascularization Techniques," *Journal of the American Heart Association* 8, no. 12 (2019): e012081, doi: 10.1161/JAHA.119.012081.

126 Paul Cuno-Booth, "Report Cites Widespread Failures at CMC Related to Troubled Heart Surgeon," New Hampshire Public Radio, Jun. 7, 2023, https://www.nhpr.org/nh-news/2023–06–07/report-cites-widespread -failures-at-cmc-related-to-troubled-heart-surgeon.

127 Rebecca Ostriker, "A Celebrated Surgeon, a Trail of Secrets and Death," *The Boston Globe*, Sep. 7, 2022, https://apps.bostonglobe.com/2022/09 /07/metro/investigations/spotlight/trail-of-secrets-and-death/yvon-baribeau -malpractice-manchester-new-hampshire.

128 Michael Berens, "How Doctors Buy Their Way Out of Trouble," Reuters, May 24, 2023, https://www.reuters.com/investigates/special-report/usa-health care-settlements.

129 Michael Nevradakis, PhD, "Big Pharma Money 'Permeates' World's Drug Regulatory Agencies, BMJ Investigation Shows," *The Defender*, Jul. 1, 2022, https://childrenshealthdefense.org/defender/big-pharma-drug-regulators.

130 Children's Health Defense Team, "Think the FDA Is Looking out for Your Health? History Tells a Different Story," *The Defender*, Oct. 6, 2021, https://childrenshealthdefense.org/defender/fda-regulatory-capture -revolving-door-jobs.

131 "FDA: User Fees Explained," U.S. Food & Drug Administration, last updated May 22, 2024, https://www.fda.gov/industry/fda-user-fee-programs/fda -user-fees-explained.

132 C. Michael White, "Why Does the FDA Get Nearly Half Its Funding from the Companies It Regulates? *The Defender*, May 14, 2021, https://childrenshealth defense.org/defender/fda-nearly-half-funding-companies-it-regulates.

133 "FY 2024 FDA Budget Summary," U.S. Food & Drug Administration, accessed Jun. 25, 2024, https://www.fda.gov/media/166050/download.

134 Arthur Allen, "Pharma-Funded FDA Gets Drugs Out Faster, but Some Work Only 'Marginally' and Most Are Pricey," *KFF Health News*, Sep. 30, 2022, https://kffhealthnews.org/news/article/pharma-fda-drugs-accelerated -approval-marginally-effective-expensive.

135 Christine Jindra, "Part II: Fast Track Announcements Have Triggered Stock Price Run-Ups and Frenzied Trading," *Cleveland.com*, Dec. 3, 2007, https: //www.cleveland.com/pdworld/2007/12/_fast_track_increases_risk.html.

136 Children's Health Defense Team, "Who Benefits When Pharma-Funded FDA Fast-Tracks Drugs and Vaccines? Not Consumers, Critics Warn," *The Defender*, Oct. 13, 2022, https://childrenshealthdefense.org/defender /fda-big-pharma-drugs-vaccines-consumers.

137 Darrell M. West, "It Is Time to Restore the US Office of Technology Assessment," Brookings, Feb. 10, 2021, https://www.brookings.edu/research /it-is-time-to-restore-the-us-office-of-technology-assessment.

138 "The Reform Work of Peter R. Breggin, MD," *Breggin.com*, Apr. 8, 2012, https://breggin.com/article-detail/post_detail/the-conscience-of-psychiatry.

139 "Books by Dr. Peter R. Breggin, MD," *Peter Breggin, MD, Psychiatric Drug Facts*, accessed Jun. 25, 2024, https://psych.breggin.com/books.

140 Peter R. Breggin, MD, *COVID-19 and the Global Predators: We Are the Prey* (Ithaca, NY: Lake Edge Press, 2021).

141 Jonathan Jones, "The 'Right' and 'Wrong' Kind of Addict: Iatrogenic Opioid Addiction in Historical Context," *Nursing Clio*, Jul. 25, 2017, https://nursing clio.org/2017/07/25/the-right-and-wrong-kind-of-addict-iatrogenic-opioid -addiction-in-historical-context.

142 Michael Kaliszewski, MD, "The Link between Substance Abuse and Suicide in Teens," *American Addiction Centers* (blog), last updated Sep. 15, 2022, https://americanaddictioncenters.org/blog/link-between-substance -abuse-suicide-in-teens.

143 "Health Disparities in Suicide," Centers for Disease Control and Prevention, last reviewed Jan. 17, 2024, https://www.cdc.gov/suicide

/disparities/?CDC_AAref_Val=https://www.cdc.gov/suicide/facts/disparities-in-suicide.html.

144 Allison McCabe, "America's Number One Prescription Sleep Aid Could Trigger 'Zombies,' Murder and other Disturbing Behavior," *AlterNet*, Jan. 15, 2014, https://www.alternet.org/2014/01/americas-number-one-prescription-sleep-aid-could-trigger-zombies-murder-and-other-disturbing.

145 "Mental Health Drug Prescriptions on the Rise," *Insurance Journal*, Apr. 22, 2021, https://www.insurancejournal.com/news/national/2021/04/22/610924.htm.

146 Ibid.

147 Nick Haslam, Jesse S. Y. Tse, and Simon De Deyne, "Concept Creep and Psychiatrization," *Frontiers in Sociology* 6 (2021), doi: 10.3389/fsoc.2021.806147.

148 Ibid.

149 Juho Honkasilta and Athanasios Koutsoklenis, "The (Un)Real Existence of ADHD—Criteria, Functions, and Forms of the Diagnostic Entity," *Frontiers in Sociology* 7, no. 814763 (2022), doi: 10.3389/fsoc.2022.814763.

150 Joanna Moncrieff, "The Serotonin Theory of Depression: A Systematic Umbrella Review of the Evidence," *Mol Psychiatry* 28, no. 8 (2023): 3243–3256, doi: 10.1038/s41380–022–01661–0.

151 Noam Shpancer, "Depression Is Not Caused by Chemical Imbalance in the Brain," *Psychology Today* (blog), Jul. 24, 2022, https://www.psychologytoday.com/us/blog/insight-therapy/202207/depression-is-not-caused-chemical-imbalance-in-the-brain.

152 Sahanika Ratnayake, "Why Has the Misleading 'Chemical Imbalance' Theory of Mental Illness Persisted for So Long?" *Slate*, Aug. 4, 2022, https://slate.com/technology/2022/08/ssris-chemical-imbalance-depression.html.

153 Ori Kapra, Ran Rotem, and Raz Gross, "The Association between Prenatal Exposure to Antidepressants and Autism: Some Research and Public Health Aspects," *Frontiers in Psychiatry* 11 (2020), doi: 10.3389/fpsyt.2020.555740.

154 Jon Rappoport, "When the Blood Boils: Vaccines and Autism," *Jon Rappoport's Blog*, Jan. 21, 2020, https://blog.nomorefakenews.com/2020/01/21/when-the-blood-boils-vaccines-and-autism.

155 Vasilios G. Masdrakis, Manolis Markianos, and David S. Baldwin, "Apathy Associated with Antidepressant Drugs: A Systematic Review," *Acta Neuropsychiatry* 35, no. 4 (2023): 1–16, doi: 10.1017/neu.2023.6.

156 Children's Health Defense Team, "Why Are 10-Year-Olds Killing Themselves?" Children's Health Defense, Nov. 14, 2019, https://childrenshealthdefense.org/news/why-are-10-year-olds-killing-themselves.

157 Rashmi Patel et al., "Do Antidepressants Increase the Risk of Mania and Bipolar Disorder in People with Depression? A Retrospective Electronic Case Register Cohort Study," *BMJ Open* 5, no. 12 (2015): e008341, doi: 10.1136/bmjopen-2015–008341.

158 Peter C. Gøtzsche, "Rapid Response: Antidepressants Increase the Risk of Suicide, Violence and Homicide at All Ages," *BMJ* 358 (2017), doi: 10. 1136/bmj.j3697.

159 Alexis Revet et al., "Antidepressants and Movement Disorders: A Postmarketing Study in the World Pharmacovigilance Database," *BMC Psychiatry* 20, no. 1 (2020): 308, doi: 10.1186/s12888–020–02711-z.

160 Peter C. Gøtzsche, "Rapid Response: Antidepressants Increase the Risk of Suicide, Violence and Homicide at All Ages," *BMJ* 358 (2017), doi: 10. 1136/bmj.j3697.

161 Peter C. Gøtzsche, "Prescription Drugs Are the Third Leading Cause of Death," *The BMJ Opinion* (blog), Jun. 16, 2016, https://blogs.bmj.com /bmj/2016/06/16/peter-c-gotzsche-prescription-drugs-are-the-third-leading -cause-of-death.

162 Children's Health Defense Team, "House of Cards Is Falling: Shake Up at Cochrane,"Children'sHealthDefense,Sep.18,2018,https://childrenshealth defense.org/news/the-house-of-cards-is-falling-the-shake-up-at-cochrane.

163 Peter C. Gøtzsche, "Prescription Drugs Are the Third Leading Cause of Death," *The BMJ Opinion* (blog), Jun. 16, 2016, https://blogs.bmj.com /bmj/2016/06/16/peter-c-gotzsche-prescription-drugs-are-the-third-leading -cause-of-death.

164 Lauren V. Moran, "Psychosis with Methylphenidate or Amphetamine in Patients with ADHD," *New England Journal of Medicine* 380, no. 12 (2019): 1128–1138, doi: 10.1056/NEJMoa1813751.

165 Ashley Welch, "Some ADHD Meds May Increase Risk of Psychosis, Study Finds," CBS News, Mar. 21, 2019, https://www.cbsnews.com/news/some -adhd-meds-may-increase-risk-of-psychosis-study-finds.

166 Cheima Bouziane et al., "White Matter by Diffusion MRI Following Methylphenidate Treatment: A Randomized Control Trial in Males with Attention-Deficit/Hyperactivity Disorder," *Radiology* 293, no. 1 (2019): 186–19, doi: 10.1148/radiol.2019182528.

167 N. S. Miller and M. S. Gold, "Benzodiazepines: Reconsidered," *Advances in Alcohol & Substance Abuse* 8, no. 3–4 (1990): 67–84, doi: 10.1300/ J251v08n03_06.

168 Dexter L. Louie, Oluwole O. Jegede, and Gretchen L. Hermes, "Chronic Use of Benzodiazepines: The Problem That Persists," *The International Journal of Psychiatry in Medicine* 58, no. 5 (2023): 426–432, doi: 10. 1177/00912174231166252.

169 Neelambika Revadigar and Vikas Gupta, *Substance-Induced Mood Disorders* (Treasure Island, Florida: StatPearls Publishing, 2023), https://www.ncbi .nlm.nih.gov/books/NBK555887.

170 Dr. Joseph Mercola, "Over-Medicating Kids Leads to More Health Problems— and More Meds," *The Defender*, Sep. 27, 2022, https://childrenshealth defense.org/defender/medicating-kids-polypharmacy-health-problems-cola.

171 Allyson J. Kemp et al., "Synergistic Effects of Psychotropics Leading to Extraordinary Weight Gain," *Cureus* 13, no. 9 (2021): e17978, doi: 10.7759/cureus.17978.

172 Ankit Gupta and Rakesh K. Chadda, "Adverse Psychiatric Effects of Non-Psychotropic Medications," *BJPsych Advances* 22, no. 5 (2016): 325–334, doi: 10.1192/apt.bp.115.015735.

173 Sidney Wolfe, MD, "How Independent Is the FDA?" *Frontline*, Nov. 13, 2003, https://www.pbs.org/wgbh/pages/frontline/shows/prescription/hazard/independent.html.

174 Jon Rappoport, "Prozac Mass Murders: The Truth Comes to Light," *Jon Rappoport's Blog*, Oct. 2, 2019, https://blog.nomorefakenews.com/2019/10/02/prozac-mass-murders-the-truth-comes-to-light.

175 "Suicidality in Children and Adolescents Being Treated with Antidepressant Medications," U.S. Food and Drug Administration, last updated Feb. 5, 2018, https://www.fda.gov/drugs/postmarket-drug-safety-information-patients-and-providers/suicidality-children-and-adolescents-being-treated-anti-depressant-medications.

176 Richard A. Friedman, MD, and Andrew C. Leon, MD, "Expanding the Black Box—Depression, Antidepressants, and the Risk of Suicide," *New England Journal of Medicine* 356, no. 23 (2007): 2343–2346, doi: 10.1056/NEJMp078015.

177 Glen I. Spielmans, Tess Spence-Sing, and Peter Parry, "Duty to Warn: Antidepressant Black Box Suicidality Warning Is Empirically Justified," *Frontiers in Psychiatry* 11, no. 18 (2020), doi: 10.3389/fpsyt.020.00018.

178 Michael P. Hengartner, "Editorial: Antidepressant Prescriptions in Children and Adolescents," *Frontiers in Psychiatry* 11, no. 600283 (2020), doi: 10.3389/fpsyt.2020.600283.

179 Leemon B. McHenry, "A Book Review—Children of the Cure: Missing Data, Lost Lives and Antidepressants," Children's Health Defense, Jun. 25, 2020, https://childrenshealthdefense.org/news/childrens-health/a-book-review-children-of-the-cure-missing-data-lost-lives-and-antidepressants.

180 "Law Project for Psychiatric Rights," *PsychRights*, accessed Jun. 25, 2024, https://psychrights.org.

181 Jim Gottstein, "Teen Screen: Pharmaceutical Company Drugging Dragnet," YouTube, 00:01:06, Sep. 22, 2012, https://www.youtube.com/watch?v=X1ULlB-VF6M&ab_channel=JimGottstein.

182 Mirelle Kass et al., "Parental Preferences for Mental Health Screening of Youths from a Multinational Survey," *JAMA Network Open* 6, no. 6 (2023): e2318892, doi:10.1001/jamanetworkopen.2023.18892.

183 Lisa Cosgrove et al., "Drivers of and Solutions for the Overuse of Antidepressant Medication in Pediatric Populations," *Frontiers in Psychiatry* 11, no. 17 (2020), doi: 10.3389/fpsyt.2020.00017.

184 Michael P. Hengartner, "Editorial: Antidepressant Prescriptions in Children and Adolescents," *Frontiers in Psychiatry* 11, no. 600283 (2020), doi: 10.3389/fpsyt.2020.600283.

185 Julia Robinson, "Number of Young Children Prescribed Antidepressants Has Risen by 41% since 2015," *The Pharmaceutical Journal*, Sep. 6, 2021, https://pharmaceutical-journal.com/article/news/number-of-young-children-prescribed-antidepressants-has-risen-by-41-since-2015.

186 Julie M. Zito, Dinci Pennap, and Daniel J. Safer, "Antidepressant Use in Medicaid-Insured Youth: Trends, Covariates, and Future Research Needs," *Frontiers in Psychiatry* 11, no. 113 (2020), doi: 10.3389/fpsyt.2020.00113.

187 Chengchen Zhang et al., "Characteristics of Youths Treated with Psychotropic Polypharmacy in the United States, 1999 to 2015," *JAMA Pediatrics* 175, no. 2 (2021): 196–198. doi: 10.1001/jamapediatrics.2020.4678.

188 Peter Simons, "Increasing Numbers of Children Prescribed Multiple Psychiatric Medications," *Mad in America*, Nov. 17, 2020, https://www.madinamerica.com/2020/11/increasing-numbers-children-prescribed-multiple-psychiatric-medications.

189 Mike McCarthy, "First Person: The Mental Health Model of 'Forced Treatment' Doesn't Work," *PublicSource*, Oct. 24, 2019, https://www.publicsource.org/mental-health-system-model-of-forced-treatment-doesnt-work.

190 Sahanika Ratnayake, "Why Has the Misleading 'Chemical Imbalance' Theory of Mental Illness Persisted for So Long?" *Slate*, Aug. 4, 2022, https://slate.com/technology/2022/08/ssris-chemical-imbalance-depression.html.

191 "Blast from the Past: Week of August 29, 2022: The Weaponization of Mental Health," *Solari Report*, Aug. 30, 2022, https://home.solari.com/blast-from-the-past-week-of-august-29–2022-the-weaponization-of-mental-health.

192 "Dr. Breggin on the Weaponization of Mental Health Edicts," interview with Sherry Strong, https://rumble.com/v1xwsr6-dr.-peter-breggin-on-the-weaponization-of-mental-health-edicts.html

193 Nazi Germany is the most obvious example of psychiatrists being used to support mass murder:

"This chapter will examine the direct and systematic involvement of psychiatry in the labeling, persecution and eventual mass murder of millions of those deemed "unfit." While the entire medical profession can and should be held accountable for the abrogation of ethics that took place during the Holocaust, the role of psychiatrists, specifically, must be explored because of their ability to conflate clinical diagnoses with the worth of an individual. The theory of eugenics allowed psychiatry to provide the scientific justification and the practical mechanisms for the 'mercy killing' of 'life unworthy of life.' The leadership and expertise of psychiatrists paved the way for a powerful merger of medicine and politics that ultimately led to the mass murder of millions under the guise of scientific and societal progress."

Susan M. Miller and Stacy Gallin, "The Transformation of Physicians from Healers to Killers: The Role of Psychiatry," In: Gallin, S., Bedzow, I. (eds) *Bioethics and the Holocaust*, The International Library of Bioethics, vol 96. Springer, Cham. https://doi.org/10.1007/978–3–031–01987–6_

194 Nadine Yousif, "One in 10 Canadians Are Hesitant to Get the Vaccine: CAMH Survey," *Toronto Star*, Apr. 13, 2021, https://www.thestar.com /news/gta/2021/04/13/one-in-10-canadians-are-hesitant-to-get-the-vaccine -camh-survey.html.

195 World Council for Health, "The World Council for Health Strongly Opposes France's Draconian New Law Criminalizing Medical Dissent," *World Council for Health*, Feb. 20, 2024, https://worldcouncilforhealth.org /news/statements/criminalizing-medical-dissent-france.

196 Gerald Posner, "Criminalizing Free Speech: The 'Pfizer Article,'" *Just the Facts*, Feb. 17, 2024, https://www.justthefacts.media/p/criminalizing-free-speech -the-pfizer.

197 [Bill No. 267, adopted by the National Assembly, in new reading, aimed at strengthening the fight against sectarian abuses and improving support for victims], Assemblée Nationale, https://www.assemblee-nationale.fr /dyn/16/textes/l16t0267_texte-adopte-seance

198 James F. Tracy, "The CIA and the Media: 50 Facts the World Needs to Know," *Global Research*, Jan. 30, 2018, https://www.globalresearch.ca/the -cia-and-the-media-50-facts-the-world-needs-to-know/5471956.

199 Rebecca Strong, "Here's What Happens When Billionaires Buy up Mainstream Media," *The Defender*, May 4, 2022, https://childrenshealth defense.org/defender/billionaires-owners-mainstream-media.

200 Jon Rappoport, *AIDS Inc.: Scandal of the Century* (San Bruno, CA: Human Energy Press, 1988).

201 Jon Rappoport, "Message to All Americans from the Medical Cartel," *Jon Rappoport's Blog*, Jan. 7, 2020, https://blog.nomorefakenews.com/2020/01/07 /message-to-americans-from-the-medical-cartel.

202 "Jon Rappoport," *Substack*, https://substack.com/@jonrappoport.

203 Jon Rappoport, "Media Won't Investigate Medically Caused Death Numbers," *Jon Rappoport's Blog*, May 16, 2018, https://blog.nomorefakenews.com /2018/05/16/media-wont-investigate-medically-caused-death-numbers.

204 Jon Rappoport, "Rejecting Rockefeller Germ Theory Once and for All," *Jon Rappoport's Blog*, Jul. 27, 2022, https://blog.nomorefakenews .com/2022/07/27/rejecting-rockefeller-germ-theory-once-and-for-all -compelling-evidence.

205 Jon Rappoport, "COVID: It's Not One Thing, It's Not One Disease," *Jon Rappoport's Blog*, Apr. 1, 2020, https://blog.nomorefakenews.com/2023/05 /12/covid-its-not-one-thing-its-not-one-disease-2.

206 Jon Rappoport, "The History of Big Pharma Any Idiot Can Understand," *Jon Rappoport,* Mar. 9, 2023, https://jonrappoport.substack.com/p/the -history-of-big-pharma-for-idiots.

207 "Understanding Parkinson's: Causes," Parkinson's Foundation, accessed Jun. 26, 2024, https://www.parkinson.org/understanding-parkinsons/causes.

208 Jon Rappoport, "Massive Lawsuit: Pesticide Causes Parkinson's; How Lawyer-Logic and Medical-Hustle Perform Stage Magic," *Jon Rappoport's Blog,* Mar. 14, 2022, https://blog.nomorefakenews.com/2022/03/14/massive -lawsuit-pesticide-causes-parkinsons-how-lawyer-logic-and-medical-hustle- perform-stage-magic.

209 Jon Rappoport, "Selling Fear—The 'Epidemic Experts' Weigh In," *Jon Rappoport's Blog,* Mar. 9, 2020, https://blog.nomorefakenews.com/2023/02 /22/selling-fear-the-epidemic-experts-weigh-in-2.

210 Jon Rappoport, "The 'Hot Zone' Theory of New Frightening Diseases," *Jon Rappoport's Blog,* May 28, 2020, https://blog.nomorefakenews.com /2020/05/28/the-hot-zone-theory-of-new-frightening-diseases.

211 Jon Rappoport, "For Alert Minds: The Art of the Covert Narrative," *Jon Rappoport's Blog,* Sep. 30, 2014, https://blog.nomorefakenews.com/2014/09 /30/for-alert-minds-the-art-of-the-covert-narrative.

212 Niamh Harris, "Formaldehyde Dumped in Liberian Water Wells, Allegedly Causing Ebola-Like Symptoms," *The People's Voice,* Oct. 16, 2014, https: //thepeoplesvoice.tv/formaldeyde-dumped-in-liberian-water-wells-allegedly -causing-ebola-like-symptoms.

213 Terrence McCoy, "The Major Liberian Newspaper Churning out Ebola Conspiracy after Conspiracy," *Washington Post,* Oct. 17, 2014, https: //archive.md/fpj1V.

214 Jon Rappoport, "The 'Hot Zone' Theory of New Frightening Diseases," *Jon Rappoport's Blog,* May 28, 2020, https://blog.nomorefakenews.com/2020/05 /28/the-hot-zone-theory-of-new-frightening-diseases.

215 "Appendix A: AIDS-Defining Conditions," Centers for Disease Control and Prevention, accessed Jun. 26, 2024, https://www.cdc.gov/mmwr/preview /mmwrhtml/rr5710a2.htm.

216 Patricia Goodson, "Questioning the HIV-AIDS Hypothesis: 30 Years of Dissent [retracted]," *Frontiers in Public Health* 2, no. 154 (2014), http: //www.davidrasnick.com/aids/ewExternalFiles/Goodson%202014.pdf.

217 Frontiers Editorial Office, "Publisher Statement on 'Questioning the HIV-AIDS Hypothesis: 30 Years of Dissent,'" *Frontiers in Public Health* 3, no. 37 (2015), doi: 10.3389/fpubh.2015.00037.

218 Frontiers Editorial Office, "Retraction: Questioning the HIV-AIDS Hypothesis: 30 Years of Dissent," *Frontiers in Public Health* 7, no. 334 (2019), doi: 10.3389/fpubh.2019.00334.

219 Jon Rappoport, "COVID: It's Not One Thing, It's Not One Disease," *Jon Rappoport's Blog*, Apr. 1, 2020, https://blog.nomorefakenews.com/2023/05 /12/covid-its-not-one-thing-its-not-one-disease-2.

220 Kit Knightly, "40 Facts You NEED to Know: The REAL Story of 'Covid,'" *Off-Guardian*, Mar. 24, 2023, https://off-guardian.org/2023/03 /24/40-facts-you-need-to-know-the-real-story-of-covid.

221 "Similarities and Differences between Flu and COVID-19," Centers for Disease Control and Prevention, last reviewed Mar. 20, 2024, https://www .cdc.gov/flu/symptoms/flu-vs-covid19.htm.

222 Leah Groth, "What Is Ground Glass Opacity (GGO)?" *Health*, updated Mar. 30, 2024, https://www.health.com/condition/infectious-diseases /coronavirus/ground-glass-opacities-covid-19.

223 "Smell and Taste Disorders," Johns Hopkins Medicine, accessed Jun. 26, 2024, https://www.hopkinsmedicine.org/health/conditions-and-diseases/smell -and-taste-disorders.

224 David James, "PCR Inventor: 'It Doesn't Tell You That You Are Sick,'" *Off-Guardian*, Oct. 5, 2020, https://off-guardian.org/2020/10/05/pcr -inventor-it-doesnt-tell-you-that-you-are-sick.

225 Kit Knightly, "WHO (Finally) Admits PCR Tests Create False Positives," *Off-Guardian*, Dec. 18, 2020, https://off-guardian.org/2020/12/18/who -finally-admits-pcr-tests-create-false-positives.

226 Kit Knightly, "WHO (Finally) Admits PCR Test Is Potentially Flawed," *Off-Guardian*, Jan. 25, 2021, https://off-guardian.org/2021/01/25/who -finally-admits-pcr-is-not-a-diagnostic-test.

227 Stand for Health Freedom, "Oregon Senators File Grand Jury Petition Alleging CDC, FDA Violated Federal Law by Inflating COVID Death Data," *The Defender*, Sep. 17, 2021, https://childrenshealthdefense.org/defender /oregon-senators-grand-jury-petition-cdc-fda-inflating-covid-death-data.

228 Kit Knightly, "40 Facts You NEED to Know: The REAL Story of 'Covid,'" *Off-Guardian*, Mar. 24, 2023, https://off-guardian.org/2023/03 /24/40-facts-you-need-to-know-the-real-story-of-covid.

229 Celia Farber, *Serious Adverse Events: An Uncensored History of AIDS* (White River Junction, VT: Chelsea Green Publishing, 2023).

230 "The Truth Barrier," *Substack*, https://celiafarber.substack.com.

231 Celeste McGovern, "Anatomy of a Science Study Censorship," Children's Health Defense, Mar. 25, 2019, https://childrenshealthdefense.org/news/ anatomy-of-a-science-study-censorship.

232 Richard Horton, "The Dawn of McScience," *The New York Review*, Mar. 11, 2004, https://www.nybooks.com/articles/2004/03/11/the-dawn-of-mcscience.

233 Mark Crispin Miller, "How 'HIV/AIDS' Foretold the 'COVID' Crisis," *News from Underground by Mark Crispin Miller*, Mar. 6, 2023, https://mark crispinmiller.substack.com/p/how-hivaids-foretold-the-covid-crisis.

234 Tim DiFerdinando, "Welcome to Rethinking AIDS!" *Rethinking AIDS*, accessed Jun. 26, 2024, https://rethinkingaids.com/index.php/introduction.

235 Catherine Austin Fitts, "The Creation of a False Epidemic with Jon Rappoport—Parts I–III," *Solari Report*, Apr. 1, 2020, https://home.solari .com/the-creation-of-a-false-epidemic-with-jon-rappoport.

236 Jon Rappoport, "Introduction to My 'Greatest COVID Hits' Series of Articles," *Jon Rappoport's Blog*, Oct. 3, 2022, https://blog.nomorefakenews. com/2022/10/03/introduction-to-my-greatest-covid-hits-series-of-articles.

237 Celia Farber, "AIDS and the AZT Scandal: SPIN's 1989 Feature, 'Sins of Omission,'" *SPIN*, Oct. 5, 2015, https://www.spin.com/2015/10/aids -and-the-azt-scandal-spin-1989-feature-sins-of-omission.

238 Celia Farber, "A Pharmacist Speaks, Remembering Fauci's AZT Putsch in the 1980s 'I Had to Dispense This Poison and Watch These Young Men Die'," *The Truth Barrier*, Jan. 2, 2022, https://celiafarber.substack .com/p/a-pharmacist-speaks-remembering-faucis.

239 Celia Farber, "Tribute to John Lauritsen, Author of 'Poison by Prescription: The AZT Story,'" *The Defender*, Apr. 26, 2022, https://childrenshealth defense.org/defender/john-lauritsen-poison-by-prescription-the-azt-story.

240 Celia Farber, "'Pre-Exposure' HIV Prophylaxis (PrEP) Is a Growing Cult in AIDS World: Activists Made Facebook Remove Ads about Class Action Lawsuits, It's Pushed on Heterosexuals and Children Are Next," *The Truth Barrier*, Jul. 22, 2023, https://celiafarber.substack.com/p/pre -exposure-hiv-prophylaxis-prep.

241 Michelle Llamas, "Truvada Lawsuits," *Drugwatch*, Jun. 13, 2023, https: //www.drugwatch.com/tenofovir-disoproxil-fumarate/lawsuits.

242 Trent Straube, "Billion-Dollar HIV PrEP Patent Lawsuit—U.S. v. Gilead—Goes to Trial," *POZ*, May 2, 2023, https://www.poz.com/article /billiondollar-hiv-prep-patent-lawsuitus-v-gilead-goes-trial.

243 ViiV Healthcare, "ViiV Healthcare Announces US FDA Approval of Apretude (Cabotegravir Extended-Release Injectable Suspension), the First and Only Long-Acting Injectable Option for HIV Prevention," press release, Dec. 21, 2021, https://viivhealthcare.com/en-us/media-center/news /press-releases/2021/december/viiv-healthcare-announces-fda-approval-of -apretude.

244 Celia Farber, "Serfs of Globalism: In the Age of Corona, What Dies First Is the Inalienable Right to Not Worry," *UncoverDC*, Mar. 5, 2020, https: //www.uncoverdc.com/2020/03/05/serfs-of-globalism-in-the-age-of -corona-what-dies-first-is-the-inalienable-right-to-not-worry.

245 Robert F. Kennedy Jr., "Why I Wrote 'The Real Anthony Fauci,'" *The Defender*, Dec. 13, 2021, https://childrenshealthdefense.org/defender/jfk-jr -why-i-wrote-the-real-anthony-fauci.

246 Celia Farber, "My Most Important Realization about What Made Fauci's Reign of Terror Possible: A New Language that Eclipsed the Scientific Tradition," *The Truth Barrier,* Mar. 5, 2023, https://celiafarber.substack.com/p/my-most-important-realization-about.

247 Jon Rappoport, "A Totalitarian Society Has Totalitarian Science," *Jon Rappoport's Blog,* Aug. 23, 2017, https://blog.nomorefakenews.com/2017/08/23/a-totalitarian-society-has-totalitarian-science-2.

248 Ibid.

249 Jon Rappoport, "The New COVID Squeeze Play, Hustle, Con; It's a Variation on the Old One, All Dressed up With Nowhere to Go—Except Fascist Tyranny," *Jon Rappoport's Blog,* Jul. 27, 2021, https://blog.nomorefakenews.com/2021/07/27/the-new-covid-squeeze-play-hustle-con.

250 Jon Rappoport, "Lockdown Civilization: Phase One and Phase Two," *Jon Rappoport's Blog,* Jan. 6, 2021, https://blog.nomorefakenews.com/2021/01/06/lockdown-civilization-phase-one-and-phase-two.

251 Brenda Baletti, PhD, "The 15-Minute City: A Climate Solution? Or Just an 'Excuse for More Control'?" *The Defender,* Feb. 8, 2023, https://childrenshealthdefense.org/defender/15-minute-city-climate-solution-control.

252 Patrick Wood, "Day 3: Technocracy in Europe and America," *Technocracy News & Trends,* Dec. 13, 2022, https://www.technocracy.news/day-3-technocracy-in-europe-and-america.

253 Twila Brase, "How Technocrats Are Taking Over the Practice of Medicine: A Wake-Up Call to the American People," *Citizens' Council on Health Care,* Jan. 2005, https://web.archive.org/web/20221212034425/https://www.cchfreedom.org/pdfreport.

254 John Laidler, "High Tech Is Watching You," *The Harvard Gazette,* Mar. 4, 2019, https://news.harvard.edu/gazette/story/2019/03/harvard-professor-says-surveillance-capitalism-is-undermining-democracy.

255 Catherine Powell and Alexandra Dent, "Data Is the New Gold, but May Threaten Democracy and Dignity," *Women Around the World* (blog), *Council on Foreign Relations,* Jan. 5, 2023, https://www.cfr.org/blog/data-new-gold-may-threaten-democracy-and-dignity-0.

256 John M. Travaline, "The Moral Dangers of Technocratic Medicine," *Linacre Quarterly* 86, no. 2–3 (2019): 231–238, doi: 10.1177/0024363919858463.

257 "Search Results for Technocracy Medicine," *Technocracy News & Trends,* https://www.technocracy.news/?s=technocracy+medicine.

258 John M. Travaline, "The Moral Dangers of Technocratic Medicine," *Linacre Quarterly* 86, no. 2–3 (2019): 231–238, doi: 10.1177/0024363919858463.

259 Brownstone Institute, "'Making a Killing': How Hospitals Profited from Deadly COVID Protocols," *The Defender,* May 19, 2023, https://childrenshealthdefense.org/defender/making-a-killing-hospitals-profit-deadly-covid-protocols.

260 Michael Nevradakis, PhD, "Exclusive: Dad Describes Hospital's COVID 'Protocols' He Believes Killed His 19-Year-Old Daughter," *The Defender*, Aug. 10, 2023, https://childrenshealthdefense.org/defender /grace-schara-covid-protocol-death.

261 John M. Travaline, "The Moral Dangers of Technocratic Medicine," *Linacre Quarterly* 86, no. 2–3 (2019): 231–238, doi: 10.1177/0024363919858463.

262 Matthew Cole, "What's Wrong with Technocracy?" *Boston Review*, Aug. 22, 2022, https://www.bostonreview.net/articles/whats-wrong-with-technocracy.

263 Children's Health Defense Team, "CHD Article on Big-Picture Look at Current Pandemic Beneficiaries Accepted by Peer-Reviewed Journal," *The Defender*, Dec. 14, 2020, https://childrenshealthdefense.org/defender /pandemic-beneficiaries-technocrats.

264 Karen Hunt, "The Forced Medication of All Citizens," *Off-Guardian*, Mar. 10, 2023, https://off-guardian.org/2023/03/10/the-forced-medication -of-all-citizens.

265 Tina E. Thomas et al., "Race, History of Abuse, and Homelessness Are Associated with Forced Medication Administration during Psychiatric Inpatient Care," *Journal of Psychiatric Practice* 26, no. 4 (2020): 294–304, doi: 10.1097/PRA.0000000000000485.

266 Simone Chérie, "The Trauma of Involuntary Treatment: Temporary Symptoms, Long-Term Suffering," *Medium*, Aug. 30, 2018, https://medium .com/antiparty/recovering-from-the-trauma-of-treatment-5f972a42c21d.

267 Brenda Baletti, PhD, "Appeal Planned after New York Supreme Court Reinstates 'Quarantine Camp' Regulation," *The Defender*, Nov. 22, 2023, https: //childrenshealthdefense.org/defender/kathleen-hochul-ny-quarantine -camp-regulation-reinstated.

268 Jon Rappoport, "Another Article Too Hot to Handle; Even Vaccine Critics Won't Run with it," *Jon Rappoport's Blog*, Feb. 23, 2022, https://blog.nomorefakenews.com/2022/02/23/another-article-too-hot -to-handle-even-vaccine-critics-wont-run-with-it.

269 Jon Rappoport, "A Cautionary Note about Most Medical Doctors," *Jon Rappoport*, Mar. 8, 2023, https://jonrappoport.substack.com/p/cautionary-note -about-most-doctors-and-vaccines.

270 Jon Rappoport, "Rejecting Rockefeller Germ Theory Once and for All," *Jon Rappoport's Blog*, Jul. 27, 2022, https://blog.nomorefakenews .com/2022/07/27/rejecting-rockefeller-germ-theory-once-and-for-all -compelling-evidence.

271 "Due Diligence and Art," *Substack*, https://sashalatypova.substack.com.

272 "Bailiwick News," *Substack*, https://bailiwicknews.substack.com.

273 Sasha Latypova, "Clarification of My Message on the Global Mass Murder Campaign," *Due Diligence and Art*, Mar. 9, 2023, https://sashalatypova .substack.com/p/clarification-of-my-message-on-the.

274 Katherine Watt, "Construction of the Kill Box: Legal History," *Bailiwick News*, May 4, 2023, https://bailiwicknews.substack.com/p/construction-of -the-kill-box-legal.

275 Katherine Watt, "PDF Compilations," *Bailiwick News*, Oct. 11, 2023, https://bailiwicknews.substack.com/p/pdf-compilations.

276 R. J. Rummel, "Democide Versus Genocide: Which Is What?" University of Hawaii, May, 1998, https://www.hawaii.edu/powerkills/GENOCIDE. HTM.

277 "Declarations of a Public Health Emergency," Administration for Strategic Preparedness & Response, accessed Jun. 26, 2024, https://aspr.hhs.gov/legal /PHE/Pages/default.aspx.

278 Team Enigma, "Intent to Harm – Evidence of the Conspiracy to Commit Mass Murder by the US DOD, HHS, Pharma Cartel," Bitchute, Dec. 2, 2022, https://www.bitchute.com/video/8ftbShzrkjl9.

279 Robert Johnson, "COVID-19 Vaccine Development Portfolio," Vaccines and Related Biological Products Advisory Committee, Oct. 22, 2020, slide 11, https://www.fda.gov/media/143560/download.

280 "BARDA COVID-19 Response," U.S. Department of Health and Human Services, accessed Jun. 27, 2024, https://medicalcountermeasures.gov/barda /barda-covid-19-response.

281 "BARDA COVID-19 Response," U.S. Department of Health and Human Services, accessed Jun. 27, 2024, https://medicalcountermeasures.gov/ barda/barda-covid-19-response.

282 Sasha Latypova, "Summary of Everything and Quick Links," *Due Diligence and Art*, Jun. 26, 2023, https://sashalatypova.substack.com/p/summary -of-everything-and-quick-links.

283 Daniel Schoeni, "Public Procurement and Disruptive Technologies: DARPA's Role in the Development of mRNA Vaccines," *Revista de la Escuela Jacobea de Posgrado* 24 (2023): 1–16, https://papers.ssrn.com/sol3 /papers.cfm?abstract_id=4495566.

284 Sasha Latypova, "The Role of the US DoD (and Their Co-Investors) in "covid countermeasures" Enterprise," *Due Diligence and Art*, Dec. 28, 2022, https://sashalatypova.substack.com/p/the-role-of-the-us-dod-and-their.

285 Robert F. Kennedy Jr., "Militarized Healthcare with Sasha Latypova," Mar. 15, 2023, in *RFK Jr. Podcast*, podcast, produced by David Whiteside, 00:59:23, https://podcasters.spotify.com/pod/show/rfkjr/episodes/Militarized -Healthcare-with-Sasha-Latypova-e20go74.

286 "COVID-19 Public Health Emergency (PHE)," U.S. Department of Health and Human Services, last updated Dec. 15, 2023, https://www.hhs .gov/coronavirus/covid-19-public-health-emergency/index.html.

287 Sasha Latypova, "Summary of Everything and Quick Links," *Due Diligence and Art*, Jun. 26, 2023, https://sashalatypova.substack.com/p /summary-of-everything-and-quick-links.

288 Jekielek J, Minick J. Pharma insider speaks out about vaccine batches. *The Epoch Times*, Jul. 20, 2023 (updated Aug. 10, 2023). https://www.theepoch times.com/us/pharma-insider-speaks-out-about-vaccine-batches-5371394.

289 Sanford Levinson, "What Is the Constitution's Role in Wartime?: Why Free Speech and Other Rights Are Not as Safe as You Might Think," *FindLaw*, Oct. 17, 2001, https://supreme.findlaw.com/legal-commentary/what-is -the-constitutions-role-in-wartime.html

290 Katherine Watt, "Bailiwick News: About," *Bailiwick News,* last updated Jan. 17, 2024, https://bailiwicknews.substack.com/about.

291 Nicholas Florko, "New Document Reveals Scope and Structure of Operation Warp Speed and Underscores Vast Military Involvement," *STAT*, Sep. 28, 2020, https://www.statnews.com/2020/09/28/operation-warp-speed -vast-military-involvement.

292 Sasha Latypova, "Summary of Everything and Quick Links," *Due Diligence and Art*, Jun. 26, 2023, https://sashalatypova.substack.com/p /summary-of-everything-and-quick-links.

293 Katherine Watt, "Construction of the Kill Box: Legal History," *Bailiwick News*, May 4, 2023, https://bailiwicknews.substack.com/p/construction-of-the -kill-box-legal.

294 Katherine Watt, "Biomedical Security State and State-Run Bioterrorism Programs: Six American Statutory Frameworks," *Bailiwick News*, Dec. 19, 2022, https://bailiwicknews.substack.com/p/biomedical-security-state-and-state.

295 "50 U.S. Code Chapter 32 - Chemical And Biological Warfare Program," Cornell Law School, accessed Jun. 27, 2024, https://www.law.cornell.edu /uscode/text/50/chapter-32.

296 "42 U.S. Code § 247d - Public Health Emergencies," Cornell Law School, accessed Jun. 27, 2024, https://www.law.cornell.edu/uscode/text/42/247d.

297 42 USC 300aa-1: Establishment," Office of the Law Revision Council, United States Code, accessed Jun. 27, 2024, https://uscode.house.gov /view.xhtml?edition=prelim&req=granuleid%3AUSC-prelim-title42 -section300aa-1&num=0.

298 Children's Health Defense Team, "NCVIA: The Legislation That Changed Everything—Conflicts of Interest Undermine Children's Health: Part II," Children's Health Defense, May 16, 2019, https://childrenshealthdefense .org/news/ncvia-the-legislation-that-changed-everything-conflicts-of -interest-undermine-childrens-health-part-ii.

299 Children's Health Defense, "CHD Responds to First Payouts from the Countermeasures Injury Compensation Program to 3 Individuals Injured by COVID Vaccines," press release, Apr. 14, 2023, https:

//childrenshealthdefense.org/press-release/chd-responds-to-first-payouts
-from-the-countermeasures-injury-compensation-program-to-3-individuals
-injured-by-covid-vaccines.

300 "21 U.S. Code § 360bbb - Expanded Access to Unapproved Therapies and Diagnostics," Cornell Law School, accessed Jun. 27, 2024, https://www.law .cornell.edu/uscode/text/21/360bbb.

301 Institute of Medicine (US) Forum on Medical and Public Health Preparedness for Catastrophic Events, *Emergency Use Authorization and the Postal Model, Workshop Summary* (Washington, DC: National Academies Press, 2010), https://www.ncbi.nlm.nih.gov/books/NBK53122.

302 "42 U.S. Code Subchapter XXVI - National All-Hazards Preparedness For Public Health Emergencies," Cornell Law School, accessed Jun. 27, 2024, https://www.law.cornell.edu/uscode/text/42/chapter-6A/subchapter-XXVI.

303 "§ 4021. Research Projects: Transactions Other than Contracts and Grants," *GovRegs*, accessed Jun. 27, 2024, https://www.govregs.com/uscode /expand/title10_subtitleA_partV_subpartE_chapter301_subchapterII _section4021#uscode_0.

304 D. Ridgway, "No-Fault Vaccine Insurance: Lessons from the National Vaccine Injury Compensation Program," *Journal of Health Politics, Policy and Law* 24, no. 1 (1999): 59–90, doi: 10.1215/03616878-24-1-59.

305 Robert F. Kennedy Jr., "Vaccines and the Liberal Mind," Children's Health Defense, Jun. 14, 2018, https://childrenshealthdefense.org/news /vaccines-and-the-liberal-mind.

306 Nora Freeman Engstrom, "A Dose of Reality for Specialized Courts: Lessons from the VICP," *University of Pennsylvania Law Review* 163 (2015): 1631–1717, https://scholarship.law.upenn.edu/cgi/viewcontent.cgi ?article=9485&context=penn_law_review.

307 Children's Health Defense, *Conflicts of Interest: Undermine Children's Health* (CHD Publishing, May 2019), https://childrenshealthdefense.org /ebook-sign-up/ebook-sign-up-conflicts-of-interest.

308 "About the National Vaccine Injury Compensation Program," Health Resources & Services Administration, last reviewed Jun. 2024, https://www .hrsa.gov/vaccine-compensation/about.

309 Nora Freeman Engstrom, "A Dose of Reality for Specialized Courts: Lessons from the VICP," *University of Pennsylvania Law Review* 163 (2015): 1631–1717, https://scholarship.law.upenn.edu/cgi/viewcontent .cgi?article=9485&context=penn_law_review.

310 *Good Morning CHD*, "Lessons of History: VICP Explained," aired Sep. 6, 2023, on CHD.TV, https://live.childrenshealthdefense.org/chd-tv/shows /good-morning-chd/lessons-of-history-vicp-explained.

311 *The National Vaccine Injury Program: Is it Working as Congress Intended?: Hearings Before the Committee on Government Reform*, 107th Cong. (2002)

(statement of the Hon. Dave Weldon) https://www.govinfo.gov/content /pkg/CHRG-107hhrg77527/html/CHRG-107hhrg77527.htm.

312 "Who Can File a Petition," Health Resources & Services Administration, last reviewed Jun. 2024, https://www.hrsa.gov/vaccine-compensation /eligible.

313 Nora Freeman Engstrom, "A Dose of Reality for Specialized Courts: Lessons from the VICP," *University of Pennsylvania Law Review* 163 (2015): 1631–1717, https://scholarship.law.upenn.edu/cgi/viewcontent.cgi?article =9485&context=penn_law_review.

314 Molly Treadway Johnson et al., *Use of Expert Testimony, Specialized Decision Makers, and Case-Management Innovations in the National Vaccine Injury Compensation Program* (Federal Judicial Center, 1998), 38, https://books .google.com/books/about/Use_of_Expert_Testimony_Specialized_Deci .html?id=iKmRAAAAMAAJ.

315 Nora Freeman Engstrom, "A Dose of Reality for Specialized Courts: Lessons from the VICP," *University of Pennsylvania Law Review* 163 (2015): 1631–1717, https://scholarship.law.upenn.edu/cgi/viewcontent .cgi?article=9485&context=penn_law_review.

316 Ibid.

317 Children's Health Defense, "CHD Responds to First Payouts from the Countermeasures Injury Compensation Program to 3 Individuals Injured by COVID Vaccines," press release, Apr. 14, 2023, https://childrenshealth defense.org/press-release/chd-responds-to-first-payouts-from-the-counter measures-injury-compensation-program-to-3-individuals-injured-by -covid-vaccines.

318 Ibid.

319 Katherine Watt, "USA v. Dr. Kirk Moore et al," *Bailiwick News*, Aug. 8, 2023, https://bailiwicknews.substack.com/p/usa-v-dr-kirk-moore-et-al.

320 Caitlin Dickerson, "Veterans Used in Secret Experiments Sue Military for Answers," NPR, Sep. 5, 2015, https://www.npr.org/2015/09/05/437555125 /veterans-used-in-secret-experiments-sue-military-for-answers.

321 "Door to Freedom," accessed Jun. 27, 2024, https://doortofreedom.org.

322 Meryl Nass, MD, "The Anthrax Vaccine Program: An Analysis of the CDC's Recommendations for Vaccine Use," *American Journal of Public Health* 92, no. 5 (2002): 715–21, doi: 10.2105/ajph.92.5.715.

323 Children's Health Defense Team, "Military's COVID Vaccine Mandate Threatens National Security, Erodes Morale," *The Defender*, May 10, 2022, https://childrenshealthdefense.org/defender/military-covid-vaccine-mandate -threatens-national-security.

324 Katherine Watt, "On the Continuing Effort to Fit a Square Peg (Legalized Manufacturing and Use of Biological Weapons) into a Round Hole (FDA Drug, Device and Biological Product Regulation),"

Bailiwick News, Jan. 3, 2024, https://bailiwicknews.substack.com/p/on-the-continuing-effort-to-fit-a.

325 Katherine Watt, "USA v. Dr. Kirk Moore et al," *Bailiwick News*, Aug. 8, 2023, https://bailiwicknews.substack.com/p/usa-v-dr-kirk-moore-et-al.

326 "Emergency Use Authorization," U.S. Food & Drug Administration, last updated May 21, 2024, https://www.fda.gov/emergency-preparedness-and-response/mcm-legal-regulatory-and-policy-framework/emergency-use-authorization.

327 Children's Health Defense Team, "Two-Tiered Medicine: Why Is Hydroxychloroquine Being Censored and Politicized?" Children's Health Defense, Jul. 30, 2020, https://childrenshealthdefense.org/news/two-tiered-medicine-why-is-hydroxychloroquine-being-censored-and-politicized.

328 Dr. Joseph Mercola, "Ivermectin Could Have Saved 'Millions' of Lives—but Doctors Were Told Not to Use It," *The Defender*, Jun. 17, 2021, https://childrenshealthdefense.org/defender/covid-ivermectin-could-have-saved-millions-lives.

329 Michael Nevradakis, PhD, "'Malicious Prosecution': Lawyers for Dr. Meryl Nass Allege Maine Medical Board Violated Nass' First Amendment Rights," *The Defender*, Jan. 22, 2024, https://childrenshealthdefense.org/defender/meryl-nass-maine-medical-board-violation-first-amendment-rights.

330 Sasha Latypova, "Summary of Everything and Quick Links," *Due Diligence and Art*, Jun. 26, 2023, https://sashalatypova.substack.com/p/summary-of-everything-and-quick-links.

331 "Safety of COVID-19 Vaccines," Centers for Disease Control and Prevention, last updated Nov. 3, 2023, https://www.cdc.gov/coronavirus/2019-ncov/vaccines/safety/safety-of-vaccines.html.

332 Brenda Baletti, PhD, "Government Contracts with COVID Vaccine Makers Let Federal Agencies Bypass Normal Regulatory Process, FOIA Documents Show," *The Defender*, Jun. 30, 2023, https://childrenshealthdefense.org/defender/hhs-barda-covid-vaccine-makers-bypass-regulation.

333 Sasha Latypova, "Reviewing the DOD Contracts for Covid 'Countermeasures,'" *Due Diligence and Art*, Jan. 13, 2023, https://sashalatypova.substack.com/p/reviewing-the-dod-contracts-for-covid.

334 Katherine Watt, "mRNA-LNP Compounds Are Cellular Genetic Dirty Bombs," *Bailiwick News*, Apr. 30, 2023, https://bailiwicknews.substack.com/p/mrna-lnp-compounds-are-cellular-genetic.

335 Suzanne Burdick, PhD, "45 Times as Many Deaths after COVID Shots in Just 2 Years Compared with All Flu Vaccine-Related Deaths since 1990, Data Show," *The Defender*, Apr. 14, 2023, https://childrenshealthdefense.org/defender/deaths-covid-shots-versus-flu-vaccines-vaers-dmed.

336 Michael Nevradakis, PhD, "Groundbreaking Analysis: COVID Vaccines Caused 300,000 Excess Deaths, $147 Billion in Damage to Economy in

2022 Alone," *The Defender*, Mar. 29, 2023, https://childrenshealthdefense .org/defender/covid-vaccine-injury-deaths-economic-damage.

337   Gavin de Becker, "Seeing Is Believing: What the Data Reveal about Deaths Following COVID Vaccine Rollouts around the World," *The Defender*, Jan. 9, 2023, https://childrenshealthdefense.org/defender/covid -vaccine-deaths-cause-unknown.

338   Katherine Watt, "Vaccine Production Facilities Are Indistinguishable from Bioweapon Production Facilities, and Vaccines Are Indistinguishable from Bioweapons," *Bailiwick News*, Apr. 13, 2023, https://bailiwicknews.substack .com/p/vaccine-production-facilities-are.

339   Katherine Watt, "93 Biochemical Weapons to Decline Whenever a Medical Mercenary Offers Them to You or Your Children," *Bailiwick News*, May 26, 2023, https://bailiwicknews.substack.com/p/93-biochemical-weapons-to -decline.

340   Katherine Watt, "Other Researchers Who Have Compiled Evidence that US Military-Public Health-Vaccination Programs Injure and Kill People," *Bailiwick News*, Apr. 9, 2024, https://bailiwicknews.substack.com/p/other -researchers-who-have-compiled.

341   Katherine Watt, "Legalized FDA Non-Regulation of Biological Products Effective May 2, 2019, by Federal Register Final Rule, Signed by Then-FDA Commissioner Scott Gottlieb," *Bailiwick News*, Dec. 19, 2023, https://baili wicknews.substack.com/p/legalized-fda-non-regulation-of-biological.

342   Ibid.

Chapter 3

1   Alexander Wilder, MD, *The Fallacy of Vaccination* (New York: The Metaphysical Publishing Company; 1899).

2   "Can Pertussis Vaccine Cause Injury & Death?" *National Vaccine Information Center*, accessed Jun. 27, 2024, https://www.nvic.org/disease-vaccine/pertussis /vaccine-injury.

3   Children's Health Defense Team, "The Changing Face of Vaccinology," Children's Health Defense, Apr. 3, 2018, https://childrenshealthdefense .org/news/the-changing-face-of-vaccinology.

4   Christopher Exley, MD, "Aluminium Adjuvants and Vaccines: A Marriage Everlasting," *Dr's Newsletter*, Jul. 11, 2023, https://drchristopherexley.substack .com/p/aluminium-adjuvants-and-vaccines.

5   Lucija Tomljenovic et al., "Significant Under-Reporting of Quadrivalent Human Papillomavirus Vaccine-Associated Serious Adverse Events in the United States: Time for Change?" *Science, Public Health Policy, and The Law* 2 (2021): 37–58, https://cf5e727d-d029fe2d3ad957f.filesusr.com/ugd/adf 864_2dede593f4a04e64ab6c0c45bc14d450.pdf.

6   National Vaccine Injury Compensation Program, "Data & Statistics," Health Resources & Services Administration, last updated Jun. 1, 2024, https://www.hrsa.gov/vaccine-compensation/data.

7   "Shocking Vaccine Admissions from Inside the W.H.O.," *The Highwire*, accessed Jun. 27, 00:31:53, https://thehighwire.com/ark-videos/shocking -vaccine-admissions-from-inside-the-w-h-o.

8   Children's Health Defense Team, "Look WHO's Talking! Vaccine Scientists Confirm Major Safety Problems," Children's Health Defense, Jan. 16, 2020, https://childrenshealthdefense.org/news/look-whos-talking-vaccine-scientists -confirm-major-safety-problems.

9   Heidi J. Larson et al., "The Vaccine-Hesitant Moment," *New England Journal of Medicine* 387, no. 1 (2022): 58–65, doi: 10.1056/NEJMra2106441.

10   Rav Arora, "Editor at 'Pro-Vaccine Publication' Experienced Serious Adverse Event after Second Pfizer Shot," *Brownstone Journal*, Sep. 12, 2023, https://brownstone.org/articles/editor-at-pro-vaccine-publication-serious -adverse-event.

11   Suzanne Burdick, PhD, "Court Again Blocks COVID Vaccine Mandate for Federal Workers," *The Defender*, Jun. 28, 2022, https://childrenshealth defense.org/defender/biden-administration-federal-covid-vaccine-mandate.

12   Children's Health Defense Team, "COVID: Spearpoint for Rolling Out a 'New Era' of High-Risk, Genetically Engineered Vaccines," Children's Health Defense, May 7, 2020, https://childrenshealthdefense.org/news /vaccine-safety/covid-19-the-spearpoint-for-rolling-out-a-new-era-of-high -risk-genetically-engineered-vaccines.

13   Ibrahim Khan, Khalid Saeed, and Idrees Khan, "Nanoparticles: Properties, Applications and Toxicities," *Arabian Journal of Chemistry* 12, no. 7 (2019): 908–931, doi: 10.1016/j.arabjc.2017.05.011.

14   Rob Verkerk, PhD, "COVID Vaccines: What's Really in Them?" *The Defender*, Dec. 9, 2021, https://childrenshealthdefense.org/defender/covid -vaccines-composition-proper-informed-consent.

15   Julie Sladden and Julian Gillespie, "The Vax-Gene Files: Have the Regulators Approved a Trojan Horse?" *Brownstone Journal*, Aug. 16, 2023, https: //brownstone.org/articles/the-vax-gene-files-have-the-regulators-approved -a-trojan-horse.

16   "COVID-19," Defense Advanced Research Projects Agency, updated Mar. 19, 2021, https://www.darpa.mil/work-with-us/covid-19.

17   Javier T. Granados-Riveron and Guillermo Aquino-Jarquin, "Engineering of the Current Nucleoside-Modified mRNA-LNP Vaccines Against SARS-CoV-2," *Biomed Pharmacotherapy* 142 (2021): 111953, doi: 10.1016/j. biopha.2021.111953.

18   Daniel Schoeni, "Public Procurement and Disruptive Technologies: DARPA's Role in the Development of mRNA Vaccines," *Revista de la Escuela*

Jacobea de Posgrado 24 (2023): 1–16, https://papers.ssrn.com/sol3/papers
.cfm?abstract_id=4495566.

19   Ulrike Granögger, "Special Report: Future Science Series: Protein Design
by Directed Evolution in modRNA Injections with Dr. Sabine Stebel,"
Solari Report, Aug. 30, 2023, https://home.solari.com/special-report-future
-science-series-protein-design-by-directed-evolution-in-modrna-injections
-with-sabine-stebel.

20   Jon Rappoport, "The COVID Vaccine and the Commercial Conquest of
the Planet: The Plan," *Jon Rappoport's Blog*, Dec. 15, 2020, https://blog
.nomorefakenews.com/2020/12/15/the-covid-vaccine-and-the-commercial
-conquest-of-the-planet-the-plan.

21   U.S. Department of Defense, "DoD Awards $1.74 Billion Agreement to
Moderna, Inc. to Secure over 65 Million Doses of COVID-19 Vaccine for
Fall Vaccinations," news release, Jul. 29, 2022, https://www.defense.gov
/News/Releases/Release/Article/3109705/dod-awards-174-billion-agreement
-to-moderna-inc-to-secure-over-65-million-doses.

22   Dr. Joseph Mercola, "COVID Vaccines Are Gene Therapy—but Big
Pharma and Big Media Don't Want You to Know That," *The Defender*, Jan.
20, 2023, https://childrenshealthdefense.org/defender/covid-vaccines-gene
-therapy-cola.

23   Jon Rappoport, "The COVID Vaccine and the Commercial Conquest of
the Planet: The Plan," *Jon Rappoport's Blog*, Dec. 15, 2020, https://blog
.nomorefakenews.com/2020/12/15/the-covid-vaccine-and-the-commercial
-conquest-of-the-planet-the-plan.

24   Annalee Armstrong, "Moderna's First Seasonal Flu Vaccine Slides into
Clinic as Pharma Giants Crowd into mRNA," *Fierce Biotech*, Jul. 7,
2021, https://www.fiercebiotech.com/biotech/moderna-s-first-flu-vaccine
-candidate-slides-into-clinic-as-pharma-giants-crowd-into-mrna.

25   Roberto Marmolani, "Opening Ceremony World Health Summit 2021,
Speech Stefan Oelrich," YouTube, Nov. 16, 2021, https://www.youtube
.com/watch?v=IKBmVwuv0Qc&ab_channel=RobertoMarmolani.

26   Dr. Joseph Mercola, "COVID Vaccines Are Gene Therapy, Contrary to
Claims by Media 'Fact-Checkers,'" *The Defender*, Aug. 22, 2023,
https://childrenshealthdefense.org/defender/covid-vaccines-gene-therapy
-media-fact-checkers-cola.

27   "750+ Studies about the Dangers of the COVID-19 Injections,"
Doctors for COVID Ethics, Jul. 4, 2022, https://doctors4covidethics.org
/750-studies-about-the-dangers-of-the-covid-19-injections.

28   Taylor Hudak, "mRNA a Serious Threat to Mankind," *Doctors for COVID
Ethics*, Jun. 25, 2022, https://doctors4covidethics.org/mrna-a-serious-threat
-to-mankind.

29 "Top 10 Medical Innovations for 2022 Unveiled," *Cleveland Clinic*, Feb. 16, 2022, https://consultqd.clevelandclinic.org/top-10-medical-innovations -for-2022-unveiled.

30 Edward Winstead, "Can mRNA Vaccines Help Treat Cancer?" *Cancer Currents Blog*, National Cancer Institute, Jan. 20, 2022, https://www.cancer .gov/news-events/cancer-currents-blog/2022/mrna-vaccines-to-treat-cancer.

31 Meg Tirrell, "Moderna Says mRNA Flu Shot Generates Better Immune Response in Study than Currently Available Vaccine," CNN, Sep. 13, 2023, https://www.cnn.com/2023/09/13/health/moderna-mrna-flu-shot/index .html.

32 Hong You et al., "The mRNA Vaccine Technology Era and the Future Control of Parasitic Infections," *Clinical Microbiology Reviews* 36, no. 1 (2023): e0024121, doi: 10.1128/cmr.00241-21.

33 "Penn Medicine Develops mRNA Vaccine against Lyme Disease-Causing Bacteria," *Penn Medicine News*, Sep. 19, 2023, https://www.pennmedicine .org/news/news-releases/2023/september/penn-medicine-develops-mrna -vaccine-against-lyme-disease.

34 Brandon Essink, MD, et al., "The Safety and Immunogenicity of Two Zika Virus mRNA Vaccine Candidates in Healthy Flavivirus Baseline Seropositive and Seronegative Adults: The Results of Two Randomised, Placebo-Controlled, Dose-Ranging, Phase 1 Clinical Trials," *Lancet Infectious Diseases* 23, no. 5 (2023): 621–633, doi: 10.1016/S1473–3099(22)00764–2.

35 "Research," Moderna, accessed Jun. 27, 2024, https://www.modernatx .com/research/product-pipeline.

36 @MakisMD, "NEW ARTICLE: Families that inject mRNA together, develop turbo cancer together?" Twitter (now X) Aug. 9, 2023, https://twitter.com /MakisMD/status/1689215783988244481.

37 Justin Hart, "Ignoring the Heart of the Matter: How Myocarditis Became the Silent Scandal of COVID-19 Vaccination," *Rational Ground by Justin Hart*, Sep. 20, 2023, https://covidreason.substack.com/p/ignoring-the-heart -of-the-matter.

38 Angus Liu, "Merck, Moderna Unveil Phase 3 Trial Details for Closely Watched mRNA Cancer Vaccine," *Fierce Biotech*, Jul. 10, 2023, https: //www.fiercebiotech.com/biotech/merck-moderna-post-phase-3-trial-closely -watched-mrna-cancer-vaccine.

39 Edward Winstead, "Can mRNA Vaccines Help Treat Cancer?" *Cancer Currents Blog*, National Cancer Institute, Jan. 20, 2022, https://www.cancer .gov/news-events/cancer-currents-blog/2022/mrna-vaccines-to-treat-cancer.

40 Jonathan Wosen, "mRNA Revolutionized the Race for a Covid-19 Vaccine. Could Cancer Be Next?" *STAT*, Nov. 21, 2022, https://www .statnews.com/2022/11/21/mrna-revolutionized-the-race-for-a-covid-19 -vaccine-could-cancer-be-next.

41   B. David Zarley, "RNA Breakthrough Offers a Potential Heart Attack Cure," *Freethink*, May 1, 2022, https://www.freethink.com/health/mrna -heart-attack-cure.

42   Jeff Hansen, "A Modified mRNA Aids Heart Attack Recovery in Mouse and Pig Models," *UAB News*, Sep. 11, 2023, https://www.uab.edu/news /research/item/13763-a-modified-mrna-aids-heart-attack-recovery-in -mouse-and-pig-models.

43   Amy Kelly, "46 Pages FOIAed Emails between CDC leaders, Dr. Fauci, Dr. Collins, and White House, NIH, HHS, Show They Knew about Vaccine-Induced Myocarditis and Thrombotic Thrombocytopenia, a Blood Clotting Disorder. Emails Over 80% Redacted," *DailyClout*, Sep. 20, 2023, https://dailyclout.io/46-pages-foiaed-from-cdc-leaders-2021-reveal -fauci-collins-white-house-nih-hhs.

44   Megan Redshaw, "More than 1.3 Million Adverse Events Following COVID Vaccines Reported to VAERS, CDC Data Show," *The Defender*, Jul. 8, 2022, https://childrenshealthdefense.org/defender/1–3-million-adverse-events -covid-vaccines-vaers-cdc-data-show.

45   Michael Nevradakis, PhD, "'Criminal': Confidential EU Documents Reveal Thousands of Deaths from Pfizer-BioNTech Shots," *The Defender*, Jun. 23, 2023, https://childrenshealthdefense.org/defender/confidential-eu-documents -deaths-pfizer-biontech-shots.

46   Dongwon Yoon, "A Nationwide Survey of mRNA COVID-19 Vaccine's Experiences on Adverse Events and Its Associated Factors," *Journal of Korean Medical Science* 38, no. 22 (2023): e170, doi: 10.3346/jkms.2023.38.e170.

47   Alexis L. Beatty, "Analysis of COVID-19 Vaccine Type and Adverse Effects Following Vaccination," *JAMA Network Open* 4, no. 12 (2021): e2140364, doi: 10.1001/jamanetworkopen.2021.40364.

48   "Investigations," *DailyClout*, accessed Jun. 27, 2024, https://dailyclout.io /category/pfizer-reports.

49   Naomi Wolf, "Foreword to the Amazon Kindle Version of the War Room/ DailyClout Pfizer Documents Analysis Volunteers' Reports," *DailyClout*, Jan. 21, 2023, https://dailyclout.io/foreword-to-the-amazon-kindle-version -of-the-war-room-dailyclout-pfizer-documents-analysis-reports.

50   Michael Palmer, MD et al., *mRNA Vaccine Toxicity* (Doctors for COVID Ethics, 2023), 133–134, https://doctors4covidethics.org/mrna-vaccine-toxicity.

51   Takehiro Nakahara et al., "Assessment of Myocardial 18F-FDG Uptake at PET/CT in Asymptomatic SARS-CoV-2-Vaccinated and Nonvaccinated Patients," *Radiology* 308, no. 3 (2023): e230743, doi: 10.1148/radiol. 230743.

52   Daniel Schroth et al., "Predictors of Persistent Symptoms after mRNA SARS-CoV-2 Vaccine-Related Myocarditis (Myovacc Registry)," *Frontiers in Cardiovascular Medicine* 10 (2023): 1204232, doi: 10.3389/ fcvm.2023.1204232.

53 John-Michael Dumais, "27% of Saudis in 'Bombshell' Study Experienced Heart Issues after mRNA Covid Shots," *The Defender*, Apr. 4, 2024, https://childrenshealthdefense.org/defender/mrna-covid-vaccine -cardiac-complications-saudi-arabia.

54 Mark Skidmore, "Retracted Article: The Role of Social Circle COVID-19 Illness and Vaccination Experiences in COVID-19 Vaccination Decisions: An Online Survey of the United States Population," *BMC Infectious Diseases* 23, no. 1 (2023): 51, doi: 10.1186/s12879–023–07998–3.

55 Mike Capuzzo, "A Disabled Vet Combed Obituaries for the Words "Suddenly" and "Unexpectedly"—Here's What He Found," *The Defender*, Jan. 4, 2024, https://childrenshealthdefense.org/defender/steve-connolly -obituaries-words-suddenly-unexpectedly-covid-vaccine.

56 John C. A. Manley, "It's Beginning to Look a Lot Like Democide," *Activist Post*, Dec. 21, 2023, https://www.activistpost.com/2023/12/its-beginning -to-look-a-lot-like-democide.html.

57 Dennis G. Rancourt et al., *COVID-19 Vaccine-Associated Mortality in the Southern Hemisphere* (Ottawa: CORRELATION Research in the Public Interest, 2023), https://correlation-canada.org/covid-19-vaccine-associated -mortality-in-the-southern-hemisphere.

58 *Covid-19 Vaccine Pharmacovigilance Report* (World Council for Health, 2022), https://worldcouncilforhealth.org/resources/covid-19-vaccine-pharma covigilance-report.

59 "Janssen COVID-19 vaccine," U.S. Food & Drug Administration, Jun. 2, 2023, https://www.fda.gov/vaccines-blood-biologics/coronavirus-covid-19 -cber-regulated-biologics/janssen-covid-19-vaccine.

60 Megan Redshaw, "'Plausible' Link between J&J Vaccine and Blood Clots, CDC Says after Confirming 28 Cases, Including 3 Deaths," *The Defender*, May 13, 2023, https://childrenshealthdefense.org/defender/link -johnson-johnson-vaccine-blood-clots-cdc-28-cases-3-deaths.

61 A Midwestern Doctor, "A Primer on Medical Gaslighting," *The Forgotten Side of Medicine*, Feb. 9, 2023, https://www.midwesterndoctor.com/p/a-primer -on-medical-gaslighting.

62 Mead et al., "Retracted: COVID-19 mRNA Vaccines: Lessons Learned from the Registrational Trials and Global Vaccination Campaign," *Cureus* 16, no. 1 (2024): e52876, doi: 10.7759/cureus.52876.

63 Brenda Baletti, PhD, "'Stunning Act of Scientific Censorship': Journal Retracts Peer-Reviewed Study Critiquing COVID-19 Vaccine," *The Defender*, Feb. 28, 2024, https://childrenshealthdefense.org/defender/cureus-retracts -study-critiquing-covid-19-vaccine-censorship.

64 Pascal Najadi and Dr. Astid Stuckelberger, "Cutting Off the Head of the Snake," Forbidden Knowledge, Oct. 22, 2023, 00:01:33–00:01:36, https://forbidden knowledgetv.net/cutting-off-the-head-of-the-snake-in-geneva-switzerland/.

65   Michael Palmer, MD et al., *mRNA Vaccine Toxicity* (Doctors for COVID Ethics, 2023), xv, 1, https://doctors4covidethics.org/mrna-vaccine-toxicity.

66   Albertsen et al., "The Role of Lipid Components in Lipid Nanoparticles for Vaccines and Gene Therapy," *Advanced Drug Delivery Reviews* 188 (2022): 114416, doi: 10.1016/j.addr.2022.114416.

67   Nathan Vardi, "Moderna's Mysterious Coronavirus Vaccine Delivery System," *Forbes*, Jul. 29, 2020, https://www.forbes.com/sites/nathanvardi /2020/07/29/modernas-mysterious-coronavirus-vaccine-delivery-system/?sh =45e3db5662d9.

68   A Mother's Anthem, "The Alarming LNP History You Haven't Been Shown—The LNNP Developer's Own Studies Dating Back 20 Years: Part 1 of 10, Introduction and Executive Brief Sent to Lawmakers & State Investigators This Summer," *A Mother's Anthem*, Sep. 7, 2023, https: //amothersanthem.substack.com/p/the-alarming-lnp-history-you-havent.

69   Nicholas A. C. Jackson et al., "The Promise of mRNA Vaccines: A Biotech and Industrial Perspective," *NPJ Vaccines* 5, no. 11 (2020), doi: 10.1038/ s41541–020–0159–8.

70   Javier T. Granados-Riveron and Guillermo Aquino-Jarquin, "Engineering of the Current Nucleoside-Modified mRNA-LNP Vaccines against SARS-CoV-2," *Biomed Pharmacotherapy* 142 (2021): 111953, doi: 10.1016/j. biopha.2021.111953.

71   Mohamed-Gabriel Alameh et al., "Lipid Nanoparticles Enhance the Efficacy of mRNA and Protein Subunit Vaccines by Inducing Robust T Follicular Helper Cell and Humoral Responses," *Immunity* 54, no. 12 (2021): 2877–2892.e7, doi: 10.1016/j.immuni.2021.11.001.

72   Seyen Moein Moghimi and Dmitri Simberg, "Pro-Inflammatory Concerns with Lipid Nanoparticles," *Molecular Therapy* 30, no. 6 (2022): 2109–2110, doi: 10.1016/j.ymthe.2022.04.011.

73   The Defender Staff, "Pharma Eyes Big Profits in Nasal Vaccines—Should We Really Spray Nanoparticles So Close to the Brain?" *The Defender,* Aug. 3, 2022, https://childrenshealthdefense.org/defender/pharma-profits -nasal-vaccines-nanoparticles-brain.

74   Alexandra Suberi, "Polymer Nanoparticles Deliver mRNA to the Lung for Mucosal Vaccination," *Science Translational Medicine* 15, no. 709 (2023), doi: 10.1126/scitranslmed.abq0603.

75   Joon Haeng Rhee, "Current and New Approaches for Mucosal Vaccine Delivery," *Mucosal Vaccines* (2020): 325–356, doi: 10.1016/B978–0–12–811924–2.00019–5.

76   Seyen Moein Moghimi and Dmitri Simberg, "Pro-Inflammatory Concerns with Lipid Nanoparticles," *Molecular Therapy* 30, no. 6 (2022): 2109–2110, doi: 10.1016/j.ymthe.2022.04.011.

77 A Mother's Anthem, "The Alarming LNP History You Haven't Been Shown—The LNNP Developer's Own Studies Dating Back 20 Years: Part 1 of 10, Introduction and Executive Brief Sent to Lawmakers & State Investigators This Summer," *A Mother's Anthem*, Sep. 7, 2023, https://amothersanthem.substack.com/p/the-alarming-lnp-history-you-havent.

78 Johan J. F. Verhoef and Thomas J. Anchordoquy, "Questioning the Use of PEGylation for Drug Delivery," *Drug Delivery and Translational Research* 3, no. 6 (2013): 499–503, doi: 10.1007/s13346–013–0176–5.

79 "What Are PEG Lipids?" *BroadPharm* (blog), Oct. 6, 2021, https://broad pharm.com/blog/what-are-peg-lipids.

80 Katherine Wylon et al., "Polyethylene Glycol as a Cause of Anaphylaxis," *Allergy, Asthma & Clinical Immunology* 12, no. 67 (2016), doi: 10.1186/s13223–016–0172–7.

81 Laura Bono, "FDA Ignores RFK Jr.'s Pleas for Vaccine Safety Oversight Concerning PEG, Suspected to Cause Anaphylaxis," press release, Dec. 14, 2020, https://childrenshealthdefense.org/press-release/fda-ignores-rfk-jr-s-pleas-for-vaccine-safety-oversight-concerning-peg-suspected-to-cause-anaphylaxis.

82 Children's Health Defense Team, "FDA Investigates Allergic Reactions to Pfizer COVID Vaccine after More Healthcare Workers Hospitalized," *The Defender*, Dec. 21, 2020, https://childrenshealthdefense.org/defender/fda-investigates-reactions-pfizer-covid-vaccine-healthcare-workers-hospitalized.

83 Children's Health Defense Team, "The Pediatric Perils of PEG: From MiraLAX to COVID Shots, FDA and CDC Ignore Safety Signals," *The Defender*, Nov. 3, 2022, https://childrenshealthdefense.org/defender/covid-vaccine-miralax-polyethylene-glycol.

84 "Empire State Consumer Project—PEG 3350: Laxatives and Children," *Empire State Consumer Project* (blog), accessed Jun. 28, 2024, https://laxative kids.weebly.com.

85 "MiraLAX," *RxList*, last updated Aug. 30, 2022, https://www.rxlist.com/miralax-drug.htm.

86 "Schering-Plough Corporation Enters Licensing Agreement to Market Prescription MiraLAX(R) Over-the-Counter," *BioSpace*, Dec. 6, 2006, https://www.biospace.com/article/releases/schering-plough-corporation-enters-licensing-agreement-to-market-prescription-miralax-r-over-the-counter.

87 Joshua G. Schier et al., "Diethylene Glycol in Health Products Sold Over-the-Counter and Imported from Asian Countries," *Journal of Medical Toxicology* 7, no. 1 (2011): 33–38, doi: 10.1007/s13181–010–0111–9.

88 Karen B. Feibus, MD, "Clinical Review, ND 22–0015, Miralax OTC (Polyethylene Glycol 3350)," *AccessData*, FDA.gov, accessed Jun. 28, 2024, 90, https://www.accessdata.fda.gov/drugsatfda_docs/nda/2006/022015s000_MedR_Part2.pdf.

89   Katherine Bortz, "Parental Concerns over MiraLAX Laxative Continue to Spur Closer Review," *Healio*, Oct. 2, 2023, originally published Apr. 3, 2017, https://www.healio.com/news/pediatrics/20170331/parental-concerns-over-miralax-laxative-continue-to-spur-closer-review.

90   Children's Hospital of Philadelphia, "NIH Project no. 1R01FD005312–01, Polyethylene Glycol Safety in Children," *Grantome*, Sep. 10, 2014, https://grantome.com/grant/NIH/R01-FD005312–01.

91   "Polyethylene Glycol Safety in Children," *ClinicalTrials.gov*, last updated Oct. 13, 2023, https://classic.clinicaltrials.gov/ct2/show/NCT05424757?term=polyethylene+glycol+safety+in+children&draw=2&rank.

92   Children's Health Defense Team, "The Pediatric Perils of PEG: From MiraLAX to COVID Shots, FDA and CDC Ignore Safety Signals," *The Defender*, Nov. 3, 2022, https://childrenshealthdefense.org/defender/covid-vaccine-miralax-polyethylene-glycol.

93   Qi Yang et al., "Analysis of Pre-existing IgG and IgM Antibodies against Polyethylene Glycol (PEG) in the General Population," *Analytical Chemistry* 88, no. 23 (2016): 11804–11812, doi: 10.1021/acs.analchem.6b03437.

94   Jamie L. Betker and Thomas J. Anchordoquy, "The Use of Lactose as an Alternative Coating for Nanoparticles," *Journal of Pharmaceutical Sciences* 109, no. 4 (2020): 1573–1580, doi: 10.1016/j.xphs.2020.01.019.

95   Cosby A. Stone Jr. et al., "Immediate Hypersensitivity to Polyethylene Glycols and Polysorbates: More Common Than We Have Recognized," *The Journal of Allergy and Clinical Immunology: In Practice* 7, no. 5 (2019): 1533–1540, doi: 10.1016/j.jaip.2018.12.003.

96   Javier T. Granados-Riveron and Guillermo Aquino-Jarquin, "Engineering of the Current Nucleoside-Modified mRNA-LNP Vaccines against SARS-CoV-2," *Biomed Pharmacotherapy* 142 (2021): 111953, doi: 10.1016/j.biopha.2021.111953.

97   "Distearoylphosphatidylcholine (DSPC)," Polysciences, accessed Jun. 28, 2024, https://www.polysciences.com/default/distearoylphosphatidylcholine.

98   Sander D. Borgsteede et al., "Other Excipients Than PEG Might Cause Serious Hypersensitivity Reactions in COVID-19 Vaccines," *Allergy* 76, no. 6 (2021): 1941–1942, doi: 10.1111/all.14774.

99   Michael Palmer, MD et al., *mRNA Vaccine Toxicity* (Doctors for COVID Ethics, 2023), https://doctors4covidethics.org/mrna-vaccine-toxicity.

100   Seyen Moein Moghimi and Dmitri Simberg, "Pro-Inflammatory Concerns with Lipid Nanoparticles," *Molecular Therapy* 30, no. 6 (2022): 2109–2110, doi: 10.1016/j.ymthe.2022.04.011.

101   "ALC-0315," Echelon Biosciences, accessed Jul. 1, 2024, https://www.echelon-inc.com/product/alc-0315.

102   "Excipients ALC-0315 and ALC-0159," *Parliamentary Question*, European Parliament, Dec. 22, 2021, https://www.europarl.europa.eu/doceo/document/P-9-2021-005690_EN.html.

103 A Mother's Anthem, "The Alarming LNP History You Haven't Been Shown—The LNNP Developer's Own Studies Dating Back 20 Years: Part 1 of 10, Introduction and Executive Brief Sent to Lawmakers & State Investigators This Summer," *A Mother's Anthem*, Sep. 7, 2023, https://amothersanthem.substack.com/p/the-alarming-lnp-history-you-havent.

104 Nehal E. Elsadek et al., "5 – Immunological Responses to PEGylated Proteins: Anti-PEG Antibodies," *Polymer-Protein Conjugates*, 2020, 103–123, doi: 10.1016/B978–0–444–64081–9.00005-X.

105 Barry W. Neun et al., "Understanding the Role of Anti-PEG Antibodies in the Complement Activation by Doxil in Vitro," *Molecules* 23, no. 7 (2018): 1700, doi: 10.3390/molecules23071700.

106 Gergely Tibor Kozma et al., "Anti-PEG Antibodies: Properties, Formation, Testing and Role in Adverse Immune Reactions to PEGylated Nano-Biopharmaceuticals," *Advanced Drug Delivery Reviews* 154–155 (2020): 163–175, doi: 10.1016/j.addr.2020.07.024.

107 Moderna Inc., "FORM S-1 Registration Statement under The Securities Act of 1933," (filed Nov. 9, 2018), 33, https://www.sec.gov/Archives/edgar/data/1682852/000119312518323562/d577473ds1.htm, accessed Jul. 1, 2024.

108 A Mother's Anthem, "Part 2 of 10—The Alarming LNP History You Haven't Been Shown—Lipid and LNP Developer's Studies Dating Back 20 years: Executive Summary Sent to Lawmakers & Investigators," *A Mother's Anthem*, Sep. 11, 2023, https://amothersanthem.substack.com/p/the-alarming-lnp-history-you-havent-ff1.

109 János Szebeni, "Complement Activation-Related Pseudoallergy: A New Class of Drug-Induced Acute Immune Toxicity," *Toxicology* 216, no. 2–3 (2005): 106–121, doi: 10.1016/j.tox.2005.07.023.

110 János Szebeni et al., "Animal Models of Complement-Mediated Hypersensitivity Reactions to Liposomes and Other Lipid-Based Nanoparticles," *Journal of Liposome Research* 17, no. 2 (2007): 107–117, doi: 10.1080/08982100701375118.

111 A Mother's Anthem, "Part 2 of 10—The Alarming LNP History You Haven't Been Shown—Lipid and LNP Developer's Studies Dating Back 20 years: Executive Summary Sent to Lawmakers & Investigators," *A Mother's Anthem*, Sep. 11, 2023, https://amothersanthem.substack.com/p/the-alarming-lnp-history-you-havent-ff1.

112 János Szebeni, "Complement Activation-Related Pseudoallergy: A Stress Reaction in Blood Triggered by Nanomedicines and Biologicals," *Molecular Immunology* 61, no. 2 (2014): 163–173, doi: 10.1016/j.molimm.2014.06.038.

113 László Dézsi et al., "Cardiopulmonary and Hemodynamic Changes in Complement Activation-Related Pseudoallergy," *Health* 5, no. 6 (2013): 1032–1038, doi: 10.4236/health.2013.56138.

114 László Dézsi et al., "A Naturally Hypersensitive Porcine Model May Help Understand the Mechanism of COVID-19 mRNA Vaccine-Induced Rare (Pseudo) Allergic Reactions: Complement Activation as a Possible Contributing Factor," *Geroscience* 44, no. 2 (2022): 597–618, doi: 10.1007/s11357–021–00495-y.

115 Yihua Yu, Marc B. Taraban, and Katharine T. Briggs, "All Vials Are Not the Same: Potential Role of Vaccine Quality in Vaccine Adverse Reactions," *Vaccine* 39, no. 45 (2021): 6565–6569, doi: 10.1016/j.vaccine.2021.09.065.

116 "Dr. Jane Ruby and Dr. Sasha Latypova Discuss Deadly 'Batches' of the Covid Vaccine. Was It Murder? Danish Batches Suggests Intentionality," Rumble, May 20, 2023, https://rumble.com/v2ox9tc—dr.-jane-ruby-and-dr.-sasha-latypova-discuss-deadly-batches-of-the-covid-v.html.

117 Sasha Latypova, "My Friend Dr. Lindsay Talked to an FDA Official about Plasmid DNA and SV40 in Vax Vials," *Due Diligence and Art*, Sep. 26, 2023, https://sashalatypova.substack.com/p/my-friend-dr-lindsay-talked-to-fda.

118 Craig-Paardekooper, "Covid Vax Variability between Lots—Independent Research by International Team," Bitchute, Dec. 15, 2021, https://www.bitchute.com/video/4HlIyBmOEJeY.

119 Jan Jekielek, "Pharma Insider Speaks Out about Vaccine Batches," *The Epoch Times*, last updated Aug. 10, 2023, https://www.theepochtimes.com/us/pharma-insider-speaks-out-about-vaccine-batches-5371394.

120 Ibid.

121 Max Schmeling et al., "Batch-Dependent Safety of the BNT162b2 mRNA COVID-19 Vaccine," *European Journal of Clinical Investigation* 53, no. 8 (2023), doi: 10.1111/eci.13998.

122 Sasha Latypova, "Photo-Report from Sweden and Iceland," *Due Diligence and Art*, Oct. 15, 2023, https://sashalatypova.substack.com/p/photo-report-from-sweden-and-iceland.

123 Michael Nevradakis, PhD, "'Bombshell' Study of Pfizer COVID Vaccine Suggests Some People Got Highly Dangerous Shots, Others Got a Placebo," *The Defender*, Jun. 29, 2023, https://childrenshealthdefense.org/defender/pfizer-biontech-covid-vaccine-placebo.

124 Rhoda Wilson, "Dr Mike Yeadon: The Variability in Serious Adverse Events by Vaccine Lot Is the 'calibration of a killing weapon,'" *The Exposé*, Jan. 11, 2022, https://expose-news.com/2022/01/11/mike-yeadon-the-variability-in-serious-adverse-events.

125 Russell L. Blaylock, "COVID Update: What Is the Truth?" *Surg Neurology International* 13, no. 167 (2022), doi: 10.25259/SNI_150_2022.

126 "ICYMI: Sen. Johnson Joins House of Representatives Panel on Injuries Caused by COVID-19 Vaccines," press release, Nov. 15, 2023, https://www.ronjohnson.senate.gov/2023/11/icymi-sen-johnson-joins-house-of-representatives-panel-on-injuries-caused-by-covid-19-vaccines.

127 Barbara Loe Fisher, "Shalala Takes Away Compensation for DPT Injured Children," *The Vaccine Reaction* 1, no. 1 (1995), https://www.nvic.org /newsletter/mar-1995/the-vaccine-reaction-march-1995.

128 Neil Z. Miller, "Vaccines and Sudden Infant Death: An Analysis of the VAERS Database 1990–2019 and Review of the Medical Literature," *Toxicology Reports* 8 (2021): 1324–1335, doi: 10.1016/j.toxrep.2021.06.020.

129 "Infant Vaccine Recalled," *Kingsport Times-News*, Mar. 4, 1990, 73, https://www.newspapers.com/article/kingsport-times-news-tennessee-dtp -scand/131597675.

130 Institute of Medicine et al., *Adverse Effects of Pertussis and Rubella Vaccines: A Report of the Committee to Review the Adverse Consequences of Pertussis and Rubella Vaccines* (Washington, DC: National Academies Press, 1991).

131 Patrice La Vigne, "NVIC.org—The Story Behind MedAlerts," National Vaccine Information Center, Aug. 20, 2013, https://www.nvic.org/news letter/aug-2013/nvic-org-the-story-behind-medalerts#_edn5.

132 National Vaccine Information Center, "NBC-TV 'Now' Show with Tom Brokaw and Katie Couric VAERS and 'Hot Lots' of DPT Vaccine," Rumble, Mar. 2, 1994, https://rumble.com/v40sz3h-nbc-tv-now-show-with-tom-brokaw -and-katie-couric-vaers-and-hot-lots-of-dpt-.html.

133 Marcel Kinsbourne, MD, "Presentation to the Committee on Government Reform, Topic: Vaccines: Finding a Balance between Public Safety and Personal Choice," *Whale*, Aug. 3, 1999, http://www.whale.to/vaccines/kins bourne.html.

134 Neil Z. Miller, "Vaccines and Sudden Infant Death: An Analysis of the VAERS Database 1990–2019 and Review of the Medical Literature," *Toxicology Reports* 8 (2021): 1324–1335, doi: 10.1016/j.toxrep.2021.06.020.

135 @LegalizeItLala_, "In 1979, eleven babies died of SIDS in one county in Tennessee," Twitter (now X) Mar. 13, 2024, https://twitter.com/LegalizeItLala_ /status/1768121702209184205.

136 Sasha Latypova, "OMG, SV40! Can We Sue Pfizer NOW?" *Due Diligence and Art*, Oct. 23, 2023, https://sashalatypova.substack.com/p/omg-sv40 -pfizer-can-be-sued-now.

137 "Adulterate," *Merriam-Webster Dictionary*, accessed Jul. 1, 2024, https://www .merriam-webster.com/dictionary/adulterate.

138 "63 Synonyms and Antonyms for Adulterate," *Thesaurus.com*, accessed Jul. 1, 2024, https://www.thesaurus.com/browse/adulterate.

139 Sasha Latypova, "OMG, SV40! Can We Sue Pfizer NOW?" *Due Diligence and Art*, Oct. 23, 2023, https://sashalatypova.substack.com/p/omg-sv40 -pfizer-can-be-sued-now.

140 Katherine Watt, "COVID-19 Injectable Bioweapons as Case Study in Legalized, Government-Operated Domestic Bioterrorism," *Bailiwick News*, Jun. 9, 2022, https://bailiwicknews.substack.com/p/covid-19-injectable-bio weapons-as.

141 Katherine Watt, "Whatever Is in the Biochemical Weapons Bearing Pfizer and Other Pharma Labels, Is There because US SecDefs and Their WHO-BIS Handlers Ordered It to Be There," *Bailiwick News*, Oct. 28, 2023, https://bailiwicknews.substack.com/p/whatever-is-in-the-biochemical-weapons.

142 Sasha Latypova, "OMG, SV40! Can We Sue Pfizer NOW?" *Due Diligence and Art*, Oct. 23, 2023, https://sashalatypova.substack.com/p/omg-sv40-pfizer-can-be-sued-now.

143 Team Enigma, "More Vials Tested—More Evidence of Product Adulteration," Bitchute, 00:10:03–00:10:15, Jul. 14, 2022, https://www.bitchute.com/video/2Uo3t2cTgjq0.

144 Michael Palmer, MD et al., *mRNA Vaccine Toxicity* (Doctors for COVID Ethics, 2023), https://doctors4covidethics.org/mrna-vaccine-toxicity.

145 David Estapé and Vivianne J. Arencibia, "Plasmids and Their Role in mRNA Technology," *Pharmaceutical Engineering*, May 18, 2023, https://ispe.org/pharmaceutical-engineering/ispeak/plasmids-and-their-role-mrna-technology.

146 Ulrike Granögger, "'Plasmidgate'—mRNA Injections Are Contaminated with Bacterial DNA," *Solari Report*, Mar. 26, 2023, https://home.solari.com/plasmidgate-mrna-injections-are-contaminated-with-bacterial-dna.

147 Team Enigma, "More Vials Tested—More Evidence of Product Adulteration," Bitchute, Jul. 14, 2022, https://www.bitchute.com/video/2Uo3t2cTgjq0.

148 Sasha Latypova, "The Real Purpose of the SV40 Promoter in Covid Injections," *Due Diligence and Art*, Sep. 1, 2023, https://sashalatypova.substack.com/p/sv40-promoter-in-covid-injections.

149 Michael Palmer, MD et al., *mRNA Vaccine Toxicity* (Doctors for COVID Ethics, 2023), https://doctors4covidethics.org/mrna-vaccine-toxicity.

150 Kevin McKernan et al., "Sequencing of Bivalent Moderna and Pfizer mRNA Vaccines Reveals Nanogram to Microgram Quantities of Expression Vector dsDNA Per Dose," *Medicinal Genomics*, Apr. 11, 2023, https://cdn-ceo-ca.s3.amazonaws.com/1i4tp3q-Sequencing%20of%20bivalent_4–11–23.pdf.

151 Julie Sladden and Julian Gillespie, "The Vax-Gene Files: An Accidental Discovery," *Brownstone Journal*, May 27, 2023, https://brownstone.org/articles/vax-gene-files-accidental-discovery.

152 Julie Sladden and Julian Gillespie, "The Vax-Gene Files: Have the Regulators Approved a Trojan Horse?" *Brownstone Journal*, Aug. 16, 2023, https://brownstone.org/articles/the-vax-gene-files-have-the-regulators-approved-a-trojan-horse.

153 Ulrike Granögger, "'Plasmidgate'—mRNA Injections Are Contaminated with Bacterial DNA," *Solari Report*, Mar. 26, 2023, https://home.solari.com/plasmidgate-mrna-injections-are-contaminated-with-bacterial-dna.

154 Ibid.

155 Trevor Hunnicutt, "Biden Plans $100 Million Drive to Combat Drug-Resistant 'Superbugs,'" Reuters, Sep. 27, 2023, https://www.reuters.com/world/us/biden-plans-100-million-drive-combat-drug-resistant-superbugs-2023-09-27.

156 Sasha Latypova, "The Real Purpose of the SV40 Promoter in Covid Injections," *Due Diligence and Art*, Sep. 1, 2023, https://sashalatypova.substack.com/p/sv40-promoter-in-covid-injections.

157 Michael Nevradakis, PhD, "'Probably the Most Important Topic of Our Time': DNA Contaminants in COVID Shots Can Trigger Cancer, Alter Human Genome," *The Defender*, Oct. 11, 2023, https://childrenshealthdefense.org/defender/covid-vaccines-dna-contamination.

158 Ibid.

159 Katherine Watt, "Whatever Is in the Biochemical Weapons Bearing Pfizer and Other Pharma Labels, Is There because US SecDefs and Their WHO-BIS Handlers Ordered It to Be There," *Bailiwick News*, Oct. 28, 2023, https://bailiwicknews.substack.com/p/whatever-is-in-the-biochemical-weapons.

160 Dawn Geske, "Moderna COVID Vaccine Recalled after Stainless Steel Particles Found in 1.6M Doses in Japan," *International Business Times*, Sep. 2, 2021, https://www.ibtimes.com/moderna-covid-vaccine-recalled-after-stainless-steel-particles-found-16m-doses-japan-3286860.

161 Reuters, "Moderna Recalls Thousands of COVID Vaccine Doses in Europe," Reuters, Apr. 8, 2022, https://www.reuters.com/business/healthcare-pharmaceuticals/moderna-recalls-thousands-covid-vaccine-doses-2022-04-08.

162 The Epoch Times, "Toxic, Metallic Compounds Found in All COVID Vaccine Samples Analyzed by German Scientists," *The Defender*, Aug. 25, 2022, https://childrenshealthdefense.org/defender/toxic-metallic-compounds-covid-vaccines-german-scientists.

163 Antonietta M. Gatti and Stefano Montanari, "New Quality-Control Investigations on Vaccines: Micro- and Nanocontamination," *International Journal of Vaccines and Vaccination* 4, no. 1 (2017): 00072, doi: 10.15406/ijvv.2017.04.00072.

164 "Hero of the Week: June 26, 2023: Dr. Antonietta M. Gatti," *Solari Report*, Jun. 26, 2023, https://home.solari.com/hero-of-the-week-june-26-2023-dr-antonietta-m-gatti.

Chapter 4

1 S. Burkhardt, "Euthanasia and Assisted Suicide: Comparison of Legal Aspects in Switzerland and Other Countries," *Medicine, Science and the Law* 46, no. 4 (2006): 287–294, doi: 10.1258/rsmmsl.46.4.287.

2 "What Do Assisted Dying, Assisted Suicide and Euthanasia Mean and What Is the Law?" BBC, Jun. 4, 2024, https://www.bbc.com/news/uk-47158287.

3   Nicole Davis, "Euthanasia and Assisted Dying Rates Are Soaring. But Where Are They Legal?" *The Guardian*, Jul. 15, 2019, https://www.th eguardian.com/news/2019/jul/15/euthanasia-and-assisted-dying-rates-are -soaring-but-where-are-they-legal.

4   Andrew McGee, "In Places Where It's Legal, How Many People Are Ending Their Lives Using Euthanasia?" *The Conversation*, updated Feb. 27, 2019, https://theconversation.com/in-places-where-its-legal-how-many-people -are-ending-their-lives-using-euthanasia-73755.

5   Lars Mehlum et al., "Euthanasia and Assisted Suicide in Patients with Personality Disorders: A Review of Current Practice and Challenges," *Borderline Personality Disorder and Emotion Dysregulation* 7, no. 15 (2020), doi: 10.1186/s40479–020–00131–9.

6   "About the Euthanasia Prevention Coalition," Euthanasia Prevention Coalition, accessed Jul. 1, 2024, https://epcc.ca/about.

7   Wesley J. Smith, "Belgian Doctor Euthanizes Patient with a Pillow," *National Review*, Sep. 9, 2023, https://www.nationalreview.com/corner /belgian-doctor-euthanizes-patient-with-a-pillow.

8   Lars Mehlum et al., "Euthanasia and Assisted Suicide in Patients with Personality Disorders: A Review of Current Practice and Challenges," *Borderline Personality Disorder and Emotion Dysregulation* 7, no. 15 (2020), doi: 10.1186/s40479–020–00131–9.

9   Christopher de Bellaigue, "Death on Demand: Has Euthanasia Gone Too Far?" *The Guardian*, Jan. 18, 2019, https://www.theguardian.com/news /2019/jan/18/death-on-demand-has-euthanasia-gone-too-far-netherlands -assisted-dying.

10 Alex Schadenberg, "Netherlands Five Year Study Shows Significant Increases in Assisted Deaths and Continued Abuse of the Law," *Euthanasia Prevention Coalition* (blog), Aug. 3, 2017, https://alexschadenberg.blogspot .com/2017/08/alex-schadenberg-executive-director.html.

11 Alex Schadenberg, "Luxembourg Euthanasia Update (56% Increase in 2020 Euthanasia Deaths)," *Euthanasia Prevention Coalition* (blog), Apr. 26, 2021, https://alexschadenberg.blogspot.com/2021/04/luxembourg-euthanasia -update-56.html.

12 "Global Human Euthanasia Services Market Size 2031:Trends & Growth Opportunities," *LinkedIn*, Jun. 21, 2024, https://www.linkedin.com/pulse /global-human-euthanasia-services-market-size-2031-trends-2zqfc/.

13 Associated Press, "Swiss See Steady Rise in Assisted Suicides," FOX News, updated Nov. 20, 2014, https://www.foxnews.com/health/swiss-see-steady -rise-in-assisted-suicides.

14 Sibilla Bondolfi, "Why Assisted Suicide Is 'Normal' in Switzerland," *SWI swissinfo.ch*, Jul. 24, 2020, https://www.swissinfo.ch/eng/why-assisted-suicide -is—normal—in-switzerland-/45924614.

15   Samuel Blouin, "'Suicide Tourism' and Understanding the Swiss Model of the Right to Die," *The Conversation*, May 23, 2018. https://theconversation .com/suicide-tourism-and-understanding-the-swiss-model-of-the-right-to -die-96698.

16   Ronny Reyes, "Inside the Swiss Assisted Suicide Clinic Where US Sisters Paid $11,000 Each to Die: Facility Has Cool White Walls, Designer Furniture—And a Death Room Where You're Hooked Up to Drip or Given Drink That Will Kill You," *Daily Mail*, updated Mar. 23, 2022, https: //www.dailymail.co.uk/news/article-10644179/Inside-Swiss-assisted -suicide-clinic-sisters-paid-11–000-die.html.

17   Mike Corder, "Dutch Euthanasia Center Sees 22% Rise in Requests in 2019," Associated Press, Feb. 7, 2020, https://apnews.com/article/bdebcf3 cd85f4e1b2e1e931dc1a51a51.

18   Christopher de Bellaigue, "Death on Demand: Has Euthanasia Gone Too Far?" *The Guardian*, Jan. 18, 2019, https://www.theguardian.com/news /2019/jan/18/death-on-demand-has-euthanasia-gone-too-far-netherlands -assisted-dying.

19   "Belgium," Patient Rights Council, accessed Jul. 1, 2024, https://www .patientsrightscouncil.org/site/belgium.

20   Richard Egan, "Don't Forget about Luxembourg, a World Leader in Assisted Dying," *BioEdge*, Sep. 18, 2023, https://web.archive.org/web /20230518065928/https://bioedge.org/end-of-life-issues/euthanasia /dont-forget-about-luxembourg-a-world-leader-in-assisted-dying.

21   "Euthanasia in the Netherlands," Alliance VITA, Nov. 24, 2017, https: //www.alliancevita.org/en/2017/11/euthanasia-in-the-netherlands.

22   Maria Cheng, "Some Dutch People Seeking Euthanasia Cite Autism or Intellectual Disabilities, Researchers Say," Associated Press, Jun. 28, 2023, https://apnews.com/article/euthanasia-autism-intellectual-disabilities -netherlands-b5c4906d0305dd97e16da363575c03ae.

23   Irene Tuffrey-Wijne et al., "'Because of His Intellectual Disability, He Couldn't Cope.' Is Euthanasia the Answer?" *Journal of Policy and Practice in Intellectual Disabilities* 16, no. 2 (2019): 113–116, doi: 10.1111/jppi.12307.

24   Kenneth Chambaere et al., "Recent Trends in Euthanasia and Other End-of-Life Practices in Belgium," *New England Journal of Medicine* 372, no. 12 (2015): 1179–1181, doi: 10.1056/NEJMc1414527.

25   Scott Kim, "How Dutch Law Got a Little Too Comfortable with Euthanasia," *The Atlantic*, Jun. 8, 2019, https://www.theatlantic.com/ideas /archive/2019/06/noa-pothoven-and-dutch-euthanasia-system/591262.

26   "Euthanasia in the Netherlands," Alliance VITA, Nov. 24, 2017, https: //www.alliancevita.org/en/2017/11/euthanasia-in-the-netherlands.

27   Irene Tuffrey-Wijne et al., "Euthanasia and Assisted Suicide for People with an Intellectual Disability and/or Autism Spectrum Disorder: An Examination

of Nine Relevant Euthanasia Cases in the Netherlands (2012–2016)," *BMC Medical Ethics* 19, no. 1 (2018): 17, doi: 10.1186/s12910–018–0257–6.

28 SMP van Veen et al., "Physician Assisted Death for Psychiatric Suffering: Experiences in The Netherlands," *Frontiers in Psychiatry* 13, no. 895387 (2022), doi: 10.3389/fpsyt.2022.895387.

29 Charles Lane, "Where the Prescription for Autism Can Be Death," *Washington Post*, Feb. 24, 2016, https://www.washingtonpost.com/opinions/where-the -prescription-for-autism-can-be-death/2016/02/24/8a00ec4c-d980–11e5 –81ae-7491b9b9e7df_story.html.

30 Scott Kim, "How Dutch Law Got a Little Too Comfortable with Euthanasia," *The Atlantic*, Jun. 8, 2019, https://www.theatlantic.com/ideas /archive/2019/06/noa-pothoven-and-dutch-euthanasia-system/591262.

31 Michael Cook, "Should Intractable Mental Illness Make You Eligible for Euthanasia?" *BioEdge*, Mar. 28, 2023, https://web.archive.org/web/2023 1204064303/https://bioedge.org/end-of-life-issues/euthanasia/should -intractable-mental-illness-make-you-eligible-for-euthanasia.

32 Irene Tuffrey-Wijne et al., "Euthanasia and Assisted Suicide for People with an Intellectual Disability and/or Autism Spectrum Disorder: An Examination of Nine Relevant Euthanasia Cases in the Netherlands (2012–2016)," *BMC Medical Ethics* 19, no. 1 (2018): 17, doi: 10.1186/s12910–018–0257–6.

33 "Unbearable Suffering without Prospect of Improvement," Regional Euthanasia Review Committees, accessed Jul. 1, 2024, https://english.euthanasie commissie.nl/due-care-criteria/unbearable-suffering-without-prospect-of -improvement.

34 Lars Mehlum et al., "Euthanasia and Assisted Suicide in Patients with Personality Disorders: A Review of Current Practice and Challenges," *Borderline Personality Disorder and Emotion Dysregulation* 7, no. 15 (2020), doi: 10.1186/s40479–020–00131–9.

35 Irene Tuffrey-Wijne et al., "'Because of His Intellectual Disability, He Couldn't Cope.' Is Euthanasia the Answer?" *Journal of Policy and Practice in Intellectual Disabilities* 16, no. 2 (2019): 113–116, doi: 10.1111/jppi.12307.

36 Irene Tuffrey-Wijne et al., "Euthanasia and Assisted Suicide for People with an Intellectual Disability and/or Autism Spectrum Disorder: An Examination of Nine Relevant Euthanasia Cases in the Netherlands (2012–2016)," *BMC Medical Ethics* 19, no. 1 (2018): 17, doi: 10.1186/s12910–018–0257–6.

37 Irene Tuffrey-Wijne et al., "Euthanasia and Assisted Suicide for People with an Intellectual Disability and/or Autism Spectrum Disorders: Investigation of 39 Dutch Case Reports (2012–2021)," *BJPsych Open* 9, no. 3 (2023): e87, doi: 10.1192/bjo2023.69.

38 Maria Cheng, "Some Dutch People Seeking Euthanasia Cite Autism or Intellectual Disabilities, Researchers Say," Associated Press, Jun. 28, 2023, https://apnews.com/article/euthanasia-autism-intellectual-disabilities -netherlands-b5c4906d0305dd97e16da363575c03ae.

39   Ibid.

40   J. Pereira, "Legalizing Euthanasia or Assisted Suicide: The Illusion of Safeguards and Controls," *Current Oncology* 18, no. 2 (2011): e38-e45, doi: 10.3747/co.v18i2.883.

41   Emma Freire, "Netherlands Considers Euthanasia for Healthy People, Doctors Say Things Are 'Getting Out Of Hand,'" *The Federalist*, Jun. 30, 2017, https://thefederalist.com/2017/06/30/netherlands-considers-euthanasia-healthy.

42   "Christian Party Leader Tells U.S. News That Dutch Doctors Kill You," *NL Times*, Jul. 21, 2017, https://nltimes.nl/2017/07/21/christian-party-leader-tells-us-news-dutch-doctors-kill.

43   Kees Van der Staaij, "In the Netherlands, the Doctor Will Kill You," *Wall Street Journal*, Jul. 20, 2017, https://www.wsj.com/articles/in-the-netherlands-the-doctor-will-kill-you-now-1500591571.

44   Emma Freire, "Netherlands Considers Euthanasia for Healthy People, Doctors Say Things Are 'Getting Out of Hand,'" *The Federalist*, Jun. 30, 2017, https://thefederalist.com/2017/06/30/netherlands-considers-euthanasia-healthy.

45   Yi-Sheng Chao et al., "International Changes in End-of-Life Practices over Time: A Systematic Review," *BMC Health Services Research* 16, no. 539 (2016), doi: 10.1186/s12913–016–1749-z.

46   Sigrid Dierickx et al., "Drugs Used for Euthanasia: A Repeated Population-Based Mortality Follow-Back Study in Flanders, Belgium, 1998–2013," *Journal of Pain and Symptom Management* 56, no. 4 (2018): 551–559, doi: 10.1016/j.jpainsymman.2018.06.015.

47   Lonny Shavelson, MD, *Medical Aid in Dying: A Guide for Patients and Their Supporters* (self pub., ACAMAID, 2022).

48   "Frequently Asked Questions," American Clinicians Academy on Medical Aid in Dying, accessed Jul. 2, 2024, https://www.acamaid.org/faq.

49   Ana Worthington, Ilora Finlay, and Claud Regnard, "Efficacy and Safety of Drugs Used for 'Assisted Dying,'" *British Medical Bulletin* 142, no. 1 (2022): 15–22, doi: 10.1093/bmb/ldac009.

50   Jonel Aleccia, "Docs in Northwest Tweak Aid-in-Dying Drugs to Prevent Prolonged Deaths," *KFF Health News*, Feb. 21, 2017, https://kffhealthnews.org/news/docs-in-northwest-tweak-aid-in-dying-drugs-to-prevent-prolonged-deaths.

51   Ana Worthington, Ilora Finlay, and Claud Regnard, "Efficacy and Safety of Drugs Used for 'Assisted Dying,'" *British Medical Bulletin* 142, no. 1 (2022): 15–22, doi: 10.1093/bmb/ldac009.

52   Lonny Shavelson and Carol Parrot, "Adding Phenobarbital to the D-DMA and DDMA Medication Protocols for Medical Aid in Dying," American Clinicians Academy on Medical Aid in Dying, Jan. 12, 2021, https://www.acamaid.org/wp-content/uploads/2021/01/Adding-Phenobarbital-to-the-D-DMA-and-DDMA-Medication-Protocols-for-Medical-Aid-in-Dying-1–1.pdf.

53   Ana Worthington, Ilora Finlay, and Claud Regnard, "Efficacy and Safety of Drugs Used for 'Assisted Dying,'" *British Medical Bulletin* 142, no. 1 (2022): 15–22, doi: 10.1093/bmb/ldac009.

54   Sigrid Dierickx et al., "Drugs Used for Euthanasia: A Repeated Population-Based Mortality Follow-Back Study in Flanders, Belgium, 1998–2013," *Journal of Pain and Symptom Management* 56, no. 4 (2018): 551–559, doi: 10.1016/j.jpainsymman.2018.06.015.

55   Ana Worthington, Ilora Finlay, and Claud Regnard, "Efficacy and Safety of Drugs Used for 'Assisted Dying,'" *British Medical Bulletin* 142, no. 1 (2022): 15–22, doi: 10.1093/bmb/ldac009.

56   "Posts with Label: Euthanasia by Deydration," *Euthanasia Prevention Coalition* (blog), accessed Jun. 26, 2024, http://alexschadenberg.blogspot.com/search/label/Euthanasia%20by%20dehydration;
      "Posts with Label: Euthanasia by Organ Donation," *Euthanasia Prevention Coalition* (blog), accessed Jun. 26, 2024, http://alexschadenberg.blogspot.com/search/label/Euthanasia%20by%20organ%20donation;
      "Posts with Label: Euthenasia by Advanced Directive," *Euthanasia Prevention Coalition* (blog), accessed Jun. 26, 2024, http://alexschadenberg.blogspot.com/search/label/euthanasia%20by%20advanced%20directive.

57   Wesley J. Smith, "Belgian Doctor Euthanizes Patient with a Pillow," *National Review*, Sep. 9, 2023, https://www.nationalreview.com/corner/belgian-doctor-euthanizes-patient-with-a-pillow.

58   Jo Fahy, "Growing Number of People Sign Up for Assisted Suicide," *SWI swissinfo.ch*, Feb. 14, 2018, https://www.swissinfo.ch/eng/a-way-out_growing-number-of-people-sign-up-for-assisted-suicide/43899702.

59   Nicole Davis, "Euthanasia and Assisted Dying Rates Are Soaring. But Where Are They Legal?" *The Guardian*, Jul. 15, 2019, https://www.theguardian.com/news/2019/jul/15/euthanasia-and-assisted-dying-rates-are-soaring-but-where-are-they-legal.

60   Maddy French, "Swiss Group to Allow Assisted Dying for Elderly Who Are Not Terminally Ill," *The Guardian*, May 26, 2014, https://www.theguardian.com/society/2014/may/26/swiss-exit-assisted-suicide-elderly-not-terminally-ill.

61   Nicole Steck et al., "Increase in Assisted Suicide in Switzerland: Did the Socioeconomic Predictors Change? Results from the Swiss National Cohort," *BMJ Open* 8, no. 4 (2018): e020922, doi: 10.1136/bmjopen-2017–020992.

62   Christine Bartsch et al., "Assisted Suicide in Switzerland: An Analysis of Death Records from Swiss Institutes of Forensic Medicine," *Deutsches Ärzteblatt International* 116, no. 33–34 (2019): 545–552, doi: 10.3238/arztebl.2019.0545.

63   Barbie Latza Nadeau, "Switzerland Approves Assisted 'Suicide Capsule,'" Yahoo! News, Dec. 6, 2021, https://news.yahoo.com/switzerland-approves-assisted-suicide-capsule-140255771.html.

64 "Kerncijfers 2020," Overheid.nl, accessed Jul. 2, 2024, https://open
.overheid.nl/documenten/ronl-cebe2a68-bf3b-418f-ba8e-53634f96f6c9/pdf.

65 Darren Boyle, "Belgian Doctors Give Healthy Woman, 24, Green Light
to Die by Euthanasia Because of 'suicidal thoughts,'" *Daily Mail*, Jun. 27,
2015, https://www.dailymail.co.uk/news/article-3141564/Belgian-doctors
-healthy-woman-green-light-die-euthanasia-suicidal-thoughts.html.

66 Ibid.

67 Steve Doughty, "Revealed: Three Children Are among Thousands to Die
from Euthanasia under Belgium's Radical Laws that Have Seen Cases
Increase Fivefold in 10 Years," *Daily Mail*, updated Jul. 24, 2018, https:
//www.dailymail.co.uk/news/article-5984023/Three-children-thousands
-die-euthanasia-Belgiums-radical-laws.html.

68 Tom Embury-Dennis, "Terminally-Ill Children Become Youngest Ever
to be Euthanised, Aged Nine and 11," *Independent*, Aug. 7, 2018, https:
//www.independent.co.uk/news/world/europe/children-euthanasia-belgium
-youngest-killed-terminally-ill-cfcee-a8481311.html.

69 Cassy Fiano-Chesser, "Report: Belgium Euthanized Record Number in
2019—Including One Child," *Live Action*, Mar. 14, 2020, https://www
.liveaction.org/news/belgium-euthanized-record-number-2019-child.

70 "Legal Situation," World Federation Right To Die Societies, Jul. 2, 2024,
https://wfrtds.org/worldmap/colombia.

71 Luke Yamaguchi, "FDA Used 'Critically Flawed' Risk-Benefit Analysis
to 'Justify' COVID Vaccines for Children," *The Defender*, Feb. 3, 2022,
https://childrenshealthdefense.org/defender/fda-risk-benefit-analysis-covid
-vaccines-children.

72 *Financial Rebellion*, episode 13, "Sucharit Bhakdi, M.D., 'What We're
Looking at Is Genocide'," hosted by Catherine Austin Fitts, Polly Tommey,
and Carolyn Betts, aired Mar. 17, 2022, on CHD.TV, https://live.children
shealthdefense.org/chd-tv/shows/financial-rebellion-with-catherine-austin
-fitts/sucharit-bhakdi-md-what-were-looking-at-is-genocide.

73 "Holland's Euthanasia Law," Patients Rights Council, Jul. 2, 2024, https:
//www.patientsrightscouncil.org/site/hollands-euthanasia-law.

74 Gali Katznelson, "Extending the Right to Die to Mature Minors in
Canada," *Bill of Health* (blog), Petrie-Flom Center at Harvard Law School,
Feb. 7, 2018, https://blog.petrieflom.law.harvard.edu/2018/02/07/extending
-the-right-to-die-to-mature-minors-in-canada.

75 Büşra Nur Bilgiç Çakmak, "Netherlands to Allow Euthanasia for Children
Age 1–12," *Anadolu Ajansi (AA)*, Oct. 14, 2020, https://www.aa.com.tr/en
/europe/netherlands-to-allow-euthanasia-for-children-age-1–12/2006465.

76 "Dutch Parliament to Discuss Possible New Laws to Help Suffering Children
to Die," *Dutch News*, Oct. 9, 2020, https://www.dutchnews.nl/2020/10
/dutch-parliament-to-discuss-possible-new-laws-to-help-suffering-children
-to-die.

77   "Health Minister Proposes Euthanasia Protocol for Children under 12," *Dutch News*, Jun. 28, 2022, https://www.dutchnews.nl/2022/06/health -minister-proposes-euthanasia-protocol-for-children-under-12.

78   Sedona Celine De Keijzer et al., "The Age Limit for Euthanasia Requests in the Netherlands: A Delphi Study among Paediatric Experts," *Journal of Medical Ethics* 49, no. 7 (2023): 458–464, doi: 10.1136/jme-2022-108448.

79   "Dutch to Widen 'Right-to-Die' to Include Terminally Ill Children," Reuters, Apr. 14, 2023, https://www.reuters.com/world/europe/dutch -widen-right-to-die-include-terminally-ill-children-2023-04-14.

80   Jonathan Van Maren, "Child Euthanasia Comes to the Netherlands," *First Things*, Oct. 15, 2020, https://www.firstthings.com/web-exclusives/2020/10 /child-euthanasia-comes-to-the-netherlands.

81   Maria Cheng, "'Disturbing': Experts Troubled by Canada's Euthanasia Laws," Associated Press, Aug. 11, 2022, https://apnews.com/article/covid -science-health-toronto-7c631558a457188d2bd2b5cfd360a867.

82   Health Canada, "Third Annual Report on Medical Assistance in Dying in Canada 2021," Government of Canada, Jul. 2022, https://www.canada .ca/en/health-canada/services/publications/health-system-services/annual -report-medical-assistance-dying-2021.html.

83   Alexandra E. Rosso, Dirk Huyer, and Alfredo Walker, "Analysis of the Medical Assistance in Dying Cases in Ontario: Understanding the Patient Demographics of Case Uptake in Ontario since the Royal Assent and Amendments of Bill C-14 in Canada," *Academic Forensic Pathology* 7, no. 2 (2017): 263–287, doi: 10.23907/2017.025.

84   Hanna Seariac, "Canada's Roadmap to Expanding Assisted Suicide to 'Mature' Minors Revealed," *Deseret News*, Mar. 1, 2023, https://www.deseret. com/2023/3/1/23617814/canada-medically-assisted-death-mature-minors.

85   Hanna Seariac, "Perspective: Inside the World of Canada's Assisted Suicide— for 'Mature Minors,'" *Deseret News*, Nov. 4, 2022, https://www.deseret .com/2022/11/4/23401482/canada-euthanasia-laws-mature-minors-medical -aid-in-dying.

86   Catherine Cullen and Alexandra Zabjek, "Federal Minister Says She's 'Shocked' by Suggestion of Assisted Deaths for Some Babies," CBC, Oct. 22, 2022, https://www.cbc.ca/news/politics/assisted-dying-carla-qualtrough -1.6625412.

87   Hanna Seariac, "Perspective: Inside the World of Canada's Assisted Suicide— for 'Mature Minors,'" *Deseret News*, Nov. 4, 2022, https://www.deseret .com/2022/11/4/23401482/canada-euthanasia-laws-mature-minors-medical -aid-in-dying.

88   "Developing, Improving, and Supporting Best Practices for the Care of Patients Considering or Completing Aid in Dying," American Clinicians Academy on Medical Aid in Dying, accessed Jul. 2, 2024, https://www .acamaid.org/introduction.

89 Sandra Normon-Eady, "Oregon's Assisted Suicide Law," Connecticut General Assembly, Jan. 22, 2002, https://www.cga.ct.gov/2002/rpt/2002 -R-0077.htm.

90 Catharine Paddock, MD, "Washington State Legalizes Assisted Suicide," *Medical News Today*, Mar. 6, 2009, https://www.medicalnewstoday.com /articles/141318#1.

91 William Breitbart, *Physician-Assisted Suicide Ruling in Montana: Struggling with Care of the Dying, Responsibility, and Freedom in Big Sky Country* (self pub., Cambridge University Press, 2010), https://www.cambridge .org/core/journals/palliative-and-supportive-care/article/physicianassisted -suicide-ruling-in-montana-struggling-with-care-of-the-dying-responsibility -and-freedom-in-big-sky-country/7B79FC7FA6C4EABED0F608C1BE2F 3EFD.

92 Joseph Austin, "Vermont Now 'Death State' with Doctor-Assisted Suicide Law, Bishop Says," *National Catholic Reporter*, May 21, 2013, https://www .ncronline.org/news/politics/vermont-now-death-state-doctor-assisted -suicide-law-bishop-says.

93 Mackenzie Maxwell, "California Euthanasia Laws: Assisted Suicide and Patient's Rights," *Legal Beagle*, Nov. 11, 2019, https://legalbeagle.com /13720898-california-euthanasia-laws-assisted-suicide-and-patients-rights .html.

94 Angela Chen, "Assisted Suicide Is Now Legal in Colorado," *The Verge*, Nov. 8, 2016, https://www.theverge.com/2016/11/8/13520908/assisted-suicide -colorado-death-dignity-right-die-election-2016.

95 Alexandra Desanctis, "Assisted-Suicide Measure Takes Effect in Washington, D.C.," *National Review*, Jul. 19, 2017, https://www.nationalreview.com /corner/assisted-suicide-takes-effect-washington-dc.

96 Carla Herreira Russo, "Hawaii Becomes the 7th State to Legalize Medically Assisted Suicide," *Huffpost*, updated Apr. 6, 2018, https://www.huffpost .com/entry/hawaii-legalizes-assisted-suicide_n_5ac6c6f5e4b0337ad1e62 1fb.

97 Associated Press, "Maine Becomes 8th State to Legalize Assisted Suicide," NBC News, Jun. 12, 2019, https://www.nbcnews.com/politics /politics-news/maine-becomes-8th-state-legalize-assisted-suicide-n1017001.

98 Monica Burke, "Physician-Assisted Suicide Comes to New Jersey. Here's Why It's Badly Misguided," The Heritage Foundation, Apr. 19, 2019, https://www.heritage.org/life/commentary/physician-assisted-suicide- comes-new-jersey-heres-why-its-badly-misguided.

99 Cedar Attanasio, "New Mexico Latest State to Adopt Medically Assisted Suicide," Associated Press, Apr. 8, 2021, https://apnews.com/article /legislature-michelle-lujan-grisham-legislation-assisted-suicide-new-mexico -62bfb8e52a96ba46c23f6ae35cabdb5a.

100 "Final Report Card on State Responses to COVID-19," *Committee to Unleash Prosperity* (blog), Apr. 11, 2022, https://committeetounleash prosperity.com/final-report-card-on-state-responses-to-covid-19.

101 Brad Polumbo, "Free States Faring Far Better Than Lockdown States in One Huge Way, New Data Show," Foundation for Economic Education, Mar. 29, 2021, https://fee.org/articles/free-states-faring-far-better-than -lockdown-states-in-one-huge-way-new-data-show.

102 Preeti Jha, "New Zealand Euthanasia: Assisted Dying to Be Legal for Terminally Ill People," BBC, Oct. 30, 2020, https://www.bbc.com/news /world-asia-54728717.

103 "Spain Passes Law Allowing Euthanasia," BBC, Mar. 18, 2021, https: //www.bbc.com/news/world-europe-56446631.

104 Euronews, "Madrid State of Emergency: Spanish Police Enforce COVID-19 Travel Restrictions," *Euronews*, Oct. 9, 2020, https://www.euronews .com/2020/10/09/court-rejects-madrid-s-partial-lockdown-because-it -harms-fundamental-rights-and-freedoms.

105 Michael X. Jin et al., "Telemedicine: Current Impact on the Future," *Cureus* 12, no. 8 (2020): e9891, doi: 10.7759/cureus.9891.

106 "Telemedicine Policy Recommendations," American Clinicians Academy on Medical Aid in Dying, Mar. 25, 2020, https://www.acamaid.org /telemedicine.

107 Anita Hannig, "Dying Virtually: Pandemic Drives Medically Assisted Deaths Online," *The Conversation*, Jun. 2, 2020, https://theconversation.com /dying-virtually-pandemic-drives-medically-assisted-deaths-online-139093.

108 Ibid.

109 Eric Wicklund, "Vermont Amends Assisted Suicide Law to Include Telemedicine," *HealthLeaders*, Apr. 28, 2022, https://www.healthleadersmedia .com/telehealth/vermont-amends-assisted-suicide-law-include-telemedicine.

110 Jonel Aleccia, "Dying Patients Protest Looming Telehealth Crackdown," Associated Press, Apr. 23, 2023, https://apnews.com/article/dea-suicide -opioid-hospice-a6d5a0d3e923760f77cd246ea1927dd1.

111 "Telehealth," *Compassion & Choices*, accessed Jul. 2, 2024, https://www .compassionandchoices.org/our-issues/telehealth.

112 Department of Justice Canada, "New Medical Assistance in Dying Legislation Becomes Law," news release, Mar. 17, 2021, https://www.canada .ca/en/department-justice/news/2021/03/new-medical-assistance-in-dying -legislation-becomes-law.html.

113 "Bill C-7: An Act to Amend the Criminal Code (Medical Assistance in Dying)," Government of Canada, modified Nov. 27, 2023, https://justice .gc.ca/eng/csj-sjc/pl/charter-charte/c7.html.

114 Hanna Seariac, "Perspective: Inside the World of Canada's Assisted Suicide— for 'Mature Minors,'" *Deseret News*, Nov. 4, 2022, https://www.deseret .com/2022/11/4/23401482/canada-euthanasia-laws-mature-minors-medical -aid-in-dying.

115 Charles Lane, "Where the Prescription for Autism Can Be Death," *Washington Post*, Feb. 24, 2016, https://www.washingtonpost.com/opinions/where-the -prescription-for-autism-can-be-death/2016/02/24/8a00ec4c-d980–11e5 –81ae-7491b9b9e7df_story.html.

116 Health Canada, *Third Annual Report on Medical Assistance in Dying in Canada 2021* (Ottawa: Health Canada, 2022), 5, https://www.canada .ca/en/health-canada/services/publications/health-system-services/annual -report-medical-assistance-dying-2021.html.

117 Ashley Carnahan, "Canada Expanding Assisted Suicide Law to the Mentally Ill," *New York Post*, Oct. 28, 2022, https://nypost.com/2022/10/28 /canada-expanding-assisted-suicide-law-to-the-mentally-ill.

118 Steve Doughty, "Dutch Euthanasia Supporter Warns UK to Be Wary of 'Slippery Slope to Random Killing of Defenceless People,'" *Daily Mail*, updated Sep. 14, 2020, https://www.dailymail.co.uk/news/article-8729235/ Dutch-euthanasia-supporter-warns-UK-wary-slippery-slope.html.

119 Martin Armstrong, "Medical Murder in Canada," *Armstrong Economics*, Nov. 17, 2022, https://www.armstrongeconomics.com/international-news /canada/medical-murder-in-canada.

120 Roger Collier, "Assisted Death Gaining Acceptance in US," *Canadian Medical Association Journal* 189, no. 3 (2017): e123, doi: 10.1503/cmaj. 109–5366.

121 Kelly Malone, "Medically Assisted Deaths Could Save Millions in Health Care Spending: Report," CBC, Jan. 23, 2017, https://www.cbc.ca/news /canada/manitoba/medically-assisted-death-could-save-millions-1.3947481.

122 Office of the Parliamentary Budget Officer, *Cost Estimate for Bill C-7 "Medical Assistance in Dying"* (Office of the Parliamentary Budget Officer, 2020), https://qsarchive-archiveqs.pbo-dpb.ca/web/default/files/Documents /Reports/RP-2021–025-M/RP-2021–025-M_en.pdf.

123 Emma Freire, "Netherlands Considers Euthanasia for Healthy People, Doctors Say Things Are 'Getting Out Of Hand,'" *The Federalist*, Jun. 30, 2017, https://thefederalist.com/2017/06/30/netherlands-considers-euthanasia -healthy.

124 Ross Douthat, "Opinion: What Medically Assisted Suicide Has Done to Canada," *Chattanooga Times Free Press*, Dec. 11, 2022, https://www.timesfree-press.com/news/2022/dec/11/opinio-medically-assisted-suicide-canada-tfp.

125 Kelly Malone, "Medically Assisted Deaths Could Save Millions in Health Care Spending: Report," CBC, Jan. 23, 2017, https://www.cbc.ca/news /canada/manitoba/medically-assisted-death-could-save-millions-1.3947481.

126 Aaron K. Trachtenberg, MD and Braden Manns, MD, "Cost Analysis of Medical Assistance in Dying in Canada," *CMAJ* 189, no. 3 (2017): e101– 105, doi: 10.1503/cmaj.160650.

127 Ibid.

128 Merinda Teller, MD, "Keep Your Organs Healthy—and Safe," *Weston A. Price Foundation*, Feb. 17, 2020, https://www.westonaprice.org/health -topics/keep-your-organs-healthy-and-safe.

129 David Shaw and Alec Morton, "Counting the Cost of Denying Assisted Dying," *Clinical Ethics* 15, no. 2 (2020): 65–70, doi: 10.1177/1477750 920907996.

130 Joachim Cohen et al., "Public Acceptance of Euthanasia in Europe: A Survey Study in 47 Countries," *Int J Public Health* 59, no. 1 (2014): 143–156, doi: 10.1007/s00038–013–0461–6.

131 Megan Brenan, "Americans' Strong Support for Euthanasia Persists," Gallup, May 31, 2018, https://news.gallup.com/poll/235145/americans -strong-support-euthanasia-persists.aspx.

132 Ibid.

133 "Strong Public Support for Right to Die," Pew Research Center, Jan. 5, 2006, https://www.pewresearch.org/politics/2006/01/05/strong-public-support -for-right-to-die.

134 Ross Douthat, "Opinion: What Medically Assisted Suicide Has Done to Canada," *Chattanooga Times Free Press*, Dec. 11, 2022, https://www.timesfree press.com/news/2022/dec/11/opinio-medically-assisted-suicide-canada-tfp.

135 Jeffery Loffredo and Whitney Webb, "Under Guise of 'Racial Justice,' Johns Hopkins Lays Out Plan to Vaccinate Ethnic Minorities and Mentally Challenged First," *The Defender*, Dec. 1, 2020, https://childrenshealth defense.org/defender/johns-hopkins-plan-to-vaccinate-ethnic-minorities -and-mentally-challenged-first.

136 *Good Morning CHD*, "Canadian Doctors Speak Out against Medically Assisted Death," aired Dec. 9, 2022, on CHD.TV, https://live.childrenshealth defense.org/chd-tv/shows/good-morning-chd/canadian-doctors-speak-out -against-medically-assisted-death.

137 Hanna Seariac, "Perspective: Inside the World of Canada's Assisted Suicide— for 'mature minors,'" *Deseret News*, Nov. 4, 2022, https://www.deseret .com/2022/11/4/23401482/canada-euthanasia-laws-mature-minors-medical -aid-in-dying.

Chapter 5

1   Paul Starr, *The Social Transformation of American Medicine: The Rise of a Sovereign Profession and the Making of a Vast Industry* (New York: Basic Books, 1982).

2   Abraham Flexner, *Medical Education in the United States and Canada* (New York: Carnegie Foundation for the Advancement of Teaching, 1910), http: //archive.carnegiefoundation.org/publications/pdfs/elibrary/Carnegie _Flexner_Report.pdf.

3   Kenneth M. Ludmerer, "Commentary: Understanding the Flexner Report," *Academic Medicine* 85, no. 2 (2010): 193–196, doi: 10.1097/ACM. 0b013e3181c8f1e7.

4   Claire Johnson, MD and Bart Green, MD, "100 Years after the Flexner Report: Reflections on its Influence on Chiropractic Education," *Journal of Chiropractic Education* 24, no. 2 (2010): 145–152, doi: 10.7899/ 1042–5055–24.2.145.

5   Ibid.

6   Jessie Wright-Mendoza, "The 1910 Report that Disadvantaged Minority Doctors," *JSTOR Daily*, May 3, 2019, https://daily.jstor.org/the-1910 -report-that-unintentionally-disadvantaged-minority-doctors.

7   Kenneth M. Ludmerer, "Commentary: Understanding the Flexner Report," *Academic Medicine* 85, no. 2 (2010): 193–196, doi: 10.1097/ACM.0b013e 3181c8f1e7.

8   Claire Johnson, MD and Bart Green, MD, "100 Years after the Flexner Report: Reflections on its Influence on Chiropractic Education," *Journal of Chiropractic Education* 24, no. 2 (2010): 145–152, doi: 10.7899/1042–5055–24.2.145.

9   Dana Ullman, "How the AMA Got Rich & Powerful: 'The AMA's Seal of Approval,'" *HuffPost*, updated Jan. 4, 2015, https://www.huffpost.com /entry/how-the-ama-got-rich-powe_b_6103720.

10   Ibid.

11   Claire Johnson, MD and Bart Green, MD, "100 Years after the Flexner Report: Reflections on Its Influence on Chiropractic Education," *Journal of Chiropractic Education* 24, no. 2 (2010): 145–152, doi: 10.7899/1042–5055–24.2.145.

12   Dana Ullman, *Homeopathy: Medicine for the 21st Century* (Berkeley: North Atlantic Books, 1988), quoted in Dana Ullman, MPH, "A Condensed History of Homeopathy," *Homeopathic Family Medicine*, Jan. 23, 2017, https://homeopathic.com/a-condensed-history-of-homeopathy.

13   ANH-USA, "Court Slows Down FDA's Homeopathy Attack," *Alliance for Natural Health*, Jan. 14, 2021, https://anh-usa.org/court-deals-blow-to -fdas-homeopathy-attack.

14   Cemre Cukaci et al., "Against All Odds—the Persistent Popularity of Homeopathy," *Wiener Klinische Wochenschrift* 132, no. 9–10 (2020): 232– 242, doi: 10.1007/s00508–020–01624-x.

15   Clare Relton, "Prevalence of Homeopathy Use by the General Population Worldwide: A Systematic Review," *Homeopathy* 106, no. 2 (2017): 69–78, doi: 10.1016/j.homp.2017.03.002.

16   Zac Carpenter, "The Royal Family and Homeopathy: A Historic Journey of Advocacy and Impact," LinkedIn, May 5, 2023, https://www.linkedin.com /pulse/royal-family-homeopathy-historic-journey-advocacy-impact-carpenter.

17   Irvine Loudon, "A Brief History of Homeopathy," *Journal of the Royal Society of Medicine* 99, no. 12 (2006): 607–610, doi: 10.1177/014107680609901206.

18   David Horobin, "'A Brief History of Homeopathy'," *Journal of the Royal Society of Medicine* 100, no. 2 (2007): 65–66, doi: 10.1177/014107680710000210.

19   William R. Barclay, "Morris Fishbein, MD—1889–1976 Editor of JAMA—1924–1950," *JAMA* 236, no. 19 (1976): 2212, doi: 10.1001/jama.1976.03270200050033.

20   Dana Ullman, "How the AMA Got Rich & Powerful: 'The AMA's Seal of Approval,'" *HuffPost*, updated Jan. 4, 2015, https://www.huffpost.com/entry/how-the-ama-got-rich-powe_b_6103720.

21   Ibid.

22   Ibid.

23   Pierre Kory, "The American Board of Internal Medicine's Longstanding War on Doctors Is Escalating," *Pierre Kory's Medical Musings*, Aug. 23, 2023, https://pierrekorymedicalmusings.com/p/the-american-board-of-internal-medicines.

24   Brenda Baletti, PhD, "Breaking: Dr. Meryl Nass Sues Maine Medical Board over Suspension, Alleges Board Violated Her First Amendment Rights," *The Defender*, Aug. 17, 2023, https://childrenshealthdefense.org/defender/meryl-nass-maine-medical-board-lawsuit-first-amendment-rights.

25   James Lyons-Weiler and Paul Thomas, "Relative Incidence of Office Visits and Cumulative Rates of Billed Diagnoses along the Axis of Vaccination [Retracted]," *Int J Environ Res Public Health* 17, no. 22 (2020): 8674, doi: 10.3390/ijerph17228674.

26   Mark Skidmore, "The Role of Social Circle COVID-19 Illness and Vaccination Experiences in COVID-19 Vaccination Decisions: An Online Survey of the United States Population [Retracted]," *BMC Infectious Diseases* 23, no. 1 (2023): 51, doi: 10.1186/s12879-023-07998-3.

27   Michael Crichton, "Aliens Cause Global Warming," (Caltech Michelin Lecture, Jan. 17, 2003), https://stephenschneider.stanford.edu/Publications/PDF_Papers/Crichton2003.pdf.

28   Michael Nevradakis, PhD, "Exclusive: 'They Euthanized Him'—Widow Recounts Husband's Fatal COVID Protocol Treatment," *The Defender*, Dec. 20, 2023, https://childrenshealthdefense.org/defender/john-springer-covid-remdesivir-ventilator-death.

29   Jacqui Deevoy, "Midazolam—The Scandal that Cannot Be Ignored," *Unity News Network*, Jul. 14, 2021, https://unitynewsnetwork.co.uk/midazolam-the-scandal-that-cannot-be-ignored.

30   "Movie of the Week: March 4, 2024: A Good Death?" *Solari Report*, Mar. 2, 2024, https://home.solari.com/movie-of-the-week-march-4-2024-a-good-death.

31   Steve Doughty, "Dutch Euthanasia Supporter Warns UK to Be Wary of Slippery Slope to Random Killing of Defenceless People'," *Daily Mail*,

updated Sep. 14, 2020, https://www.dailymail.co.uk/news/article-8729235
/Dutch-euthanasia-supporter-warns-UK-wary-slippery-slope.html.

32   Steve Cook, "The UK Midazolam Murders—Update on Legal Progress,"
The Liberty Beacon, Nov. 8, 2021, https://www.thelibertybeacon.com/the
-uk-midazolam-murders-update-on-legal-progress.

33   Rachel Rettner, "How Does Execution Drug Midazolam Work?" *Live
Science*, Jun. 29, 2015, https://www.livescience.com/51384-execution-drug
-midazolam-effect.html.

34   Rhoda Wilson, "UK Lawyer Clare Wills Harrison: Midazolam Orders and
the Liverpool Care Pathway," *Europe Reloaded*, Jan. 25, 2022, https://www
.europereloaded.com/uk-lawyer-clare-wills-harrison-midazolam-orders
-and-the-liverpool-care-pathway.

35   Steve Cook, "The UK Midazolam Murders—Update on Legal Progress,"
The Liberty Beacon, Nov. 8, 2021, https://www.thelibertybeacon.com/the
-uk-midazolam-murders-update-on-legal-progress.

36   Peter Halligan, "Eugenics—UK Style," *Peter's Newsletter*, Aug. 23, 2022,
https://peterhalligan.substack.com/p/eugenics-uk-style.

37   Children's Health Defense Team, "Tens of Thousands of Lives Could Have
Been Saved If Research on Covid Treatments Hadn't Been Suppressed,
Doctors and Economists Say," *The Defender*, May 13, 2021, https:
//childrenshealthdefense.org/defender/tens-of-thousands-lives-could
-have-been-saved-covid-treatments.

38   Children's Health Defense Team, "Two-Tiered Medicine: Why Is
Hydroxychloroquine Being Censored and Politicized?" Children's Health
Defense, Jul. 30, 2020, https://childrenshealthdefense.org/news/two-tiered
-medicine-why-is-hydroxychloroquine-being-censored-and-politicized.

39   Meryl Nass, MD, "Pfizer, FDA Documents Contradict Official COVID
Vaccine Safety Narrative—Is This Fraud?" *The Defender*, Mar. 15, 2022,
https://childrenshealthdefense.org/defender/pfizer-fda-documents-contradict
-covid-vaccine-safety-narrative.

40   Michelle Rogers, "Fact Check: Hospitals Get Paid More If Patients Listed
as COVID-19, on Ventilators," *USA Today*, updated Apr. 27, 2020, https:
//www.usatoday.com/story/news/factcheck/2020/04/24/fact-check
-medicare-hospitals-paid-more-covid-19-patients-coronavirus/30006
38001.

41   Michael Nevradakis, PhD, "Exclusive: Dad Describes Hospital's COVID
'Protocols' He Believes Killed His 19-Year-Old Daughter," *The Defender*,
Aug. 10, 2023, https://childrenshealthdefense.org/defender/grace-schara
-covid-protocol-death.

42   Children's Health Defense Team, "The Emperor Has No Clothes: COVID
Math Simply Doesn't Add Up," *The Defender*, Sep. 23, 2021, https://childrens
healthdefense.org/defender/covid-health-data-mainstream-media-vaccine
-risks.

43   AJ DePriest and TN Liberty Network, "Summary Brief—Follow The Money: Blood Money in U.S. Healthcare," TN Liberty Network, revised Jul. 8, 2022, https://childrenshealthdefense.org/wp-content/uploads/Follow -the-Money_Blood-Money-in-US-Hospitals_BRIEF_9-Jul_2022.pdf.

44   Andrew Lohse, "Press Release: 'Medical Murder' Outpaces Heart Disease and Cancer, Becoming America's #1 Cause of Death," *DailyClout*, Sep. 22, 2023, https://dailyclout.io/press-release-medical-murder-outpaces-heart -disease-and-cancer-becoming-americas-1-cause-of-death.

45   Ibid.

46   *The People's Study*, CHD.TV, accessed Jul. 2, 2024, https://live.childrens healthdefense.org/chd-tv/bus.

47   "About CHBMP," The Covid-19 Humanity Betrayal Memory Project, accessed Jul. 26, 2024, https://chbmp.org/about.

48   "The 25 Commonalities," COVID-19 Humanity Betrayal Memory Project, accessed Jul. 2, 2024, https://chbmp.org/commonalities.

49   Sasha Latypova, "Highland Hospital, Rochester NY, Attempted to Kill My Family Member with Covid Protocol in August 2023," *Due Diligence and Art*, Sep. 9, 2023, https://sashalatypova.substack.com/p /highland-hospital-rochester-ny-attempted.

50   Ibid.

51   "The 25 Commonalities," COVID-19 Humanity Betrayal Memory Project, accessed Jul. 2, 2024, https://chbmp.org/commonalities.

52   Jagdish Khubchandani et al., "COVID-19 Vaccine Refusal among Nurses Worldwide: Review of Trends and Predictors," *Vaccines* (Basel) 10, no. 2 (2022): 230, doi: 10.3390/vaccines10020230.

53   Hannah Mitchell, "30% of Nurses Would Quit over Vaccine Mandate, Ohio Union Says," *Becker's Hospital Review*, Sep. 2, 2021, https://www .beckershospitalreview.com/workforce/30-of-nurses-would-quit-over-vaccine -mandate-ohio-union-says.html.

54   Sarai Rodriguez, "Over One Third of Physicians Disagree with Covid Vaccine Mandate," *RevCycle Intelligence*, Nov. 8, 2021, https://revcycleintelligence .com/news/over-one-third-of-physicians-disagree-with-covid-vaccine-mandate.

55   Kelly Gooch, "Vaccination-Related Employee Departures At 55 Hospitals, Health Systems," *Becker's Hospital Review*, updated Feb. 17, 2023, https: //www.beckershospitalreview.com/workforce/vaccination-requirements -spur-employee-terminations-resignations-numbers-from-6-health-systems .html?oly_enc_id=5489D8214856H7J.

56   Nick Murray, "Gov. Mills' Mandate Pushed 10% of Healthcare Workers out of the Industry," *Maine Policy Institute* (blog), Nov. 2, 2021, https://mainepolicy .org/gov-mills-mandate-pushed-10-of-healthcare-workers-out-of-the -industry.

57   Ibid.

58 Mark Hagland, "McKinsey Report: Nursing Shortage Will Become Dire by 2025," *Healthcare Innovation*, May 17, 2022, https://www.hcinnovation group.com/policy-value-based-care/staffing-professional-development /news/21268125/mckinsey-report-nursing-shortage-will-become-dire-by -2025.

59 "1 in 4 Healthcare Workers Quitting over Vaccine Mandates Will Leave the Profession," Resume Builder, last updated Nov. 23, 2021, https://www .resumebuilder.com/1-in-4-healthcare-workers-quitting-over-vaccine-mandates -will-leave-the-profession/

60 "AONL Longitudinal Nursing Leadership Insight Study," American Organization for Nursing Leadership, January 2024, https://www.aonl.org /resources/nursing-leadership-survey.

61 "Travel Nursing Trends: A Look into the Future of Travel Nursing," Trusted Nurse Staffing, Jan. 4, 2023, https://www.trustednursestaffing.com/trends -in-travel-nursing.

62 Ibid.

63 Cristal Mackay, "Digging into the Data: Travel Nurse Demographics," *Aya Healthcare* (blog), May 20, 2022, https://www.ayahealthcare.com/blog /digging-into-the-data-travel-nurse-demographics.

64 Emma Curchin, "How Does Travel Nurse Pay Compare to Permanent Staff Nurses?" Center for Economic and Policy Research, Jun. 15, 2023, https: //cepr.net/how-does-travel-nurse-pay-compare-to-permanent-staff-nurses.

65 Robert Painter, "Who's Responsible for the Negligence of Temporary or Travel Nurses?" Painter Law Firm, Jan. 26, 2022, https://painterfirm.com/medmal /Whos-responsible-for-the-negligence-of-temporary-or-travel-nurses.

66 Emma Curchin, "How Does Travel Nurse Pay Compare to Permanent Staff Nurses?" Center for Economic and Policy Research, Jun. 15, 2023, https: //cepr.net/how-does-travel-nurse-pay-compare-to-permanent-staff-nurses.

67 Ibid.

68 Judith Garber, "The Rising Danger of Private Equity in Healthcare," Lown Institute, Jan. 23, 2024, https://lowninstitute.org/the-rising-danger-of -private-equity-in-healthcare.

69 Common Dreams, "Patients Face Greater Risks at Hospitals Acquired by Private Equity Firms, Study Reveals," *The Defender*, Jan. 4, 2024, https: //childrenshealthdefense.org/defender/hospitals-private-equity-firms -patients-injuries-infections-cd.

70 Dr. Joseph Mercola, "Lawsuits Pile Up Alleging Remdesivir Killed COVID Patients," *The Defender*, Mar. 14, 2023, https://childrenshealthdefense.org /defender/lawsuits-remdesivir-covid-cola.

71 Dr. Joseph Mercola, "Dr. Pierre Kory: How and Why Pharma Killed Ivermectin," *The Defender*, Nov. 7, 2022. https://childrenshealthdefense .org/defender/war-on-ivermectin-pierre-kory-big-pharma-cola.

72   Brian Shilhavy, *Medical Kidnapping: A Threat to Every Family in America* (Bastrop, TX: Sophia Media, 2016), https://healthytraditions.com/pages /medical-kidnapping.

73   Pharmaphorum Editor, "A History of the Pharmaceutical Industry," *Pharmaphorum*, Sep. 1, 2020, https://pharmaphorum.com/r-d/a_history _of_the_pharmaceutical_industry.

74   Denis G. Arnold, PhD, Oscar Jerome Stewart, PhD, and Tammy Beck, PhD, "Financial Penalties Imposed on Large Pharmaceutical Firms for Illegal Activities," *JAMA* 17, no. 324 (2020): 1995–1997, doi: 10.1001 /jama.2020.18740.

75   David Badcott, and Stephan Sahm, "The Dominance of Big Pharma: Unhealthy Relationships?" *Medicine, Health Care and Philosophy* 16, no. 2 (2013): 245–247, doi: 10.1007/s11019–012–9387–7.

76   Ajai R. Singh, MD, "Modern Medicine: Towards Prevention, Cure, Well-Being and Longevity," *Mens Sana Monographs* 8, no. 1 (2010): 17–29, doi: 10.4103/0973–1229.58817.

77   Patrick Radden Keefe, *Empire of Pain* (New York: Doubleday, 2021).

78   Andrew W. D. Adyama, "The Two Arthur Sacklers," *The Harvard Crimson*, Oct. 17, 2019, https://www.thecrimson.com/article/2019/10/17/two-arthur -sacklers.

79   Joanna Walters, "Meet the Sacklers: The Family Feuding over Blame for the Opioid Crisis," *The Guardian*, Feb. 13, 2018, https://www.theguardian.com /us-news/2018/feb/13/meet-the-sacklers-the-family-feuding-over-blame -for-the-opioid-crisis.

80   Andrew W. D. Adyama, "The Two Arthur Sacklers," *The Harvard Crimson*, Oct. 17, 2019, https://www.thecrimson.com/article/2019/10/17/two-arthur -sacklers.

81   Robert Bud, "Antibiotics, Big Business, and Consumers. The Context of Government Investigations into the Postwar American Drug Industry," *Technology and Culture* 46, no. 2 (2005): 329–349, doi: 10.1353/tech. 2005.0066.

82   "A Brief History of Benzodiazepines," Benzodiazepine Information Coalition, accessed Jul. 2, 2024, https://www.benzoinfo.com/a-brief-history-of -benzodiazepines.

83   David W. McFadden, Elizabeth Calvario, and Cynthia Graves, "The Devil Is in the Details: The Pharmaceutical Industry's Use of Gifts to Physicians as Marketing Strategy," *Journal of Surgical Research* 140, no. 1 (2007): 1–5, doi: 10.1016/j.jss.2006.10.010.

84   Ibid.

85   David Badcott and Stephan Sahm, "The Dominance of Big Pharma: Unhealthy Relationships?" *Medicine, Health Care and Philosophy* 16, no. 2 (2013): 245–247, doi: 10.1007/s11019–012–9387–7.

86 A Midwestern Doctor, "Why Are Vaccine Injured Patients Silenced?" *The Forgotten Side of Medicine*, Sep. 23, 2023, https://www.midwesterndoctor.com/p/why-are-vaccine-injured-patients.

87 David W. McFadden, Elizabeth Calvario, and Cynthia Graves, "The Devil Is in the Details: The Pharmaceutical Industry's Use of Gifts to Physicians as Marketing Strategy," *Journal of Surgical Research* 140, no. 1 (2007): 1–5, doi: 10.1016/j.jss.2006.10.010.

88 Art Van Zee, "The Promotion and Marketing of OxyContin: Commercial Triumph, Public Health Tragedy," *American Journal of Public Health* 99, no. 2 (2009): 221–227, doi: 10.2105/AJPH.2007.131714.

89 Maria Chutchian, "New York Bankruptcy Judge Who Oversaw Purdue Pharma Case to Retire," Reuters, Sep. 29, 2021, https://www.reuters.com/legal/transactional/new-york-bankruptcy-judge-who-oversaw-purdue-pharma-case-retire-2021-09-28.

90 Devan Cole and Ariane de Vogue, "Supreme Court Blocks $6 Billion Opioid Settlement That Would Have Given the Sackler Family Immunity," CNN, Aug. 10, 2023, https://www.cnn.com/2023/08/10/politics/supreme-court-purdue-pharma-opioid-settlement/index.html.

91 Art Van Zee, "The Promotion and Marketing of OxyContin: Commercial Triumph, Public Health Tragedy," *American Journal of Public Health* 99, no. 2 (2009): 221–227, doi: 10.2105/AJPH.2007.131714.

92 "Panelists for 'Talking Our Way Out': Dr. Art Van Zee," Emory & Henry College, accessed Jul. 3, 2024, https://www.ehc.edu/live/blurbs/1319-panelists-for-talking-our-way-out.

93 Art Van Zee, "The Promotion and Marketing of OxyContin: Commercial Triumph, Public Health Tragedy," *American Journal of Public Health* 99, no. 2 (2009): 221–227, doi: 10.2105/AJPH.2007.131714.

94 "11 Dietary Principles: Avoid Refined and Denatured Foods," Weston A. Price Foundation, accessed Jun. 25, 2024, https://www.westonaprice.org/wp-content/uploads/11Principles-chapter1.pdf.

95 "Dr. Weston A. Price Movietone," Weston A. Price Foundation, accessed Jun. 25, 2024, https://www.westonaprice.org/about-us/dr-weston-a-price-movietone.

96 Weston A. Price, *Nutrition and Physical Degeneration*, 8th ed. (La Mesa, CA: Price-Pottenger Nutrition Foundation, 2009).

97 Hilda Labrada Gore, "Episode 155: The 'Isaac Newton of Nutrition' with Chris Masterjohn," Oct. 8, 2018, in *Wise Traditions Podcast*, produced by Hilda Labrada Gore, 00:36:34, https://www.westonaprice.org/podcast/155-the-isaac-newton-of-nutrition.

98 Hilda Labrada Gore, "Episode 30: Nutrient Density (principle #3) with Sally Fallon Morell," Jul. 4, 2016, in *Wise Traditions Podcast*, produced by Hilda Labrada Gore, 00:28:16, https://www.westonaprice.org/podcast/30-nutrient-density-principle-3.

99   "11 Dietary Principles: Avoid Refined and Denatured Foods," Weston A. Price Foundation, accessed Jun. 25, 2024, https://www.westonaprice.org /wp-content/uploads/11Principles-chapter1.pdf.

100 Jim Earles, "Sugar-Free Blues: Everything You Wanted to Know about Artificial Sweeteners," *Wise Traditions in Food, Farming and the Healing Arts*, Feb. 19, 2004, https://www.westonaprice.org/health-topics/modern-foods /sugar-free-blues-everything-you-wanted-to-know-about-artificial-sweeteners.

101 Chris Knobbe, MD, "The Omega-6 Apocalypse," *Wise Traditions in Food, Farming and the Healing Arts*, Jul. 12, 2023, https://www.westonaprice.org /health-topics/the-omega-6-apocalypse.

102 Sally Fallon and Mary G. Enig, PhD, "Be Kind to Your Grains . . . and Your Grains Will Be Kind to You," Weston A. Price Foundation, Jan. 1, 2000, https://www.westonaprice.org/health-topics/food-features/be-kind-to -your-grains-and-your-grains-will-be-kind-to-you.

103 "Dirty Secrets of the Food Processing Industry," Weston A. Price Foundation, Dec. 26, 2005, https://www.westonaprice.org/health-topics/modern-foods /dirty-secrets-of-the-food-processing-industry.

104 "A Campaign for Real Milk: Nature's Perfect Food," Weston A. Price Foundation, accessed Jun. 25, 2024, https://www.westonaprice.org/wp -content/uploads/RealMilkBrochure.pdf

105 "21 CFR Part 74—Listing of Color Additives Subject to Certification. Subpart A—Foods," Code of Federal Regulations, 42 FR 15654, Mar. 22, 1977, https://www.ecfr.gov/current/title-21/chapter-I/subchapter-A/part-74.

106 Merinda Teller, MPH, PhD, "MSG: Three Little Letters That Spell Big Fat Trouble," *Wise Traditions in Food, Farming and the Healing Arts*, Apr. 18, 2017, https://www.westonaprice.org/health-topics/modern-diseases/msg-three -little-letters-spell-big-fat-trouble.

107 Sally Fallon and Mary G. Enig, PhD, "Caustic Commentary, Fall 2006: Additives Are Additive," Weston A. Price Foundation, Sep. 30, 2006, https: //www.westonaprice.org/health-topics/caustic-commentary/caustic -commentary-fall-2006/#gsc.tab=0.

108 Sally Fallon Morell, "Dissecting Those New Fake Burgers," *Wise Traditions in Food, Farming and the Healing Arts*, Nov. 5, 2019, https://www.westonaprice .org/health-topics/dissecting-those-new-fake-burgers.

109 Kaayla Daniel, "What Can I Drink Instead of Soy Milk?" Weston A. Price Foundation, Mar. 3, 2012, https://www.westonaprice.org/health-topics/soy -alert/what-can-i-drink-instead-of-soy-milk.

110 Jessica Fu, "If Food Is Medicine, Why Isn't It Taught at Medical Schools?" *The Counter*, Oct. 14, 2019, https://thecounter.org/medical-schools-lack -nutritional-education.

111 Robin L. Danek et al., "Perceptions of Nutrition Education in the Current Medical School Curriculum," *Family Medicine* 49, no. 10 (2017): 803–806, PMID: 29190407.

112 Sylvia Onusic, "Great Pioneers in Nutrition of the Twentieth Century," *Wise Traditions in Food, Farming and the Healing Arts*, Jul. 1, 2015, https://www.westonaprice.org/health-topics/nutrition-greats/great-pioneers-in-nutrition-of-the-twentieth-century.

113 Chris Knobbe, MD, "The Omega-6 Apocalypse," *Wise Traditions in Food, Farming and the Healing Arts*, Jul. 12, 2023, https://www.westonaprice.org/health-topics/the-omega-6-apocalypse.

114 P. K. Nguyen, S. Lin, and P. Heidenreich, "A Systematic Comparison of Sugar Content in Low-Fat vs. Regular Versions of Food," *Nutr Diabetes* 25, no. 6 (2016): e193, doi: 10.1038/nutd.2015.43.

115 Hilda Labrada Gore, "Episode 425: How Vegetable Oils Destroy Our Health with Chris Knobbe," Jun. 19, 2023, in *Wise Traditions Podcast*, podcast, produced by Hilda Labrada Gore, 00:41:23, https://www.westonaprice.org/podcast/how-vegetable-oils-destroy-our-health.

116 Uffe Ravnskov, *The Cholesterol Myths: Exposing the Fallacy That Saturated Fat and Cholesterol Cause Heart Disease* (Washington DC: New Trends Publishing, 2003), 265.

117 Sally Fallon Morell, "The Salt of the Earth," *Wise Traditions in Food, Farming and the Healing Arts*, Jul. 4, 2011, https://www.westonaprice.org/health-topics/abcs-of-nutrition/the-salt-of-the-earth.

118 Timothy M. Smith, "The FDA's New Guidance on Sodium Could Be Lifesaving. Here's Why," American Medical Association, Jan. 17, 2022, https://www.ama-assn.org/delivering-care/hypertension/fda-s-new-guidance-sodium-could-be-lifesaving-here-s-why.

119 Morton Satin, "Salt and Our Health," *Wise Traditions in Food, Farming and the Healing Arts*, Mar. 26, 2012, https://www.westonaprice.org/health-topics/abcs-of-nutrition/salt-and-our-health.

120 Sally Fallon Morell, "The Salt of the Earth," *Wise Traditions in Food, Farming and the Healing Arts*, Jul. 4, 2011, https://www.westonaprice.org/health-topics/abcs-of-nutrition/the-salt-of-the-earth.

121 Ibid.

122 Morton Satin, "Salt and Our Health," *Wise Traditions in Food, Farming and the Healing Arts*, Mar. 26, 2012, https://www.westonaprice.org/health-topics/abcs-of-nutrition/salt-and-our-health.

123 Sally Fallon Morell, "The Salt of the Earth," *Wise Traditions in Food, Farming and the Healing Arts*, Jul. 4, 2011, https://www.westonaprice.org/health-topics/abcs-of-nutrition/the-salt-of-the-earth.

124 "Why Some Think Lowering Salt Intake May Do More Harm Than Good," ABC News, Apr. 7, 2015, https://abcnews.go.com/Health/lowering-salt-intake-harm-good/story?id=30139911.

125 Melinda Wenner Moyer, "It's Time to End the War on Salt," *Scientific American*, Jul. 8, 2011, https://www.scientificamerican.com/article/its-time-to-end-the-war-on-salt.

126 Sally Fallon Morell, "The Salt of the Earth," *Wise Traditions in Food, Farming and the Healing Arts*, Jul. 4, 2011, https://www.westonaprice.org/health-topics/abcs-of-nutrition/the-salt-of-the-earth.

127 Ana Fernandez et al., "Evidence-Based Medicine: Is It a Bridge Too Far?" *Health Research Policy and Systems* 13, no. 66 (2015), doi: 10.1186/s12961-015-0057-0.

128 A Midwestern Doctor, "Why Can Doctors Not Diagnose Medical Injuries?" *The Forgotten Side of Medicine*, May 2, 2022, https://www.midwesterndoctor.com/p/why-can-doctors-not-diagnose-medical.

129 Ana Fernandez et al., "Evidence-Based Medicine: Is It a Bridge Too Far?" *Health Research Policy and Systems* 13, no. 66 (2015), doi: 10.1186/s12961-015-0057-0.

130 Ariel L. Zimerman, "Evidence-Based Medicine: A Short History of a Modern Medical Movement," *Virtual Mentor* 15, no. 1 (2013): 71–76, doi: 10.1001/virtualmentor.2013.15.1.mhst1–1301.

131 Ibid.

132 Ibid.

133 Joshua J. Goldman and Tiffany L. Shih, "The Limitations of Evidence-Based Medicine: Applying Population-Based Recommendations to Individual Patients," *Virtual Mentor* 13, no. 1 (2011): 26–30, doi: 10.1001/virtualmentor.2011.13.1.jdsc1–1101.

134 Achilleas Thoma and Felmont F. Eaves II, "A Brief History of Evidence-Based Medicine (EBM) and the Contributions of Dr David Sackett," *Aesthetic Surgery Journal* 35, no. 8 (2015): NP261-NP263, doi: 10.1093/asj/sjv130.

135 Joshua J. Goldman and Tiffany L. Shih, "The Limitations of Evidence-Based Medicine: Applying Population-Based Recommendations to Individual Patients," *Virtual Mentor* 13, no. 1 (2011): 26–30, doi: 10.1001/virtualmentor.2011.13.1.jdsc1–1101.

136 Christopher Worsham and Anupam B. Jena, "The Art of Evidence-Based Medicine," *Harvard Business Review*, Jan. 30, 2019, https://hbr.org/2019/01/the-art-of-evidence-based-medicine.

137 Ibid.

138 Ibid.

139 Ibid.

140 Brian Hurwitz, "Legal and Political Considerations of Clinical Practice Guidelines," *BMJ* 318, no. 7184 (1999): 661–664, doi: 10.1136/bmj.318.7184.661.

141 Joshua J. Goldman and Tiffany L. Shih, "The Limitations of Evidence-Based Medicine: Applying Population-Based Recommendations to Individual Patients," *Virtual Mentor* 13, no. 1 (2011): 26–30, doi: 10.1001/virtualmentor.2011.13.1.jdsc1–1101.

142 Sandra Vento, Francesca Cainelli, and Alfredo Vallone, "Defensive Medicine: It Is Time to Finally Slow Down an Epidemic," *World J Clin Cases* 6, no. 11 (2018): 406–409, doi: 10.12998/wjcc.v6.i11.406.

143 Ibid.

144 SeverlyPrecocious, "Shower Thought: Medical School Does Not Select for Thinkers, It Selects for Robots," Reddit, May 4, 2019, https://www.reddit .com/r/medicalschool/comments/bkhn94/comment/emiw30n.

145 A Midwestern Doctor, "Why Can Doctors Not Diagnose Medical Injuries," *The Forgotten Side of Medicine,* May 2, 2022, https://www.midwesterndoctor .com/p/why-can-doctors-not-diagnose-medical.

146 R. Brian Haynes, "What Kind of Evidence Is It That Evidence-Based Medicine Advocates Want Health Care Providers and Consumers to Pay Attention To?" *BMC Health Services Research* 2, no. 3 (2002), doi: 10.1186/ 1472–6963–2–3.

147 Stefan Sauerland and Christoph Seiler, "Role of Systematic Reviews and Meta-Analysis in Evidence-Based Medicine," *World Journal of Surgery* 29, no. 5 (2005): 582–587, doi: 10.1007/s00268–005–7917–7.

148 Patricia B. Burns, Rod J. Rohrich, and Kevin Chung, "The Levels of Evidence and Their Role in Evidence-Based Medicine," *Plastic and Reconstructive Surgery* 128, no. 1 (2011): 305–310, doi: 10.1097/PRS.0b013e318219c171.

149 Ana Fernandez et al., "Evidence-Based Medicine: Is It a Bridge Too Far?" *Health Research Policy and Systems* 13, no. 66 (2015), doi: 10.1186/s12961 –015–0057–0.

150 R. Brian Haynes, "What Kind of Evidence Is It That Evidence-Based Medicine Advocates Want Health Care Providers and Consumers to Pay Attention To?" *BMC Health Services Research* 2, no. 3 (2002), doi: 10.1186 /1472–6963–2–3.

151 Elise Gamertsfelder and Leeza Osipenko, "What If Trial Participants Knew Their Contributions Were for Naught?" *Medpage Today*, Oct. 28, 2023, https: //www.medpagetoday.com/opinion/second-opinions/107038?xid=nl _secondopinion_2023–10–31&eun=g1366012d0r.

152 Nicholas J. DeVito, Seb Bacon, and Ben Goldacre, "Compliance with Legal Requirement to Report Clinical Trial Results on ClinicalTrials.gov: A Cohort Study," *Lancet* 395, no. 10221 (2020): 361–369, doi: 10.1016/ S0140–6736(19)33220–9.

153 Charles Piller, "Failure to Report: A STAT Investigation of Clinical Trials Reporting," *STAT*, Dec. 13, 2015, https://www.statnews.com/2015/12/13 /clinical-trials-investigation.

154 Ibid.

155 Robert J. Snyder, "Lack of Transparency in Publishing Negative Clinical Trial Results," *Clinics in Podiatric Medicine and Surgery* 37, no. 2 (2020): 385–389, doi: 10.1016/j.cpm.2019.12.013.

156 ClinicalTrials.gov, "Adverse Event Information: Introduction, Results Database Train-the-Trainer Workshop," Aug. 2021, National Library of Medicine, https://www.nlm.nih.gov/oet/ed/ct/2021_prs_ttt/4_adverse/2021 _08_m4_file_00_AdverseEvents_Intro_final.pdf.

157 Joshua J. Goldman and Tiffany L. Shih, "The Limitations of Evidence-Based Medicine: Applying Population-Based Recommendations to Individual Patients," *Virtual Mentor* 13, no. 1 (2011): 26–30, doi: 10.1001/virtualme ntor.2011.13.1.jdsc1–1101.

158 Robert F. Kennedy Jr., "New Study: Vaccine Manufacturers and FDA Regulators Used Statistical Gimmicks to Hide Risks of HPV Vaccines," Children's Health Defense, Aug. 11, 2017, https://childrenshealthdefense. org/news/new-study-vaccine-manufacturers-fda-regulators-used-statistical -gimmicks-hide-risks-hpv-vaccines.

159 S. Bala Bhaskar, "Concealing Research Outcomes: Missing Data, Negative Results and Missed Publications," *Indian Journal of Anaesthesia* 61, no. 6 (2017): 453–455, doi: 10.4103/ija.IJA_361_17.

160 Stephen L. George and Mark Buyse, "Data Fraud in Clinical Trials," *Clinical Investigation (Lond)* 5, no. 2 (2015): 161–173, doi: 10.4155/cli.14.116.

161 Daniele Fanelli, "How Many Scientists Fabricate and Falsify Research? A Systematic Review and Meta-Analysis of Survey Data," *PLoS One* 4, no. 5 (2009): e5738, doi: 10.1371/journal.pone.0005738.

162 Yu Xie, Kai Wang, Yan Kong, "Prevalence of Research Misconduct and Questionable Research Practices: A Systematic Review and Meta-Analysis," *Science and Engineering Ethics* 27, no. 4 (2021): 41, doi: 10.1007/s11948 –021–00314–9.

163 Richard Van Noorden, "Medicine Is Plagued by Untrustworthy Clinical Trials. How Many Studies Are Faked or Flawed?" *Nature*, Jul. 18, 2023, https://www.nature.com/articles/d41586–023–02299-w.

164 Amy Beck, Ann Bianchi, and Darlene Showalter, "Evidence-Based Practice Model to Increase Human Papillomavirus Vaccine Uptake: A Stepwise Approach," *Nursing for Women's Health* 25, no. 6 (2021): 430–436, doi: 10.1016/j.nwh.2021.09.006.

165 Claire P. Rees et al., "Will HPV Vaccination Prevent Cervical Cancer?" *Journal of the Royal Society of Medicine* 113, no. 2 (2020): 64–78, doi: 10. 1177/0141076819899308.

166 Children's Health Defense Team, "Bombshell Study Questioning HPV Vaccine Efficacy Appears as UK's Cervical Cancer Rates Rise in Young," Children's Health Defense, Jan. 23, 2020, https://childrenshealthdefense .org/news/bombshell-study-questioning-hpv-vaccine-efficacy-appears-as -the-uks-cervical-cancer-rates-rise-in-young.

167 Children's Health Defense Team, "HPV Vaccine's Likely Contribution to Sweden's Spike in Cervical Cancer," Children's Health Defense, May 15, 2018, https://childrenshealthdefense.org/news/hpv-vaccines-likely -contribution-to-swedens-spike-in-cervical-cancer.

168 Robert F. Kennedy Jr., "Australian Data: Cancer Epidemic in Gardasil Girls," Children's Health Defense, Jul. 14, 2020, https://childrenshealth defense.org/news/australian-data-cancer-epidemic-in-gardasil-girls.

169 The Defender Staff, "Utah Woman Is First to Sue Merck Alleging Gardasil HPV Vaccine Caused Cervical Cancer," *The Defender*, Apr. 26, 2023, https://childrenshealthdefense.org/defender/caroline-cantera-merck -gardasil-hpv-vaccine-lawsuit-cervical-cancer.

170 Brenda Baletti, PhD, "Injured by Merck's Gardasil HPV Vaccine? You or Your Child May Have a Case," *The Defender*, Jun. 20, 202, https://childrens healthdefense.org/defender/injured-merck-gardasil-hpv-vaccine-case.

171 "Word of the Year 2022," *Merriam-Webster Dictionary*, accessed Nov. 26, 2023, https://www.merriam-webster.com/wordplay/word-of-the-year-2022.

172 CJ Hopkins, "The Year of the Gaslighter," *CJ Hopkins*, Dec. 18, 2022, https://cjhopkins.substack.com/p/the-year-of-the-gaslighter.

173 "Word of the Year 2023," *Merriam-Webster Dictionary*, accessed Nov. 26, 2023, https://www.merriam-webster.com/wordplay/word-of-the-year.

174 Cynthia Vinney, "How to Spot Medical Gaslighting and What to Do about It," *Verywell Mind*, May 18, 2023, https://www.verywellmind.com /what-is-medical-gaslighting-6831284.

175 Children's Health Defense Team, "Laughing All the Way to the Bank: Vaccine Makers and Liability Protection—Conflicts of Interest Undermine Children's Health: Part III," Children's Health Defense, May 23, 2019, https://childrenshealthdefense.org/news/laughing-all-the-way-to-the-bank -vaccine-makers-and-liability-protection-conflicts-of-interest-undermine -childrens-health-part-iii.

176 Mary S. Holland, "Liability for Vaccine Injury: The United States, the European Union, and the Developing World," *Emory Law Journal* 67, no. 3 (2018): 415, https://scholarlycommons.law.emory.edu/elj/vol67/iss3/3.

177 David Charbonneau, PhD, "The 'Dark Truth' behind America's 'Vaccine Court,'" *The Defender*, May 9, 2022, https://childrenshealthdefense.org /defender/truth-americas-vaccine-court.

178 Children's Health Defense Team, "Gaslighting Autism Families: CDC, Media Continue to Obscure Decades of Vaccine-Related Harm," *The Defender*, Dec. 17, 2021, https://childrenshealthdefense.org/defender/autism -cdc-media-vaccine-related-harm.

179 JB Handley, "Vaccines and Autism—Is the Science Really Settled?" Children's Health Defense, Aug. 11, 2020, https://childrenshealthdefense .org/news/vaccines-and-autism-is-the-science-really-settled.

180 A Midwestern Doctor, "A Primer on Medical Gaslighting," *The Forgotten Side of Medicine*, Feb. 9, 2023, https://www.midwesterndoctor .com/p/a-primer-on-medical-gaslighting.

181 Children's Health Defense Team, "Why You Shouldn't Take Mainstream Media's Latest Wave of Polio Alarmism at Face Value," *The Defender*, Sep.14,2022,https://childrenshealthdefense.org/defender/mainstream-media-polio-vaccine-alarmism.

182 Tessa Lena, "A Story about Polio, Pesticides and the Meaning of Science," *The Defender*, Mar. 7, 2022, https://childrenshealthdefense.org/defender/polio-pesticides-ddt-science.

183 Rodney Dodson, "What You Didn't Know about Polio," *Medium*, May 3, 2021, https://rodneydodson000.medium.com/what-you-didnt-know-about-polio-26d20cba98e5.

184 Forrest Maready, *The Moth in the Iron Lung: A Biography of Polio* (self pub., CreateSpace Independent Publishing Platform, 2018), https://www.amazon.com/Moth-Iron-Lung-Biography-Polio/dp/1717583679#detailBullets_feature_div.

185 Children's Health Defense Team, "The Intertwined History of Myelitis and Vaccines," Children's Health Defense, Sep. 25, 2020, https://childrenshealthdefense.org/news/the-intertwined-history-of-myelitis-and-vaccines.

186 Stephen E. Mawdsley, "Polio Provocation: Solving a Mystery with the Help of History," *The Lancet* 384, no. 9940 (2014): 300–301, doi: 10.1016/s0140-6736(14)61251-4.

187 H. V. Wyatt, "Diagnosis of Acute Flaccid Paralysis: Injection Injury or Polio?" *The National Medical Journal of India* 16, no. 3 (2003): 156–158, PMID: 12929860.

188 Children's Health Defense Team, "Read the Fine Print, Part Two—Nearly 400 Adverse Reactions Listed in Vaccine Package Inserts," Children's Health Defense, Aug. 14, 2020, https://childrenshealthdefense.org/news/read-the-fine-print-part-two-nearly-400-adverse-reactions-listed-in-vaccine-package-inserts.

189 Marco Cáceres, "Polio Wasn't Vanquished, It Was Redefined," *The Vaccine Reaction*, Jul. 9, 2015, https://thevaccinereaction.org/2015/07/polio-wasnt-vanquished-it-was-redefined/#_edn1.

190 Rodney Dodson, "What You Didn't Know about Polio," *Medium*, May 3, 2021, https://rodneydodson000.medium.com/what-you-didnt-know-about-polio-26d20cba98e5.

191 "Bernard Greenberg, PhD," UNC University Libraries, Health Sciences Library, accessed Jul. 3, 2024, https://web.archive.org/web/20240215172134/https://hsl.lib.unc.edu/gillings/history-deans-4.

192 Rodney Dodson, "What You Didn't Know about Polio," *Medium*, May 3, 2021, https://rodneydodson000.medium.com/what-you-didnt-know-about-polio-26d20cba98e5.

193 Cynthia Vinney, "How to Spot Medical Gaslighting and What to Do about It," *Verywell Mind*, May 18, 2023, https://www.verywellmind.com /what-is-medical-gaslighting-6831284.

194 A Midwestern Doctor, "A Primer on Medical Gaslighting," *The Forgotten Side of Medicine*, Feb. 9, 2023, https://www.midwesterndoctor .com/p/a-primer-on-medical-gaslighting.

195 Caitjan Gainty, "Medical Gaslighting: When Conditions Turn out Not to Be 'All in the Mind,'" *The Conversation*, Sep. 18, 2023, https://theconversation .com/medical-gaslighting-when-conditions-turn-out-not-to-be-all-in-the -mind-209611.

196 Callum Wells, "'I Know Three People Who Died from It!' M.I.A. Claims Being Labelled an Anti-Vaxxer Is 'Frustrating' as She Slams 'Society' for 'Gaslighting' Her . . . Despite Declaring She'd Rather 'Choose Death,'" *Daily Mail*, Oct. 14, 2022, https://www.dailymail.co.uk/tvshowbiz/article -11317153/M-says-labelled-anti-vaxxer-frustrating-slams-society-gaslighting -her.html.

197 "What Is the UK Covid-19 Inquiry," UK Covid-19 Inquiry, accessed Jul. 3, 2024, https://covid19.public-inquiry.uk.

198 Douglas Dickie, "Scotland's Covid Vaccine Injured Call for End of 'Gaslighting' from Medical Profession as They Prepare for UK Inquiry," *Scottish Daily Express*, Sep. 13, 2023, https://www.scottishdailyexpress .co.uk/news/scottish-news/scotlands-covid-vaccine-injured-call-30925072.

199 Ibid.

200 Jack Evans, "'It's Killing Us': Family's Horror Arm-Wrestle with Vax Scheme," *News.com.au*, Nov. 20, 2022, https://www.news.com.au/lifestyle/health /health-problems/its-killing-us-familys-horror-armwrestle-with-vax-scheme /news-story/ec80e2b805a56197e33d17a5c2c572da.

201 @SenatorRennick, "The gaslighting by the TGA in regards to the actual injury rate . . . " Twitter (now X), Aug. 11, 2023, https://twitter.com/Senator Rennick/status/1690128982782672897.

202 A Midwestern Doctor, "Why Are Vaccine Injured Patients Silenced?" *The Forgotten Side of Medicine*, Sep. 23, 2023, https://www.midwesterndoctor .com/p/why-are-vaccine-injured-patients.

203 Paul Bennett et al., "Living with Vaccine-Induced Immune Thrombocytopenia and Thrombosis: A Qualitative Study," *BMJ Open* 13, no. 7 (2023): e072658, doi: 10.1136/bmjopen-2023–072658.

204 Paul Bennett et al., "Qualitative Study of the Consequences of Vaccine-Induced Immune Thrombocytopenia and Thrombosis: The Experiences of Family Members," *BMJ Open* 13, no. 12 (2023): e080363, doi: 10.1136/ bmjopen-2023–080363.

205 A Midwestern Doctor, "Why Can Doctors Not Diagnose Medical Injuries," *The Forgotten Side of Medicine*, May 2, 2022, https://www.midwesterndoctor .com/p/why-can-doctors-not-diagnose-medical.

206 A Midwestern Doctor, "A Primer on Medical Gaslighting," *The Forgotten Side of Medicine*, Feb. 9, 2023, https://www.midwesterndoctor.com /p/a-primer-on-medical-gaslighting.

207 Megan Redshaw, "U.S. Sen. Johnson Holds News Conference with Families Injured by COVID Vaccines, Ignored by Medical Community," *The Defender*, Jun. 29, 2021, https://childrenshealthdefense.org/defender/sen -johnson-ken-ruettgers-press-conference-families-injured-covid-vaccines.

208 Paul D. Thacker, "Researcher Blows the Whistle on Data Integrity Issues in Pfizer's Vaccine Trial," *BMJ* 375, no. 2635 (2021), doi: 10.1136/bmj. n2635.

209 Barbara Gehrett, MD, Chris Flowers, MD, and Loree Britt, "Report 93: Pfizer's 'Post-Marketing Surveillance Report' Reveals that Pfizer Manipulated Data and Wrongly Tabulated Adverse Events, Which Concealed Them," *Daily Clout*, Nov. 28, 2023, https://dailyclout.io/pfizers-post-marketing -surveillance-reveals-pfizer-manipulated-data.

210 A Midwestern Doctor, "Why Can Doctors Not Diagnose Medical Injuries," *The Forgotten Side of Medicine*, May 2, 2022, https://www.midwesterndoctor .com/p/why-can-doctors-not-diagnose-medical.

211 Ibid.

212 Alexander Wilder, MD, *The Fallacy of Vaccination* (New York: The Metaphysical Publishing Company, 1899), 6, https://archive.org/details /101229606.nlm.nih.gov.

213 Katherine Watt, "Construction of the Kill Box: Legal History," *Bailiwick News,* May 4, 2023, https://bailiwicknews.substack.com/p/construction-of -the-kill-box-legal.

214 Kit Knightly, "'Pandemic Treaty' Will Hand WHO Keys to Global Government," *Off-Guardian*, Apr. 19, 2022, https://off-guardian.org/2022/04/19/pandemic -treaty-will-hand-who-keys-to-global-government.

215 "International Health Regulations," World Health Organization, accessed Jul. 3, 2024, https://www.who.int/health-topics/international-health-regulations.

216 Meryl Nass, MD, "What Are the International Health Regulations?" *Door to Freedom*, updated Oct. 30, 2023, https://doortofreedom.org/2023/10/30 /what-are-the-international-health-regulations.

217 Michael Nevradakis, PhD, "As Dec. 1 Deadline to Reject Controversial Who Proposals Looms, Activists Mount Global Opposition Efforts," *The Defender*, Nov. 28, 2023, https://childrenshealthdefense.org/defender/december-deadline -who-pandemic-treaty-ihr-amendments.

218 "About Us," *Door to Freedom*, accessed Jul. 3, 2024, https://doortofreedom .org/about.

219 "Meryl's COVID Newsletter," *Meryl's COVID Newsletter*, accessed Jul. 3, 2024, https://merylnass.substack.com.

220 *Good Morning CHD*, "Decoding the WHO's New Pandemic "Agreement" + New World Dis-Order with James Corbett," aired Nov. 20, 2023, on CHD. TV, https://live.childrenshealthdefense.org/chd-tv/shows/good-morning -chd/decoding-the-whos-new-pandemic-agreement—new-world-disorder -with-james-corbett.

221 Meryl Nass, MD, "An Update on the Pandemic Preparedness Agenda— This Affects YOU," *Door to Freedom*, accessed Jul. 3, 2024, https: //doortofreedom.org/2023/07/27/an-update-on-the-pandemic-preparedness -agenda-this-affects-you.

222 Michael Nevradakis, PhD, "As Dec. 1 Deadline to Reject Controversial Who Proposals Looms, Activists Mount Global Opposition Efforts," *The Defender*, Nov. 28, 2023, https://childrenshealthdefense.org/defender /december-deadline-who-pandemic-treaty-ihr-amendments.

223 Meryl Nass, MD, "Don't Read This if You Already Read the Brownstone Version on the WHO Treaty. I Added 1,000 Words, Tightened It Up, BEST VERSION," *Meryl's COVID Newsletter*, Aug. 20, 2023, https://merylnass .substack.com/p/dont-read-this-if-you-already-read.

224 Meryl Nass, MD, "What Are the International Health Regulations?" *Door to Freedom*, updated Oct. 30, 2023, https://doortofreedom.org/2023/10/30 /what-are-the-international-health-regulations.

225 Kit Knightly, "'Pandemic Treaty' Will Hand WHO Keys to Global Government," *Off-Guardian*, Apr. 19, 2022, https://off-guardian.org/2022 /04/19/pandemic-treaty-will-hand-who-keys-to-global-government.

226 Meryl Nass, MD, "Taking a Good Look at Pandemic Preparedness," *Door to Freedom*, accessed Jul. 3, 2024, https://doortofreedom.org/taking -a-good-look-at-pandemic-preparedness.

227 WHO Media Team, "Tripartite and UNEP Support OHHLEP's Definition of 'One Health,'" news release, Dec. 1, 2021, https://www.who.int/news /item/01–12–2021-tripartite-and-unep-support-ohhlep-s-definition-of -one-health.

228 Elze Van Hamelen, "'One Health'—Where Biosecurity Meets Agenda 2030," *Global Research*, Aug. 30, 2022, https://www.globalresearch.ca /one-health-where-biosecurity-meets-agenda-2030/5791710.

229 Meryl Nass, MD, "Taking a Good Look at Pandemic Preparedness," *Door to Freedom*, accessed Jul. 3, 2024, https://doortofreedom.org/taking -a-good-look-at-pandemic-preparedness.

230 Meryl Nass, MD, "Don't Read This If You Already Read the Brownstone Version on the WHO Treaty. I Added 1,000 Words, Tightened It Up, BEST VERSION," *Meryl's COVID Newsletter*, Aug. 20, 2023, https://merylnass .substack.com/p/dont-read-this-if-you-already-read.

231 Alliance for Natural Health International, "Think WHO Shouldn't Make Decisions about Your Health? You Must Be a Conspiracy Theorist," *The*

Defender, Feb. 12, 2024, https://childrenshealthdefense.org/defender/who-ihr-pandemic-treaty-medical-ethics-conspiracy-theorist.

232 "Proposal for Negotiating Text of the WHO Pandemic Agreement," World Health Organization, Oct. 30, 2023, https://apps.who.int/gb/inb/pdf_files/inb7/A_INB7_3-en.pdf.

233 "FACT SHEET: White House launches Office of Pandemic Preparedness and Response Policy," The White House, Jul. 21, 2023, https://www.whitehouse.gov/briefing-room/statements-releases/2023/07/21/fact-sheet-white-house-launches-office-of-pandemic-preparedness-and-response-policy.

234 "Dr. Paul Friedrichs Will Head New White House Pandemic Response Office," American College of Surgeons, Jul. 25, 2023, https://www.facs.org/for-medical-professionals/news-publications/news-and-articles/acs-brief/july-25–2023-issue/dr-paul-friedrichs-will-head-new-white-house-pandemic-response-office.

235 Michael Nevradakis, PhD, "'Health Program or Military Program'? White House Taps Military Official to Lead New Pandemic Policy Office," *The Defender,* Jul. 26, 2023, https://childrenshealthdefense.org/defender/white-house-paul-friedrichs-pandemic-policy-office.

236 Meryl Nass, MD, "USG Announces New Pandemic Preparedness and Response Office: Is It a Health Program or a Military Program?" *Meryl's COVID Newsletter,* Jul. 24, 2023, https://merylnass.substack.com/p/usg-announces-new-pandemic-preparedness.

237 "ZERO DRAFT: Political Declaration of the United Nations General Assembly High-Level Meeting on Pandemic Prevention, Preparedness and Response," United Nations, accessed Jul. 3, 2024, https://www.un.org/pga/77/wp-content/uploads/sites/105/2023/06/Zero-draft-PPPR-Political-Declaration-5-June.pdf.

238 Michael Nevradakis, PhD, "'Health Program or Military Program'? White House Taps Military Official to Lead New Pandemic Policy Office," *The Defender,* Jul. 26, 2023, https://childrenshealthdefense.org/defender/white-house-paul-friedrichs-pandemic-policy-office.

239 The Defender Staff, "A COVID Silver Lining? More Parents Than Ever Questioning 'Routine' Childhood Vaccines," *The Defender,* Aug. 12, 2022, https://childrenshealthdefense.org/defender/parents-questioning-routine-childhood-vaccines-covid.

240 Caroline Lewis, "Rockland County Records First U.S. Case of Polio since 2013," *Gothamist,* Jul. 21, 2022, https://gothamist.com/news/rockland-county-records-first-us-case-of-polio-since-2013.

241 Mami Taniuchi et al., "Kinetics of Poliovirus Shedding Following Oral Vaccination as Measured by Quantitative Reverse Transcription-PCR versus Culture," *Journal of Clinical Microbiology* 53, no. 1 (2015): 206–211, doi: 10.1128/JCM.02406–14.

242 Nsikan Akpan, "New York Declares Polio a State Disaster as Wastewater Surveillance Spots Virus in Nassau County," *Gothamist*, Sep. 9, 2022, https://gothamist.com/news/new-york-declares-polio-a-state-disaster-as -wastewater-surveillance-spots-virus-in-nassau-county.

243 Jessica Rendall, "Polio in New York: 'We Simply Cannot Roll the Dice'," CNET, Sep. 13, 2022, https://www.cnet.com/health/medical /polio-in-new-york-we-simply-cannot-roll-the-dice.

244 Matilda Hill, Ananda S. Bandyopadhyay, and Andrew J. Pollard, "Emergence of Vaccine-Derived Poliovirus in High-Income Settings in the Absence of Oral Polio Vaccine Use," *Lancet* 400, no. 10354 (2022): 713–715, doi: 10.1016/S0140–6736(22)01582–3.

245 Ivana Kottasová, "Around 1 Million Children in London Offered Polio Boosters after Virus Is Detected in Sewage," CNN, Aug. 10, 2022, https: //www.cnn.com/2022/08/10/europe/polio-vaccine-children-london-intl /index.html.

246 Associated Press, "Polio in US, UK and Israel Reveals Rare Risk of Oral Vaccine," *U.S. News*, Aug. 21, 2022, https://www.usnews.com/news/world /articles/2022–08–21/polio-in-us-uk-and-israel-reveals-rare-risk-of-oral -vaccine.

247 Jerusalem Post Staff, "Polio in Israel: Virus Found Outside Jerusalem for First Time," *The Jerusalem Post*, Mar. 17, 2022, https://www.jpost.com /breaking-news/article-701596.

248 Judy Siegel-Itzkovich, "Israel's Polio Outbreak Finally Under Control— Public Health Chief," *The Jerusalem Post*, Jul. 7, 2022, https://www.jpost .com/health-and-wellness/article-711466.

249 Children's Health Defense Team, "COVID Restrictions May Be Winding Down, but Global Control Is Ramping Up," *The Defender*, Mar. 10, 2022, https://childrenshealthdefense.org/defender/covid-restrictions-winding -down-global-control-ramping-up.

250 T. Hovi et al., "Poliovirus Surveillance by Examining Sewage Specimens. Quantitative Recovery of Virus after Introduction into Sewerage at Remote Upstream Location," *Epidemiology & Infection* 127, no. 1 (2001): 101–6, doi: 10.1017/s0950268801005787.

251 Lawrence Goodridge, "Sewage Surveillance: How Scientists Track and Identify Diseases Like COVID-19 before They Spread," *The Conversation*, Oct. 26, 2020, https://theconversation.com/sewage-surveillance-how-scien-tists-track-and-identify-diseases-like-covid-19-before-they-spread-148307.

252 Erika Ito et al., "Detection of Rotavirus Vaccine Strains in Oysters and Sewage and Their Relationship with the Gastroenteritis Epidemic," *Applied and Environmental Microbiology* 87, no. 10 (2021): e02547–20, doi: 10.1128/AEM.02547–20.

253 Amy Rosenberg, "Wastewater Surveillance for COVID-19: It's Complicated," *Tufts Now*, Jun. 21, 2022, https://now.tufts.edu/2022/06/21 /wastewater-surveillance-covid-19-its-complicated.

254 Life in the Lab Staff, "A Brief History of Wastewater Testing and Pathogen Detection," *ThermoFisher Scientific* (blog), Oct. 5, 2021, https://www.thermo fisher.com/blog/life-in-the-lab/a-brief-history-of-wastewater-testing-and -pathogen-detection.

255 "COVID-19 Testing PCR—A Critical Appraisal," Children's Health Defense, Sep. 14, 2020, https://childrenshealthdefense.org/news/covid-19-testing-pcr -a-critical-appraisal.

256 Este Geraghty, "How GIS Brings Wastewater Surveillance to Life," *Industry Blogs,* Esri, May 21, 2022, https://www.esri.com/en-us/industries/blog /articles/how-gis-brings-wastewater-surveillance-to-life.

257 Warish Ahmed et al., "Minimizing Errors in RT-PCR Detection and Quantification of SARS-CoV-2 RNA for Wastewater Surveillance," *Science of the Total Environment* 805 (2022): 149877, doi: 10.1016/j. scitotenv.2021.149877.

258 Marisa Eisenberg, Andrew Brouwer, and Joseph Eisenberg, "Sewage Surveillance Is the Next Frontier in the Fight against Polio," *The Conversation*, Oct. 19, 2018, https://theconversation.com/sewage-surveillance -is-the-next-frontier-in-the-fight-against-polio-105012.

259 Denise Chow, "CDC to Ramp up Wastewater Surveillance," NBC News, Feb. 4, 2022, https://www.nbcnews.com/science/science-news/cdc-ramp -wastewater-surveillance-rcna14892.

260 "Wastewater Surveillance: A New Frontier for Public Health," Centers for Disease Control and Prevention, updated Apr. 15, 2024, https://www.cdc .gov/advanced-molecular-detection/php/success-stories/wastewater-surveillance .html?CDC_AAref_Val=https://www.cdc.gov/amd/whats-new/wastewater -surveillance.html.

261 Lawrence Goodridge, "Sewage Surveillance: How Scientists Track and Identify Diseases like COVID-19 before They Spread," *The Conversation*, Oct. 26, 2020, https://theconversation.com/sewage-surveillance-how-scien tists-track-and-identify-diseases-like-covid-19-before-they-spread-148307.

262 Warish Ahmed et al., "Minimizing Errors in RT-PCR Detection and Quantification of SARS-CoV-2 RNA for Wastewater Surveillance," *Science of the Total Environment* 805 (2022): 149877, doi: 10.1016/j. scitotenv.2021.149877.

263 Sarah Arnold, "White House Finally Makes an Admission about the 6 Feet Social Distancing Rule," *Townhall*, Aug. 17, 2022, https://townhall.com /tipsheet/saraharnold/2022/08/17/white-house-covid-czar-admits-6foot -social-distancing-rule-not-the-right-way-n2611892.

264 Marta Massano et al., "Wastewater Surveillance for Different Classes of Pharmaceutical Drugs: Focus on Psychotropic Drugs and Their Metabolites," *Toxicologie Analytique et Clinique* 34, no. 3 Suppl (2022): S70-S71, doi: 10.1016/j.toxac.2022.06.096.

265 Allen G. Ross et al., "Can We 'WaSH' Infectious Diseases out of Slums?" *International Journal of Infectious Diseases* 92 (2020): 130–132, doi: 10.1016 /j.ijid.2020.01.014.

Chapter 6

1 Tony Rogers, "[Movie Review] The Zone of Interest," *Utobian*, Dec. 21, 2023, https://tobyrogers.substack.com/p/movie-review-the-zone-of-interest.

2 Bob Herman, "The U.S. Is the Drug Industry's Gold Mine," Axios, Sep. 30, 2021, https://www.axios.com/2021/09/30/drug-prices-pharma-revenue -usa-international.

3 "Vaccine Capitalism: Five Ways Big Pharma Makes So Much Money," *Corporate Watch*, Mar. 18, 2021, https://corporatewatch.org/five-ways-big -pharma-makes-so-much-money.

4 Bob Herman, "The U.S. Is the Drug Industry's Gold Mine," Axios, Sep. 30, 2021, https://www.axios.com/2021/09/30/drug-prices-pharma-revenue-usa -international.

5 Michael Erman and Patrick Wingrove, "Exclusive: Drugmakers Set to Raise US Prices on at Least 500 Drugs in January," Reuters, Dec. 29, 2023 https://www.reuters.com/business/healthcare-pharmaceuticals/drugmakers -set-raise-us-prices-least-500-drugs-january-2023–12–29.

6 "About," Pharmacopoeia, accessed Jul. 3, 2024, http://www.pharmacopoeia -art.net/about.

7 "'Cradle to Grave', In Sickness and In Health," Pharmacopoeia, accessed Jul. 3, 2024, http://www.pharmacopoeia-art.net/articles/in-sickness-and -in-health.

8 World Health Organization, "The European Commission and WHO Launch Landmark Digital Health Initiative to Strengthen Global Health Security," news release, Jun. 5, 2023, https://www.who.int/news/item/05– 06–2023-the-european-commission-and-who-launch-landmark-digital -health-initiative-to-strengthen-global-health-security.

9 "Poll Shows Americans Are Fed up with Pharmaceutical Industry," Harvard T.H. Chan School of Public Health, 2019, https://www.hsph .harvard.edu/news/hsph-in-the-news/poll-shows-americans-are-fed -up-with-pharmaceutical-industry.

10 Lydia Saad, "Retail, Pharmaceutical Industries Slip in Public Esteem," Gallup, Sep. 13, 2023, https://news.gallup.com/poll/510641/retail-pharmaceutical -industries-slip-public-esteem.aspx.

11 Fred D. Ledley et al., "Profitability of Large Pharmaceutical Companies Compared with Other Large Public Companies," *JAMA* 323, no. 9 (2020): 834–843, doi: 10.1001/jama.2020.0442.

12 Andrew Bloomenthal, "Gross Profit Margin: Formula and What It Tells You," Investopedia, Jun. 12, 2023, https://www.investopedia.com/terms/g/gross_profit_margin.asp.

13 Megan Brenan, "U.S. Business Sector Average Rating Worst Since 2008," Gallup, Sep. 9, 2022, https://news.gallup.com/poll/400835/business-sector-average-rating-worst-2008.aspx.

14 Lydia Saad, "Retail, Pharmaceutical Industries Slip in Public Esteem," Gallup, Sep. 13, 2023, https://news.gallup.com/poll/510641/retail-pharmaceutical-industries-slip-public-esteem.aspx.

15 Annalisa Merelli, "Guess How Much Big Pharma Paid in US Taxes on $110 Billion of Profit in 2022," *Quartz*, May 17, 2023, https://qz.com/guess-how-much-big-pharma-paid-in-us-taxes-on-110-bill-1850441135.

16 Matej Mikulic, "Domestic and International Revenue of the U.S. Pharmaceutical Industry between 1975 and 2022," Statista, Aug. 25, 2023, https://www.statista.com/statistics/275560/domestic-and-international-revenue-of-the-us-pharmaceutical-industry.

17 Brian Buntz, "GSK, Pfizer and J&J among the Most-Fined Drug Companies, According to Study," Pharmaceutical Processing World, Nov. 18, 2020, https://www.pharmaceuticalprocessingworld.com/gsk-pfizer-and-jj-among-the-most-fined-drug-companies-according-to-study.

18 Nick Dearden, "Big Pharma Is Still Making Absurd Profits off of the Pandemic," *Jacobin*, May 9, 2022, https://jacobin.com/2022/05/pharmaceutical-industry-pfizer-covid-vaccines-patents.

19 "Big Pharma: How Did They Perform in 2022 and What's Ahead in 2023?" DCAT Value Chain Insights, Feb. 16, 2023, https://www.dcatvci.org/features/big-pharma-how-did-they-perform-in-2022-and-whats-ahead-in-2023.

20 Kevin Dunleavy, "The Top 20 Pharma Companies by 2022 Revenue," *Fierce Pharma*, Apr. 18, 2023, https://www.fiercepharma.com/pharma/top-20-pharma-companies-2022-revenue.

21 Taylor Giorno, "Top 5 Largest US Pharma Firms' Net Earnings Topped $81.9 Billion Last Year: Watchdog," *The Hill*, Jul. 23, 2023, https://thehill.com/policy/healthcare/4116604-five-largest-us-pharma-firms-net-earnings-topped-81-9-billion-last-year-watchdog.

22 Kevin Dunleavy, "The Top 20 Pharma Companies by 2022 Revenue," *Fierce Pharma*, Apr. 18, 2023, https://www.fiercepharma.com/pharma/top-20-pharma-companies-2022-revenue.

23 Jennifer Kates, Cynthia Cox, and Josh Michaud, "How Much Could COVID-19 Vaccines Cost the U.S. after Commercialization?" KFF, Mar. 10, 2023, https://www.kff.org/coronavirus-covid-19/issue-brief/how-much-could-covid-19-vaccines-cost-the-u-s-after-commercialization.

24 Ibid.

25 *Cross-Border Rx: Pharmaceutical Manufacturers and U.S. International Tax Policy, Before the Committee on Finance*, 118th Cong. (2023) (prepared statement of Brad W. Setser, Whitney Shepherdson Senior Fellow, Council on Foreign Relations), https://www.cfr.org/report/cross-border-rx-pharmaceutical -manufacturers-and-us-international-tax-policy.

26 Annalisa Merelli, "Guess How Much Big Pharma Paid in US Taxes on $110 Billion of Profit in 2022," *Quartz*, May 17, 2023, https://qz.com /guess-how-much-big-pharma-paid-in-us-taxes-on-110-bill-1850441135.

27 Samanth Subramanian, "In the Push for New Vaccines, Taxpayers Keep Paying and Paying," *Quartz*, May 12, 2021, https://qz.com/2006390 /taxpayers-are-paying-twice-or-more-for-the-covid-19-vaccine.

28 "AAP Schedule of Well-Child Care Visits," *Healthychildren.org*, last updated Jul. 24, 2023, https://www.healthychildren.org/English/family-life/health-management/Pages/Well-Child-Care-A-Check-Up-for-Success.aspx.

29 Children's Health Defense, *Profiles of the Vaccine-Injured: "A Lifetime Price to Pay"* (New York: Skyhorse Publishing, 2022), https://www.skyhorsepublishing .com/9781510776593/profiles-of-the-vaccine-injured.

30 John-Michael Dumais, "'A Very Dangerous Medical Experiment': CDC Expands Vaccine Schedules for Kids, Pregnant Women and Most Adults," *The Defender*, updated Jan. 31, 2024, https://childrenshealthdefense.org /defender/cdc-expand-vaccine-schedule-children-pregnant-women.

31 Robert F. Kennedy Jr. and Brian Hooker, *Vax-Unvax: Let the Science Speak* (New York: Skyhorse Publishing, 2023), https://www.skyhorsepublishing .com/9781510766969/vax-unvax.

32 Brian Shilhavy, "Murderous Medical Doctors: How Pediatricians Kill Babies with Multiple Vaccines in One Office Visit," *Vaccine Impact*, Sep. 29, 2023, https://vaccineimpact.com/2023/murderous-medical-doctors-how -pediatricians-kill-babies-with-multiple-vaccines-in-one-office-visit.

33 Children's Health Defense Team, "Read the Fine Print, Part Two—Nearly 400 Adverse Reactions Listed in Vaccine Package Inserts," Children's Health Defense, Aug. 14, 2020, https://childrenshealthdefense.org/news/read-the -fine-print-part-two-nearly-400-adverse-reactions-listed-in-vaccine-package -inserts.

34 James Lyons-Weiler and Paul Thomas, "Retracted: Relative Incidence of Office Visits and Cumulative Rates of Billed Diagnoses along the Axis of Vaccination," *International Journal of Environmental Research and Public Health* 17, no. 22 (2020): 8674, doi: 10.3390/ijerph17228674.

35 Mandy A. Allison et al., "Financing of Vaccine Delivery in Primary Care Practices," *Academic Pediatrics* 17, no. 7 (2017): 770–777, doi: 10.1016/j. acap.2017.06.001.

36   "Current Partners," American Academy of Pediatrics, last updated May. 8, 2024, https://www.aap.org/en/philanthropy/corporate-and-organizational -partners/current-partners.

37   "Managing Costs Associated with Vaccinating," American Academy of Pediatrics, last updated Aug. 30, 2021, https://www.aap.org/en/patient -care/immunizations/implementing-immunization-administration-in -your-practice/managing-costs-associated-with-vaccinating.

38   "Prescribing Information: M-M-R-II," Merck, 2023, https://www.merck .com/product/usa/pi_circulars/m/mmr_ii/mmr_ii_pi.pdf.

39   Children's Health Defense Team, "A Six-in-One Vaccine Associated with Sudden Infant Death . . . ," Children's Health Defense, Mar. 5, 2020, https://childrenshealthdefense.org/news/a-six-in-one-vaccine-associated -with-sudden-infant-death.

40   James Lyons-Weiler and Russell L. Blaylock, "Revisiting Excess Diagnoses of Illnesses and Conditions in Children Whose Parents Provided Informed Permission to Vaccinate Them," *International Journal of Vaccine Theory, Practice, and Research* 2, no. 2 (2022), doi: 10.56098/ijvtpr.v2i2.59.

41   Institute for Pure and Applied Knowledge, "New Study Supports Conclusion of Retracted 2020 Study Showing Unvaxxed Kids Healthier Than Vaxxed," *The Defender*, Sep. 27, 2022, https://childrenshealthdefense.org/defender /study-unvaccinated-healthier-vaccinated-kids.

42   Ibid.

43   "Safer Medication Use in Pregnancy," Centers for Disease Control and Prevention, last rev. Apr. 10, 2023, https://web.archive.org/web/20230929 041423/https://www.cdc.gov/pregnancy/meds/treatingfortwo/infographic _large.html.

44   Press Association, "Thalidomide Scandal: 60-Year Timeline," *The Guardian*, Sep. 1, 2012, https://www.theguardian.com/society/2012/sep/01/thalidomide -scandal-timeline.

45   Ruth B. Merkatz, "Inclusion of Women in Clinical Trials: A Historical Overview of Scientific, Ethical, and Legal Issues," *Journal of Obstetric, Gynecologic & Neonatal Nursing* 27, no. 1 (1998): 78–84, doi: 10.1111/ j.1552–6909.1998.tb02594.x.

46   Ibid.

47   "Safer Medication Use in Pregnancy," Centers for Disease Control and Prevention, last rev. Apr. 10, 2023, https://web.archive.org/web/202309290 41423/https://www.cdc.gov/pregnancy/meds/treatingfortwo/infographic _large.html.

48   Susan E. Andrade et al., "Use of Prescription Medications with a Potential for Fetal Harm among Pregnant Women," *Pharmacoepidemiology & Drug Safety* 15, no. 8 (2006): 546–554, doi: 10.1002/pds.1235.

49   "Birth Defects from Drugs," Birth Injury Help Center, accessed Jul. 3, 2024, https://www.birthinjuryhelpcenter.org/medication-birth-defects.html.

50   Atsushi Sato et al., "Influence of Prenatal Drug Exposure, Maternal Inflammation, and Parental Aging on the Development of Autism Spectrum Disorder," *Frontiers in Psychiatry* 13 (2022), doi: 10.3389/fpsyt.2022.821455.

51   Jill Jin, MD, MPH, "Safety of Medications Used during Pregnancy," *JAMA* 328, no. 5 (2022): 486, doi: 10.1001/jama.2022.8974.

52   Yu-Chien Chang et al., "Prevalence, Trends, and Characteristics of Polypharmacy among US Pregnant Women Aged 15 To 44 Years: NHANES 1999 to 2016," *Medicine (Baltimore)* 102, no. 22 (2023): e33828, doi: 10.1097/MD.0000000000033828.

53   Martina Ayad, MD and Maged M. Costantine, MD, "Epidemiology of Medications Use in Pregnancy," *Seminars in Perinatology* 39, no. 3 (2015): 508–11, doi: 10.1053/j.semperi.2015.08.002.

54   Sara K. Head, PhD, MPH et al., "Behaviors Related to Medication Safety and Use During Pregnancy," *Journal of Women's Health (Larchmont)* 32, no. 1 (2023): 47–56, doi: 10.1089/jwh.2022.0205.

55   Allen A. Mitchell, MD et al., "Medication Use During Pregnancy, with Particular Focus on Prescription Drugs: 1976–2008," *American Journal of Obstetrics & Gynecology* 205, no. 1 (2011): 51.e1–8, doi: 10.1016/j.ajog.2011.02.029.

56   Leanna Skarnulis, "Toxins and Pregnancy," WebMD, Oct. 1, 2008, https://www.webmd.com/baby/features/pregnancy-and-toxins#1.

57   Children's Health Defense Team, "That Was Then, This Is Now: Open Season on Vaccinating Pregnant Women," Children's Health Defense, Apr. 23, 2018, https://childrenshealthdefense.org/news/that-was-then-this-is-now-open-season-on-vaccinating-pregnant-women.

58   F. M. Vassoler, E. M. Byrnes, and R. C. Pierce, "The Impact of Exposure to Addictive Drugs on Future Generations: Physiological and Behavioral Effects," *Neuropharmacology* 76, part B (2014): 269–275, doi: 10.1016/j.neuropharm.2013.06.016.

59   "Are Prescription Drugs Safe to Take When Pregnant?" National Institute on Drug Abuse Research Report, Apr. 13, 2021, https://nida.nih.gov/publications/research-reports/misuse-prescription-drugs/are-prescription-drugs-safe-to-take-when-pregnant.

60   "Neonatal Opioid Withdrawal Syndrome (Formerly Known as Neonatal Abstinence Syndrome)," *Cleveland Clinic*, reviewed Jun. 12, 2022, https://my.clevelandclinic.org/health/diseases/23226-neonatal-abstinence-syndrome.

61   Danya M. Qato and Aakash B. Gandhi, "Opioid and Benzodiazepine Dispensing and Co-Dispensing Patterns among Commercially Insured

Pregnant Women in the United States, 2007–2015," *BMC Pregnancy and Childbirth* 21, no. 1 (2021): 350, doi: 10.1186/s12884–021–03787–5.

62   Ashleigh Garrison, "Antianxiety Drugs—Often More Deadly than Opioids—Are Fueling the Next Drug Crisis in US," CNBC, Aug. 3, 2018, https://www.cnbc.com/2018/08/02/antianxiety-drugs-fuel-the-next -deadly-drug-crisis-in-us.html.

63   "Benzodiazepine Drugs Market," *Fact.MR*, Sep. 2022, https://www.factmr .com/report/4432/benzodiazepine-drugs-market.

64   CMI, "Leading Companies—Benzodiazepine Drugs Industry," *Coherent Market Insights* (blog), Mar. 2023, https://www.coherentmarketinsights. com/blog/insights/leading-companies-benzodiazepine-drugs-industry-352.

65   F. M. Vassoler, E. M. Byrnes, and R. C. Pierce, "The Impact of Exposure to Addictive Drugs on Future Generations: Physiological and Behavioral Effects," *Neuropharmacology* 76, part B (2014): 269–275, doi: 10.1016/j. neuropharm.2013.06.016.

66   Chittaranjan Andrade, MD, "Gestational Exposure to Benzodiazepines and Z-Hypnotics and the Risk of Major Congenital Malformations, Ectopic Pregnancy, and Other Adverse Pregnancy Outcomes," *Journal of Clinical Psychiatry* 84, no. 2 (2023): 23f14874, doi: 10.4088/JCP.23f14874.

67   Babette Bais, "Prevalence of Benzodiazepines and Benzodiazepine-Related Drugs Exposure before, during and after Pregnancy: A Systematic Review and Meta-Analysis," *Journal of Affective Disorders* 269 (2020): 18–27, doi: 10.1016/j.jad.2020.03.014.

68   Babette Bais, "Prevalence of Benzodiazepines and Benzodiazepine-Related Drugs Exposure before, during and after Pregnancy: A Systematic Review and Meta-Analysis," *Journal of Affective Disorders* 269 (2020): 18–27, doi: 10.1016/j.jad.2020.03.014.

69   Nora D. Volkow et al., "Self-Reported Medical and Nonmedical Cannabis Use among Pregnant Women in the United States," *JAMA* 322, no. 2 (2019): 167–169, doi: 10.1001/jama.2019.7982.

70   Committee on Obstetric Practice, "Committee Opinion No. 722: Marijuana Use during Pregnancy and Lactation," *Obstetrics & Gynecology* 130, no. 4 (2017): e205-e209, doi: 10.1097/AOG.0000000000002354.

71   Ed Susman, "Cannabis May Interfere with Pregnancy—Retrospective Study Finds Exposure Linked with Higher Rates of Stillbirth, Hypertensive Disorders," *Medpage Today*, Feb. 10, 2023, https://www.medpagetoday. com/meetingcoverage/smfm/103063.

72   Rikki Schlott, "From Crime and Homelessness to Schizophrenia and Suicide: Mothers Share How Pot Stole Their Sons," *New York Post*, updated May 12, 2023, https://nypost.com/2023/05/11/moms-how-pot-left-our -sons-schizophrenic-homeless-and-dead.

73 F. M. Vassoler, E. M. Byrnes, and R. C. Pierce, "The Impact of Exposure to Addictive Drugs on Future Generations: Physiological and Behavioral Effects," *Neuropharmacology* 76, part B (2014): 269–275, doi: 10.1016/j. neuropharm.2013.06.016.

74 "State Medical Cannabis Laws," National Conference of State Legislatures, updated Jun. 4, 2024, https://www.ncsl.org/health/state-medical-cannabis -laws.

75 John C. Hagan III, MD, "Big Tobacco, Big Opioid, Big Weed: The Successful Commercialization of Habituation & Addiction," *Missouri Medicine* 115, no. 6 (2018): 476–480, PMID: 30643323.

76 Kelly C. Young-Wolff et al., "Rates of Prenatal Cannabis Use among Pregnant Women before and during the COVID-19 Pandemic," *JAMA* 326, no. 17 (2021): 1745–1747, doi: 10.1001/jama.2021.16328.

77 Iris Dorbian, "Global Cannabis Sales to Skyrocket to $57 Billion in 2026, Says Top Market Research Firm," *Forbes*, Sep. 13, 2022, https://www.forbes .com/sites/irisdorbian/2022/09/13/global-cannabis-sales-to-skyrocket-to -57-billion-in-2026-says-new-report/?sh=6f3a0e677b07.

78 Katie Jones, "The Big Pharma Takeover of Medical Cannabis," *Visual Capitalist*, Aug. 12, 2019, https://www.visualcapitalist.com/the-big-pharma -takeover-of-medical-cannabis.

79 Dario Sabaghi, "Pfizer Bets on Medical Cannabis with $6.7 Billion Acquisition," *Forbes*, Dec. 21, 2021, https://www.forbes.com/sites /dariosabaghi/2021/12/20/pfizer-to-enter-the-medical-cannabis-industry -with-67-billion-acquisition/?sh=15b5622e6072.

80 Ben Stevens, "Pharma Industry Showing 'Growing Interest' in Cannabis as M&A Activity Increases," *Business of Cannabis*, Aug. 22, 2023, https: //businessofcannabis.com/pharma-industry-showing-growing-interest-in -cannabis-as-ma-activity-increases.

81 Stephen Murphy, "Big Pharma Eyes Cannabis: The Next Frontier in High-Growth Investment," *Business of Cannabis*, Sep. 21, 2023, https: //businessofcannabis.com/big-pharma-eyes-cannabis-the-next-frontier-in -high-growth-investment.

82 Matej Mikulic, "Top 10 Medical Cannabis Patent Holders in the U.S. as of 2019, by Number of Patents," Statista, Oct. 28, 2020, https://www.statista.com /statistics/1038262/medical-cannabis-companies-us-by-number-of-patents.

83 Reuters, "US Health Officials Look to Move Marijuana to Lower-Risk Drug Category," Reuters, Aug. 30, 2023, https://www.reuters.com/business /healthcare-pharmaceuticals/hhs-official-calls-move-marijuana-lower-risk -drug-category-bloomberg-news-2023-08-30.

84 David Hodes, "Big Pharma Watching, Waiting and Building Cannabis Patent Portfolio," *Cannabis Science and Technology*, Jul. 10, 2019, https://www.cannabissciencetech.com/view/big-pharma-watching -waiting-and-building-cannabis-patent-portfolio.

85   Office of the Surgeon General, "U.S. Surgeon General's Advisory: Marijuana Use and the Developing Brain," U.S. Department of Health and Human Services, last rev. Aug. 29, 2019, https://www.hhs.gov/surgeongeneral /reports-and-publications/addiction-and-substance-misuse/advisory-on -marijuana-use-and-developing-brain/index.html.

86   Kelly C. Young-Wolff et al., "California Cannabis Markets—Why Industry-Friendly Regulation Is Not Good Public Health," *JAMA Health Forum* 3, no. 7 (2022): e222018, doi: 10.1001/jamahealthforum.2022.2018.

87   Children's Health Defense Team, "Prenatal Ultrasound—Not So Sound after All," Children's Health Defense, Aug. 20, 2019, https://childrens healthdefense.org/child-health-topics/known-culprits/ultrasound /prenatal-ultrasound-not-so-sound-after-all.

88   Daniel F. O'Keeffe and Alfred Abuhamad, "Obstetric Ultrasound Utilization in the United States: Data from Various Health Plans," *Seminars in Perinatology* 37, no. 5 (2013): 292–294, doi: 10.1053/j.semperi.2013. 06.003.

89   Michael Bachner, "Israeli Device Would Let Pregnant Women Take Ultrasound Scans on Phone," *The Times of Israel*, May 3, 2018, https://www.times ofisrael.com/israeli-device-would-let-pregnant-women-take-ultrasound -scans-on-phone.

90   "BioInitiative 2012: A Rationale for Biologically-Based Exposure Standards for Low-Intensity Electromagnetic Radiation," *Bioinitiative Report*, accessed Jul. 3, 2024, https://bioinitiative.org.

91   "Point of Care Ultrasound vs Traditional Ultrasound: What's the Difference?" GE HealthCare, Jan. 18, 2023, https://www.gehealthcare.com/insights /article/-of-care-ultrasound-vs-traditional-ultrasound-what's-the-difference.

92   "Non-Ionizing Radiation from Wireless Technology," U.S. Environmental Protection Agency, last updated Sep. 19, 2023, https://www.epa.gov /radtown/non-ionizing-radiation-wireless-technology.

93   J. D. Campbell, R. W. Elford, and B. F. Brant, "Case-Control Study of Prenatal Ultrasonography Exposure in Children with Delayed Speech," *Canadian Medical Association Journal* 149, no. 10 (1993): 1435–1440, PMID: 8221427.

94   Jim West, *50 Human Studies: A New Bibliography Reveals Extreme Risk for Prenatal Ultrasound* (self pub., 2015), https://harvoa.org/chs/pr/dusbk1 .htm.

95   R. Morgan Griffin, "3D and 4D Ultrasounds," WebMD, reviewed Jun. 9, 2023, https://www.webmd.com/baby/3d-4d-ultrasound.

96   Stephanie Grady, "It's a Booming Baby Business, But Are 3D/4D Ultrasounds Really Worth the Risk?" *Fox6 Milwaukee*, Feb. 17, 2015, https://www.fox6now.com/news/its-a-booming-baby-business-but-are -3d-4d-ultrasounds-really-worth-the-risk.

97   R. Morgan Griffin, "3D and 4D Ultrasounds," WebMD, reviewed Jun. 9, 2023, https://www.webmd.com/baby/3d-4d-ultrasound.

98   "5 Steps to Launching a Profitable Elective Ultrasound Business," *Ultrasound Trainers* (blog), Apr. 24, 2023, https://ultrasoundtrainers.com /blogs/5-steps-to-launching-a-profitable-elective-ultrasound-business.

99   Stephanie Grady, "It's a Booming Baby Business, But Are 3D/4D Ultrasounds Really Worth the Risk?" Fox6 Milwaukee, Feb. 17, 2015, https://www.fox6 now.com/news/its-a-booming-baby-business-but-are-3d-4d-ultrasounds -really-worth-the-risk.

100 Patrick Lindsay et al., "Portable Point of Care Ultrasound (PPOCUS): An Emerging Technology for Improving Patient Safety," *Anesthia Patient Safety Foundation Newsletter* 35, no. 1 (2020): 15–17, https://www.apsf.org/article /portable-point-of-care-ultrasound-ppocus-an-emerging-technology-for -improving-patient-safety.

101 Andrea Park, "Exo Officially Enters Hand-Held Ultrasound Race with Launch of AI-Powered Iris System," *Fierce Biotech*, Sep. 29, 2023, https: //www.fiercebiotech.com/medtech/exo-officially-enters-handheld-ultrasound -race-launch-ai-powered-iris-system.

102 Ibid.

103 Adam Hsieh et al., "Handheld Point-of-Care Ultrasound: Safety Considerations for Creating Guidelines," *Journal of Intensive Care Medicine* 37, no. 9 (2022): 1146–1151, doi: 10.1177/08850666221076041.

104 Ketan Patel, "Overlooked Cyber Risk: Why IT Must Improve POCUS Workflow," *Exo* (blog), Sep. 6, 2022, https://www.exo.inc/article/overlooked -cyber-risk-why-it-must-improve-pocus-workflow.

105 Michael Bachner, "Israeli Device Would Let Pregnant Women Take Ultrasound Scans on Phone," *The Times of Israel*, May 3, 2018, https://www .timesofisrael.com/israeli-device-would-let-pregnant-women-take-ultrasound -scans-on-phone.

106 Ibid.

107 Ermioni Tsarna et al., "Associations of Maternal Cell-Phone Use during Pregnancy with Pregnancy Duration and Fetal Growth in 4 Birth Cohorts," *American Journal of Epidemiology* 188, no. 7 (2019): 1270–1280, doi: 10.1093/aje/kwz092.

108 Laura Birks et al., "Maternal Cell Phone Use during Pregnancy and Child Behavioral Problems in Five Birth Cohorts," *Environment International* 104 (2017): 122–131, doi: 10.1016/j.envint.2017.03.024.

109 Laura Dorwart, "What Is a Psychopath?" *Verywell Health*, Nov. 10, 2023, https://www.verywellhealth.com/psychopath-5235293.

110 Catherine Austin Fitts, "Solari Core Concepts: Tapeworm Economics," *Solari Report*, Feb. 22, 2018, https://pension.solari.com/solari-core-concepts /#Tapeworm.

111 "Revolving Door: Overview," *Open Secrets*, accessed Jul. 3, 2024, https://www.opensecrets.org/revolving.

112 "Vaccine Capitalism: Five Ways Big Pharma Makes So Much Money," *Corporate Watch*, Mar. 18, 2021, https://corporatewatch.org/five-ways-big-pharma-makes-so-much-money.

113 "Seven Decades of Firsts with Seven CDC Directors: Julie Gerberding, MD, MPH," Centers for Disease Control and Prevention, Jul. 12, 2016, https://www.cdc.gov/grand-rounds/pp/2016/20160712-pdf-seven-directors-gerberding-508.pdf.

114 Erika Edwards, "Beginning to Look 'Pretty Intense': Former CDC Head Who Led U.S. SARS Response Speaks about Coronavirus," NBC News, Jan. 28, 2020, https://www.nbcnews.com/health/health-news/beginning-look-pretty-intense-former-cdc-head-who-led-u-n1124531.

115 Robert F. Kennedy Jr., "Merck's Vaccine Division President Julie Gerberding Sells $9.1 Million in Shares—Is She Jumping Ship?" Children's Health Defense, Feb. 5, 2020, https://childrenshealthdefense.org/news/mercks-vaccine-division-president-julie-gerberding-sells-9–1-million-in-shares-is-she-jumping-ship.

116 "Gardasil Lawsuit," Wisner Baum, https://www.wisnerbaum.com/prescription-drugs/gardasil-lawsuit/, accessed Sept. 30, 2024.

117 "Julie Louise Gerberding, MD, MPH," Center for Health Incentives and Behavioral Economics, accessed Jul. 3, 2024, https://chibe.upenn.edu/expert/julie-louise-gerberding-md-mph.

118 "Merck & Co. Inc., Most Recent Insider Transactions," MarketWatch, https://www.marketwatch.com/investing/stock/mrk/company-profile?pid=66190237, accessed Sept. 30, 2024.

119 North Bethesda, MD, "Dr. Julie Gerberding Named Chief Executive Officer of the Foundation for the National Institutes of Health," Foundation for the National Institutes of Health, Mar. 1, 2022, https://fnih.org/press-release/dr-julie-gerberding-named-chief-executive-officer-of-the-foundation-for-the-national-institutes-of-health.

120 Michael Carome, MD, "Outrage of the Month: Revolving Door to FDA Commissioner's Office Sows Distrust in Agency," Public Citizen, Oct. 2019, https://www.citizen.org/article/outrage-of-the-month-revolving-door-to-fda-commissioners-office-sows-distrust-in-agency.

121 Karen Hobert Flynn, "For Big Pharma, the Revolving Door Keeps Spinning," *Congress Blog, The Hill*, Jul. 11, 2019, https://thehill.com/blogs/congress-blog/politics/452654-for-big-pharma-the-revolving-door-keeps-spinning.

122 Kimani Hayes, "Gottlieb Says That 'at Some Point,' COVID-19 Vaccines Could Be Considered a 'Three-Dose Vaccine,'" CBS News, Nov. 21, 2021, https://www.cbsnews.com/news/scott-gottlieb-covid-19-vaccine-three-doses.

123 Kimani Hayes, "Gottlieb Says Kids Could Start Getting COVID-19 Vaccine as Soon as November 4 or 5," CBS News, Oct. 25, 2021, https://www.cbsnews.com/news/covid-vaccine-kids-ages-5–11-november -gottlieb-face-the-nation.

124 The Epoch Times, "Pfizer Board Member Urged Twitter to Censor Posts on Natural Immunity, Low COVID Risk for Kids," *The Defender*, Jan. 10, 2023, https://childrenshealthdefense.org/defender/pfizer-scott-gottlieb -censor-twitter-natural-immunity-covid-risk-et.

125 John-Michael Dumais, "Texas Sues Pfizer for 'False' and 'Deceptive' Marketing of COVID Vaccines," *The Defender*, Dec. 1, 2023, https://childrenshealth defense.org/defender/texas-sues-pfizer-false-deceptive-covid-vaccines.

126 Brownstone Institute, "The First Amendment, Brought to You by Pfizer," *Brownstone Journal*, Jan. 9, 2024, https://brownstone.org/articles/the -first-amendment-brought-to-you-by-pfizer.

127 "Vaccines and Related Biological Products Advisory Committee," U.S. Food & Drug Administration, updated Apr. 26, 2019, https://www .fda.gov/advisory-committees/blood-vaccines-and-other-biologics/vaccines -and-related-biological-products-advisory-committee.

128 "Program Overview," Open Payments, accessed Jul. 4, 2024, https://open paymentsdata.cms.gov/about.

129 Children's Health Defense Team, "17 Pharma Henchmen Who Voted to Experiment on Your Kids—And How to Shun Them," *The Defender*, Nov. 1, 2021, https://childrenshealthdefense.org/defender/fda-pfizer-covid -kids-pharma.

130 Megan Redshaw, "'This Is Politics, Not Science': White House, CDC Prepare to Vaccinate 5- to 11-Year-Olds Prior to FDA Authorization," *The Defender*, Oct. 20, 2021, https://childrenshealthdefense.org/defender /white-house-cdc-plans-guidance-covid-vaccines-kids.

131 Toby Rogers, "Some Thoughts on Today's VRBPAC Meeting . . . " *uTobian*, Oct. 27, 2021, https://tobyrogers.substack.com/p/some-thoughts -on-todays-vrbpac-meeting.

132 "Funding to Universities by the Bill & Melinda Gates Foundation," University Philanthropy, accessed Jul. 4, 2024, https://www.university philanthropy.com/bill-and-melinda-gates-foundation-funding.

133 "ACIP Membership Roster," Centers for Disease Control and Prevention, last rev. Mar. 28, 2024, https://www.cdc.gov/vaccines/acip/members/bios .html.

134 Children's Health Defense Team, "14 ACIP Members Who Voted to Jab Your Young Children—And Their Big Ties to Big Pharma," *The Defender*, Nov. 24, 2021, https://childrenshealthdefense.org/defender /cdc-acip-pfizer-pediatric-covid-vaccine-big-pharma.

135 Toby Rogers, "Some Thoughts on Today's ACIP Meeting," *uTobian*, Nov. 2, 2021, https://tobyrogers.substack.com/p/some-thoughts-on-todays-acip-meeting.

136 "Brown's COVID-19 Vaccination Guidance Has Shifted from a Requirement to a Strong Recommendation," Brown University, last updated Jul. 6, 2023, https://healthy.brown.edu/vaccinations.

137 Marla J. Gold, MD, "Changes to COVID Vaccination Requirements," *Drexel News*, Apr. 12, 2023, https://drexel.edu/news/archive/2023/april/campus-update-changes-to-covid-vaccination-requirements.

138 "Harvard's Fall 2021 Vaccine Requirements for All Harvard Affiliates," Harvard International Office, Jul. 8, 2021, https://www.hio.harvard.edu/news/harvards-fall-2021-vaccine-requirements-all-harvard-affiliates.

139 Samuel L. Stanley Jr., MD, "April 15, 2022: Important Update on COVID-19 Directives," Michigan State University, Office of the President, Apr. 15, 2022, https://web.archive.org/web/20220415154929/https://president.msu.edu/communications/messages-statements/2022_community_letters/2022–04–15-Important-update-on-COVID-19-directives.html.

140 "COVID-19 Vaccine Information," The Ohio State University, accessed Jul. 4, 2024, https://safeandhealthy.osu.edu/vaccine.

141 Persis Drell, Lloyd Minor, MD, and Russell Furr, "Compliance with New Federal Vaccination Requirement," Stanford COVID-19 Health Alerts, Oct. 6, 2021, https://healthalerts.stanford.edu/covid-19/2021/10/06/compliance-with-new-federal-vaccination-requirement.

142 Zshekinah Collier, "Maryland Universities Drop COVID-19 Vaccine Requirements, Masks Optional," WYPR, Aug. 10, 2022, https://www.wypr.org/wypr-news/2022–08–10/maryland-universities-drop-covid-19-vaccine-requirements-masks-optional.

143 Michelle Ma, "UW Announces COVID-19 Vaccine Requirement for All Employees," news release, Jun. 3, 2021, https://www.washington.edu/news/2021/06/03/uw-announces-covid-19-vaccine-requirement-for-all-employees.

144 "COVID-19 Vaccination Requirement: University Issues Additional Guidance," *My VU News*, Jun. 9, 2021, https://news.vanderbilt.edu/2021/06/09/covid-19-vaccination-requirement-university-issues-additional-guidance.

145 Communications and External Relations, "Wake Forest COVID-19 Vaccine Policy," Wake Forest University Campus Health, accessed Jul. 4, 2024, https://campushealth.wfu.edu/2021/05/wake-forest-covid-19-vaccine-policy.

146 Berkeley Lovelace Jr., "White House Says about 900,000 Kids Ages 5 to 11 Got a Covid Vaccine in the First Week after its Approval," CNBC, Nov. 10, 2021, https://www.cnbc.com/2021/11/10/white-house-says-about-900000-kids-ages-5-to-11-got-a-covid-vaccine-in-first-week.html.

147 Rosemary Gibson and Janardan Prasad Singh, *China Rx: Exposing the Risks of America's Dependence on China for Medicine* (Guilford, CT: Prometheus Books, 2018), https://www.prometheusbooks.com/9781633886414/china-rx.

148 Anna Nishino, "The Great Medicines Migration: How China Took Control of Key Global Pharmaceutical Supplies," *Nikkei Asia*, Apr. 5, 2022, https://asia.nikkei.com/static/vdata/infographics/chinavaccine-3.

149 Niels Graham, "The US Is Relying More on China for Pharmaceuticals—And Vice Versa," *Atlantic Council* (blog), Apr. 20, 2023, https://www.atlantic council.org/blogs/econographics/the-us-is-relying-more-on-china-for -pharmaceuticals-and-vice-versa.

150 Ken Roberts, "In 2022, China Dominates U.S. Exports of Immunological Drugs, Plasma and Vaccines," *Forbes*, Oct. 26, 2022, https://www.forbes .com/sites/kenroberts/2022/10/26/in-2022-china-now-dominates-us -exports-of-plasma-and-vaccines/?sh=4963fd1f5af7.

151 David Smith D, "'It's Gamified': Inside America's Blood Plasma Donation Industry," *The Guardian*, Mar. 2, 2023, https://www.theguardian.com /books/2023/mar/02/blood-money-book-kathleen-mclaughlin.

152 Al Roth, "Plasma and Plasma Products (Such as Antibodies) Are a Big Business (and the U.S. Dominates the International Market)," *Market Design* (blog), May 18, 2020, http://marketdesigner.blogspot.com/2020/05 /plasma-and-plasma-products-such-as.html.

153 Sasha Latypova, "Why Are They Doing It?" *Due Diligence and Art*, Feb. 3, 2023, https://archive.ph/Lpr7x.

154 "The Solari Report," *Solari Report*, accessed Jul. 4, 2024, https://home. solari.com.

155 Catherine Austin Fitts, *Financial Rebellion*, hosted by Catherine Austin Fitts, Polly Tommey, and Carolyn Betts (2024; CHD.TV), https://live.childrenshealthdefense.org/chd-tv/shows/financial-rebellion-with -catherine-austin-fitts.

156 Catherine Austin Fitts, "1st Quarter 2022 Wrap Up: Introduction," *Solari Report*, Jul. 2022, https://planetequity2022.solari.com/introduction.

157 Mark Skidmore and Catherine Austin Fitts, "Summary Report on 'Unsupported Journal Voucher Adjustments' in the Financial Statements of the Office of the Inspector General for the Department of Defense and the Department of Housing and Urban Development," *Solari Report*, accessed Jul. 4, 2024, https://missingmoney.solari.com/wp-content/uploads /2018/08/Unsupported_Adjustments_Report_Final_4.pdf.

158 "The Missing Money," *Solari Report*, accessed Jul. 4, 2024, https://missing money.solari.com.

159 Catherine Austin Fitts, "Now Available! *mRNA Vaccine Toxicity* by Doctors for COVID Ethics with Afterword by Catherine Austin Fitts," *Solari Report*, Jul. 31, 2023, https://home.solari.com/now-available-mrna-vaccine -toxicity-by-doctors-for-covid-ethics-with-afterword-by-catherine-austin -fitts.

160 Steven H. Woolf et al., "The New Crisis of Increasing All-Cause Mortality in US Children and Adolescents," *JAMA Network* 329, no. 12 (2023): 975–976, doi: 10.1001/jama.2023.3517.

161 Annalise Merelli, "Life Expectancy for Men in U.S. Falls to 73 years—Six Years Less than for Women, per Study," *STAT*, Nov. 13, 2023, https://www .statnews.com/2023/11/13/life-expectancy-men-women.

162 "What's Behind 'Shocking' U.S. Life Expectancy Decline—And What to Do about It," Harvard School of Public Health, Apr. 13, 2023, https: //www.hsph.harvard.edu/news/hsph-in-the-news/whats-behind-shocking -u-s-life-expectancy-decline-and-what-to-do-about-it.

163 Annalise Merelli, "Life Expectancy for Men in U.S. Falls to 73 years—Six Years Less than for Women, per Study," *STAT*, Nov. 13, 2023, https://www .statnews.com/2023/11/13/life-expectancy-men-women.

164 Charlotte Morabito, "Here's Why American Men Die Younger Than Women on Average and How to Fix It," CNBC, last updated Mar. 3, 2023, https://www.cnbc.com/2023/03/01/why-american-men-die-younger -than-women-on-average-and-how-to-fix-it.html.

165 Selena Simmons-Duffin, "'Live Free and Die?' The Sad State of U.S. Life Expectancy," NPR, Mar. 25, 2023, https://www.npr.org/sections /health-shots/2023/03/25/1164819944/live-free-and-die-the-sad -state-of-u-s-life-expectancy.

166 Steven H. Woolf, MD, MPH et al., "The New Crisis of Increasing All-Cause Mortality in US Children and Adolescents," *JAMA Network* 329, no. 12 (2023): 975–976, doi: 10.1001/jama.2023.3517.

167 Selena Simmons-Duffin, "'Live Free and Die?' The Sad State of U.S. Life Expectancy," NPR, Mar. 25, 2023, https://www.npr.org/sections /health-shots/2023/03/25/1164819944/live-free-and-die-the-sad-state -of-u-s-life-expectancy.

168 Shameek Rakshit, Matthew McGough, and Krutika Amin, "How Does U.S. Life Expectancy Compare to Other Countries?" Peterson-KFF Health System Tracker, Jan. 30, 2024, https://www.healthsystemtracker.org /chart-collection/u-s-life-expectancy-compare-countries.

169 Orli Belman, "Immigration Boosts U.S. Life Expectancy, According to USC/Princeton Study," USC Leonard Davis School of Gerontology, Sep. 30, 2021, https://gero.usc.edu/2021/09/30/immigration-boosts-u-s -life-expectancy-according-to-usc-princeton-study.

170 Mike Schneider and the Associated Press, "U.S. Population Increase in 2023 Was Driven by the Most Immigrants since 2001—And Immigration Will Be the 'Main Source of Growth in the Future,'" *Fortune*, Dec. 20, 2023, https://fortune.com/2023/12/20/u-s-population-increase-in-2023-was -driven-by-the-most-immigrants-since-2001-and-immigration-will-be-the -main-source-of-growth-in-the-future.

171 Suzanne Burdick, PhD, "U.S. Military Runs COVID Vaccines, Former Pharma Exec Tells RFK Jr.," *The Defender*, Mar. 30, 2023, https://childrens healthdefense.org/defender/military-covid-vaccines-rfk-jr-podcast.

172 Catherine Austin Fitts, "Hero of the Week: April 10, 2023: Sasha Latypova & Robert F. Kennedy, Jr.," *Solari Report*, Apr. 10, 2023, https://home.solari .com/hero-of-the-week-april-10–2023-sasha-latypova-robert-f-kennedy-jr.

173 Editorial Team, "FASAB Statement 56: Understanding New Government Financial Accounting Loopholes," *Solari Report*, Dec. 29, 2018, https: //missingmoney.solari.com/fasab-statement-56-understanding-new -government-financial-accounting-loopholes.

174 Matt Taibbi, "Has the Government Legalized Secret Defense Spending?" *Rolling Stone*, Jan. 16, 2019, https://www.rollingstone.com/politics/politics -features/secret-government-spending-779959.

175 Sasha Latypova, "Why Are They Doing It?" *Due Diligence and Art*, Feb. 3, 2023, https://archive.ph/Lpr7x.

176 *The Solari Report*, "Special Report: The Great Steal: Is WHO Fronting for Mr. Global's Land Rush? with Sasha Latypova," hosted by Catherine Austin Fitts, aired Aug. 31, 2023, on *Solari Report*, https://home.solari.com/special -report-the-great-steal-is-who-fronting-for-mr-globals-land-rush-with -sasha-latypova.

177 Michael Nehls, *The Indoctrinated Brain: How to Successfully Fend Off the Global Attack on Your Mental Freedom* (New York: Skyhorse Publishing, 2023), https://www.skyhorsepublishing.com/9781510778368/the-indoctrinated -brain.

178 *Future Science Series*, "The Indoctrinated Brain with Dr. Michael Nehls," hosted by Ulrike Granögger, aired Jan. 8, 2024, on *Solari Report*, https://home.solari .com/future-science-series-the-indoctrinated-brain-with-dr-michael-nehls.

179 Taylor Wendt, "Hippocampus: What to Know," WebMD, Sep. 1, 2022, https://www.webmd.com/brain/hippocampus-what-to-know.

180 Michael Nehls, "Unified Theory of Alzheimer's Disease (UTAD): Implications for Prevention and Curative Therapy," *Journal of Molecular Psychiatry* 4 (2016): 3, doi: 10.1186/s40303–016–0018–8.

181 Catherine Austin Fitts and Carolyn Betts, Esq, "Financial Transaction Freedom: What Is It, What Threatens It, and How Do I Take Action to Secure It?" *Solari Report*, May 30, 2023, https://home.solari.com /financial-transaction-freedom.

182 R-CALF USA, "Peeling Back the Layers of Biden's 30x30 Land Grab— Margaret Byfield," YouTube, Aug. 26, 2021, https://www.youtube.com /watch?v=htgqG_Eo4AM&ab_channel=R-CALFUSA.

183 *The Solari Report*, "Special Report: The Great Steal: Is WHO Fronting for Mr. Global's Land Rush? with Sasha Latypova," hosted by Catherine Austin Fitts, aired Aug. 31, 2023, on *Solari Report*, https://home.solari.com

/special-report-the-great-steal-is-who-fronting-for-mr-globals-land-rush
-with-sasha-latypova.

184 Children's Health Defense Team, "Vaccine-Induced Myocarditis Injuring
Record Number of Young People, Will Shots Also Bankrupt Families?"
The Defender, Jan. 31, 2022, https://childrenshealthdefense.org/defender
/vaccine-induced-myocarditis-injuring-young-people.

185 Charlotte Morabito, "Here's Why American Men Die Younger than
Women on Average and How to Fix It," CNBC, last updated Mar. 3, 2023,
https://www.cnbc.com/2023/03/01/why-american-men-die-younger
-than-women-on-average-and-how-to-fix-it.html.

186 Terry Turner, "49+ US Medical Bankruptcy Statistics for 2023," RetireGuide,
updated Oct. 20, 2023, https://www.retireguide.com/retirement-planning
/risks/medical-bankruptcy-statistics.

187 *Medical Debt Burden in the United States* (Consumer Financial Protection
Bureau, Feb. 2022), https://s3.amazonaws.com/files.consumerfinance.gov/f
/documents/cfpb_medical-debt-burden-in-the-united-states_report
_2022–03.pdf.

188 Terry Turner, "49+ US Medical Bankruptcy Statistics for 2023," RetireGuide,
updated Oct. 20, 2023, https://www.retireguide.com/retirement-planning
/risks/medical-bankruptcy-statistics.

189 "Consumers Increasingly Using Crowdfunding Campaigns to Pay Medical
Bills," Consumers for Quality Care, Nov. 1, 2023, https://consumers4quality
care.org/consumers-increasingly-using-crowdfunding-campaigns-to-pay
-medical-bills.

190 Rachel Bluth, "GoFundMe CEO: 'Gigantic Gaps' in Health System
Showing Up in Crowdfunding," *KFF Health News*, Jan. 16, 2019, https:
//kffhealthnews.org/news/gofundme-ceo-gigantic-gaps-in-health-system
-showing-up-in-crowdfunding.

191 Dan Goldberg, "State Lawmakers Find America's Medical Debt Problem
'Can No Longer Be Ignored,'" *Politico*, Sep. 7, 2023, https://www.politico
.com/news/2023/09/07/state-lawmakers-america-medical-debt-00114330.

192 Neil Bennett et al., "Who Had Medical Debt in the United States?" United
States Census Bureau, Apr. 7, 2021, https://www.census.gov/library/stories
/2021/04/who-had-medical-debt-in-united-states.html.

193 Preeti Vankar, "Percentage of Adults in the United States with Medical
Debt Who Owed Select Amounts as of 2021," *Statista*, Jun. 20, 2022,
https://www.statista.com/statistics/1248996/share-of-adults-medical-debt
-amount-owed-us.

194 Terry Turner, "49+ US Medical Bankruptcy Statistics for 2023," RetireGuide,
updated Oct. 20, 2023, https://www.retireguide.com/retirement-planning
/risks/medical-bankruptcy-statistics.

195 "Medical Debt Burden in the United States," Consumer Financial Protection Bureau, Feb. 2022, https://s3.amazonaws.com/files.consumer finance.gov/f/documents/cfpb_medical-debt-burden-in-the-united-states _report_2022–03.pdf.

196 Jesse Bedayn, "States Confront Medical Debt That's Bankrupting Millions," Associated Press, Apr. 12, 2023, https://apnews.com/article/medical-debt -legislation-2a4f2fab7e2c58a68ac4541b8309c7aa.

197 Toby Rogers, "What We Are Up Against," *uTobian*, Dec. 28, 2023, https: //tobyrogers.substack.com/p/what-we-are-up-against.

198 Catherine Austin Fitts, "The Injection Fraud—It's Not a Vaccine," *Solari Report*, May 27, 2020, https://home.solari.com/deep-state-tactics-101-the -covid-injection-fraud-its-not-a-vaccine.

199 Catherine Austin Fitts and Ricardo Oskam, "3rd Quarter 2022 Wrap Up: Building Wealth," *Solari Report*, 2023, https://buildingwealth.solari.com /solaris-building-wealth.

200 Catherine Austin Fitts, "The Injection Fraud—It's Not a Vaccine," *Solari Report*, May 27, 2020, https://home.solari.com/deep-state-tactics-101-the -covid-injection-fraud-its-not-a-vaccine.

Chapter 7

1 NPR Staff, "Ike's Warning of Military Expansion, 50 Years Later," NPR, Jan. 17, 2011, https://www.npr.org/2011/01/17/132942244/ikes-warning -of-military-expansion-50-years-later.

2 Arnold S. Relman, "The New Medical-Industrial Complex," *New England Journal of Medicine* 303, no. 17 (1980): 963–970, doi: 10.1056/ NEJM198010233031703.

3 Toby Rogers, "What We Are Up Against," *uTobian*, Dec. 28, 2023, https: //tobyrogers.substack.com/p/what-we-are-up-against.

4 Yashaswini Singh and Christopher Whaley, "Private Equity Is Buying Up Health Care, but the Real Problem Is Why Doctors Are Selling," *The Hill*, Dec. 21, 2023, https://thehill.com/opinion/healthcare/4365741-private -equity-is-buying-up-health-care-but-the-real-problem-is-why-doctors-are -selling.

5 Whitney Webb, "This Biden Proposal Could Make the US a 'Digital Dictatorship,'" *Unlimited Hangout*, May 5, 2021, https://unlimitedhangout .com/2021/05/investigative-reports/this-biden-proposal-could-make -the-us-a-digital-dictatorship.

6 "The FEMMIT Complex—Why Finance, Energy, Medicine, Military, and Information Technology Industries Are Influence Leaders," Foresight University, accessed Jul. 5, 2024, https://foresightguide.com/the-femmit -complex-why-finance-energy-medicine-military-and-it-industries-are -influence-leaders.

7   *The People's Study*, CHD.TV, accessed Jul. 2, 2024, https://live.childrens healthdefense.org/chd-tv/browse-all/chd-bus-collection.

8   Megan Brenan and Jeffery M. Jones, "Ethics Ratings of Nearly All Professions Down in U.S," Gallup, Jan. 22, 2024, https://news.gallup.com /poll/608903/ethics-ratings-nearly-professions-down.aspx.

9   Kyla Thomas and Jill Darling, "Education Is Now a Bigger Factor Than Race in Desire for COVID-19 Vaccine," USC Schaeffer, Mar. 2, 2021, https://healthpolicy.usc.edu/evidence-base/education-is-now-a-bigger -factor-than-race-in-desire-for-covid-19-vaccine.

10   Bobbie Anne Flower Cox, "Conspiracy Theorists Were Right about Climate Lockdowns," *Brownstone Journal*, Jan. 22, 2024, https://brownstone.org /articles/conspiracy-theorists-were-right-about-climate-lockdowns.

11   Mary Kekatos, "Exemptions for Routine Childhood Vaccination at Highest Level Ever: CDC Report," ABC News, Nov. 9, 2023, https://abcnews.go .com/Health/routine-childhood-vaccination-rates-kindergartners-lower -pre-pandemic/story?id=104753746.

12   Ibid.

13   Sylvia Hui, "Millions in the UK Are Being Urged to Get Vaccinations during a Surge in Measles Cases," Associated Press, Jan. 22, 2024, https://apnews.com/article/uk-measles-outbreak-mmr-vaccination-rates -5c060634a19b8bf272d0f304e9e0f42d.

14   Children's Health Defense Team, "Vaccine Failures: The Glaring Problem Officials Are Ignoring. Part I: Measles Vaccination," Children's Health Defense, Jan. 7, 2020, https://childrenshealthdefense.org/news/vaccine -failure-the-glaring-problem-officials-are-ignoring-part-i-measles-vaccination.

15   Brenda Baletti, PhD, "MMR Vaccine Debate Heats Up as Media Claim 'Vaccine Hesitancy' to Blame for Recent Outbreaks," *The Defender*, Jan. 25, 2024, https://childrenshealthdefense.org/defender/mmr-vaccine-media-measles -outbreaks.

16   Carlo Martuscelli and Hanne Cokelaere, "EU Countries Destroy €4B Worth of COVID Vaccines," Politico, Dec. 18, 2023, https://www.politico.eu /article/europe-bonfire-covid-vaccines-coronavirus-waste-europe-analysis.

17   Aussie17, "BREAKING: Consumer's Association in Malaysia (PPIM) Demands Immediate Withdrawal of mRNA Vaccines following Alarming Safety Concerns," *PharmaFiles*, Dec. 28, 2023, https://www.aussie17 .com/p/breaking-consumers-association-of.

18   M. Nathaniel Mead et al., "Retracted: COVID-19 mRNA Vaccines: Lessons Learned from the Registrational Trials and Global Vaccination Campaign," *Cureus* 16, no. 1 (2024): e52876, doi: 10.7759/cureus.52876.

19   Attorney Bobbie Anne Cox, "Quarantine Lawsuit Summary," *Attorney Bobbie Anne Cox . . . Knowledge Is Power!* Dec. 17, 2023, https://attorneycox .substack.com/p/quarantine-lawsuit-summary.

20   Whitney Webb, "This Biden Proposal Could Make the US a 'Digital Dictatorship,'" *Unlimited Hangout*, May 5, 2021, https://unlimitedhang out.com/2021/05/investigative-reports/this-biden-proposal-could-make -the-us-a-digital-dictatorship.

21   "Frequently Asked Questions," ARPA-H, accessed Jul. 5, 2024, https: //arpa-h.gov/about/faqs.

22   Joyce Frieden, "New Federal Health Agency Sets Bold Goals," *Medpage Today*, last updated Apr. 1, 2024, https://www.medpagetoday.com/publich ealthpolicy/washington-watch/108401.

23   Whitney Webb, "This Biden Proposal Could Make the US a 'Digital Dictatorship,'" *Unlimited Hangout*, May 5, 2021, https://unlimitedhangout .com/2021/05/investigative-reports/this-biden-proposal-could-make -the-us-a-digital-dictatorship.

24   Christina D. Bethell et al., "A National and State Profile of Leading Health Problems and Health Care Quality for US Children: Key Insurance Disparities and Across-State Variations," *Academic Pediatrics* 11, no. 3S (2011): S22-S33, doi: 10.1016/j.acap.2010.08.011.

25   James Lyons-Weiler and Paul Thomas, "Relative Incidence of Office Visits and Cumulative Rates of Billed Diagnoses along the Axis of Vaccination [Retracted]," *International Journal of Environmental Research and Public Health* 17, no. 22 (2020): 8674, doi: 10.3390/ijerph17228674.

26   Hilda Labrada Gore, "The GAPS Diet: A Powerful Healing Protocol with Natasha Campbell-McBride," Oct. 31, 2022, in *Wise Traditions Podcast*, 42:47, https://www.westonaprice.org/podcast/the-gaps-diet-a-powerful -healing-protocol.

27   Eugene Reznik, "A Review of a Ketogenic Diet in the Treatment of Autism Spectrum Disorder," Loma Linda University, Jun. 2024, https://scholars repository.llu.edu/etd/1713.

28   Celia Farber, "Meat Cures Depression and 'Mental Illness,'" *The Truth Barrier*, Jan. 27, 2024, https://celiafarber.substack.com/p/meat-cures-depression -and-mental.

29   "What Is the Wim Hof Method?" *WimHofMethod*, accessed Jul. 5, 2024, https://www.wimhofmethod.com.

30   "Coherence Healing," *Unlimited Dr. Joe Dispenza*, accessed Jul. 5, 2024, https://drjoedispenza.com/coherence-healing.

31   Ryan Miller, "A Totalitarian State Can Only Rule a Desperately Poor Society," Foundation for Economic Education, Sep. 5, 2016, https://fee .org/articles/a-totalitarian-state-can-only-rule-a-desperately-poor-society.

32   Sasha Latypova, "Catherine Austin Fitts Explains the Cabal's Land and Real Estate Stealing Tactics," *Due Diligence and Art*, Aug. 30, 2023, https: //sashalatypova.substack.com/p/catherine-austin-fitts-explains-the.

33   Catherine Austin Fitts, "Turtle Forth," *Solari Report*, Nov. 23, 2014, https: //home.solari.com/turtle-forth.